Engineering Materials

This series provides topical information on innovative, structural and functional materials and composites with applications in optical, electrical, mechanical, civil, aeronautical, medical, bio- and nano-engineering. The individual volumes are complete, comprehensive monographs covering the structure, properties, manufacturing process and applications of these materials. This multidisciplinary series is devoted to professionals, students and all those interested in the latest developments in the Materials Science field, that look for a carefully selected collection of high quality review articles on their respective field of expertise.

Indexed at Compendex (2021) and Scopus (2022)

Ilya A. Verbenko · Ivan A. Parinov ·
Ekaterina V. Glazunova · Svetlana I. Dudkina ·
Konstantin P. Andryushin · Dmitry V. Volkov ·
Larisa A. Reznichenko

Environmentally Friendly Smart Materials with Special Electrical and Magnetic Properties

Springer

Ilya A. Verbenko
Research Institute of Physics
Southern Federal University
Rostov-on-Don, Russia, Rostov Region

Ekaterina V. Glazunova
Research Institute of Physics
Southern Federal University
Rostov-on-Don, Russia, Rostov Region

Konstantin P. Andryushin
Research Institute of Physics
Southern Federal University
Rostov-on-Don, Russia, Rostov Region

Larisa A. Reznichenko
Research Institute of Physics
Southern Federal University
Rostov-on-Don, Russia, Rostov Region

Ivan A. Parinov
I. I. Vorovich Mathematics, Mechanics
and Computer Sciences Institute
Southern Federal University
Rostov-on-Don, Russia, Rostov Region

Svetlana I. Dudkina
Research Institute of Physics
Southern Federal University
Rostov-on-Don, Russia, Rostov Region

Dmitry V. Volkov
Research Institute of Physics
Southern Federal University
Rostov-on-Don, Russia, Rostov Region

ISSN 1612-1317 ISSN 1868-1212 (electronic)
Engineering Materials
ISBN 978-3-031-90386-1 ISBN 978-3-031-90387-8 (eBook)
https://doi.org/10.1007/978-3-031-90387-8

This Springer imprint is published by the registered company Springer Nature Switzerland AG
The registered company address is: Gewerbestrasse 11, 6330 Cham, Switzerland

Preface

The basis of most modern ferroelectric piezoceramic materials (FPCMs), used in production and in numerous applications, is currently solid solutions of lead-containing complex oxides: $PbTiO_3$–$PbZrO_3$ (PZT), $Pb(Nb_{2/3}Mg_{1/3})O_3$–$PbTiO_3$ (PMN-PT) and so on. The most common technology for their manufacture involves solid-phase synthesis and sintering at high temperatures. However, the presence of very toxic Pb in these compounds has predefined the need to transition to environmentally friendly methods of manufacture, designed compositions of FPCMs and products made from them; the task of designing and developing new, lead-free similar materials and composites of PZT becomes urgent.

Environmentally friendly FPCMs and the methodologies employed for their manufacture take a very important place in modern ferro- and piezoelectric material science. The importance of these active materials has been pointed, for example by DIRECTIVE 2002/95/EC OF THE EUROPEAN PARLIAMENT AND OF THE COUNCIL of 27 January 2003 on the restriction of the use of certain hazardous substances in electronic equipment.

The authors of the book are the representatives of the so-called Rostov Scientific School on Ferro-piezoelectricity. Investigations of the three-component systems, based on lead zirconate titanate (PZT), started in the Rostov State University (now Southern Federal University, Russia) at the end of the 1960s. Almost immediately, there were performed intensive studies of four-, five-component solid solutions and then the studies of six-component systems, based on PZT at the end 1990s. The well-known brand is Piezoelectric Ceramic Rostovskaya (PCR), originally presenting itself PZT-type ceramics. From the past, Rostov scientists designed and manufactured more than 100 systems of PCR, as based on PZT composition as on the base of other ferro-piezoelectric solid solutions. Many materials, composites and devices on their base were developed, researched and produced in Rostov-on-Don. Rostov scientists have published more 5000 journal papers and books in these topics and received more than 300 Soviet, Russian and international patents (see, for example, monographs [1–13] and references there in). The authors of the book from the Research Institute of Physics, Southern Federal University, Rostov-on-Don, Russia, have been engaged in the research activities related to the search, development, and study of lead-free ferro-piezoelectric materials and environmentally friendly technologies for over five decades. However, very significant parts of these publications are not readily accessible in open access, that restricts their citation. Due to the Russian studies in the ferro-piezoelectricity are insufficiently known in world science in a whole.

The book is devoted to R&D of modern lead-free environmentally friendly ferroelectric piezoceramic materials and the obtained attainments of the authors and their colleagues.

The book is comprised of six chapters.

In the initial chapter, a comprehensive literature overview is performed on environmentally friendly ferroelectric piezoelectric ceramic materials and the methodologies employed for their preparation. The majority of the most promising lead-free FPCMs are based on alkali metal niobates ($KNbO_3$, $NaNbO_3$, $LiNbO_3$), including those modified by various cations, such as Pb^{2+}, Cd^{2+}, Mg^{2+}, Ca^{2+}, Sr^{2+}, Ba^{2+}, Bi^{3+}, La^{3+}, Ti^{4+} and Ge^{4+}. These modifications are employed to improve the manufacturability of the materials and their final properties. The influence of structural inhomogeneities on electrical and physical properties is the subject of investigation. A strategy for the synthesis of a new generation of active FPCMs is formulated. The potential for lead-free FPCMs, applied widely in the most advanced fields, including biomedicine, information technology, robotics, precision engineering, nondestructive testing, special measuring equipment, military and civilian sonar, and ultrasonic power technologies, is demonstrated.

Chapter 2, a comprehensive analysis and generalization of the extensive experimental data are discussed, and the classification of the utilized modifiers, the order and method of their introduction, and the influence of different modifiers on the ceramic (T_{sint}, ρ), dielectric ($\varepsilon_{33}^{T}/\varepsilon_0$, T_C), piezoelectric (d_{31}, K_p), and elastic (V_R, E_y) characteristics of solid solutions of the systems $NaNbO_3-KNbO_3$ and $NaNbO_3-LiNbO_3$ are evaluated. This study demonstrates the advantages of glass-ceramic technology in obtaining solid solutions of the system $NaNbO_3-LiNbO_3$. The impact of α-, β-, and γ-glasses on $T_{sint.}$, $\bar{D}$, and σ of solid solutions of the system $NaNbO_3-LiNbO_3$ is assessed. It is demonstrated that the selection of the optimal qualitative and quantitative composition of modifying additives and glasses allows one to produce materials with promising applications in various types of radio-electronic devices.

Chapter 3 elaborates modification schemes and the construction of quasi-chemical reactions of defect formation and considers modification formulae for the solid solutions of (Na, Li)NbO_3 and $LiNbO_3$ systems. The stoichiometric modification of the material by the addition of donor and acceptor modifiers is evaluated to ascertain their influence on the structural, dielectric, ferroelectric, and mechanical characteristics of the materials. The regularities of changes in the above properties at super stoichiometric modification of one (Ag)-, two (Cd)-, three (Al, Cr, Fe, La)-, four (Ti, Si, Mn)-, five (V)-, and six (U, W)-valence elements of sodium niobate-based solid solutions are investigated. The analysis of defect formation and types of ordering-disorder leads to the proposal of a "model with packing disorder" that describes the crystallochemical situation in solid solutions, based on (Na, Li)NbO_3. Furthermore, the influence of non-stoichiometry on the electrical conductivity of the obtained materials is studied.

Chapter 4 presents the studies of the (Na, Li)NbO_3 system, which is modified by various oxides and their combinations. The following oxides are considered: MnO_2, CuO, NiO, MnO_2 + CuO, and Bi_2O_3 + Fe_2O_3. Phase diagrams of solid solutions of the modified system (Na, Li)NbO_3 are constructed, and the dependencies of perovskite cell parameters and homogeneous deformation parameter δ on the modifier concentration are presented. The impact of the quantity of the modifier additive on the grain structure of ceramics and the uniformity of the modifier distribution is elucidated. The conditions of formation of electrophysical characteristics, including: $\varepsilon_{33}^T/\varepsilon_0$, $\tan\delta$, d_{33}, $|d_{31}|$, g_{33}, g_{31}, K_p, Q_M, V_1^E, are analyzed for modified solid solutions. The behavior of ceramics in high electric fields is investigated. Consequently, the analysis of the dependencies of the relative reversible dielectric constant on the electric field strength for unpolarized ($\varepsilon/\varepsilon_0$) and polarized ($\varepsilon_{33}^T/\varepsilon_0$) ceramics of the studied compositions permits to perform a classification of the studied ceramics, based on (Na, Li)NbO_3 as the group of ferro-rigid FPCMs.

Chapter 5 presents studies of multielement compositions, based on solid solutions sodium-potassium niobates of compositions $(Na_{0.46}K_{0.54})_{1-x}Li_x(Nb_{0.9}Ta_{0.1})_{1-x}Sb_xO_3$ ($x = 0.02 - 0.06$). The compositions are modified super stoichiometrically and stoichiometrically with MnO_2, CuO, NiO, Bi_2O_3, Fe_2O_3, and combined additives. The results of studies, investigating the kinetics of the formation (synthesis and sintering) of ceramics, based on alkali and alkaline-earth metal niobates, are presented. It is demonstrated that sintering invariably occurs in conjunction with the liquid phase, which exerts a profound impact on the fundamental stages of diffusion, polymorphic transformations, recrystallization, dissociation, and other processes. The influence of each additive, the alloying process, the preparation environment, and the morphological characteristics of the modified samples on the structure and behavior of the dielectric, piezoelectric, and ferroelectric properties of multielement alkaline niobates is extensively researched.

Chapter 6 investigates the thermal-frequency characteristics of modified solid solutions, based on alkali metal niobates in the presence of a wide range of external impacts, including temperature and the frequency of alternating electric fields. The causes and mechanisms of oxygen non-stoichiometry in complex compounds with perovskite-type structure are described in detail. The pivotal role of oxygen non-stoichiometry in the genesis of thermal-frequency behavior for such structures is demonstrated. Attempts are made to determine the nature of relaxation processes, occurring in multielement solid solutions, based on sodium-potassium niobates.

This self-sufficient book, presenting modern manufacture technologies and study of environmentally friendly smart materials, is intended to a wide range of students, engineers and specialists interested and participating in R&D of the modern problems of the ferro- and piezoelectrics. The book gives new knowledge in the pointed scientific-technical areas and allows one to understand the corresponding level of modern investigations.[1]

Rostov-on-Don, Russia
February 2025

Ilya A. Verbenko DrSc
iaverbenko@sfedu.ru

Ivan A. Parinov DrSc
iparinov@sfedu.ru

Ekaterina V. Glazunova PhD
glazunova@sfedu.ru

Svetlana I. Dudkina PhD
sidudkina@sfedu.ru

Konstantin P. Andryushin DrSc
kpandryushin@sfedu.ru

Dmitry V. Volkov PhD
dvvolkov@sfedu.ru

Larisa A. Reznichenko DrSc
lareznichenko@sfedu.ru

References

1. Parinov, I.A. (ed.): Advanced Nano- and Piezoelectric Materials and their Applications. Nova Science Publishers, New York (2014)
2. Akopyan, V.A., Soloviev, A.N., Parinov, I.A., Shevtsov, S.N.: Definition of Constants for Piezoceramic Materials. Nova Science Publishers, New York (2010)
3. Danciger, A.Y., et al: Multicomponent systems of ferroelectric complex oxides: physics, crystal chemistry, technology. In: Design Aspects of Ferro-piezoelectric Materials. Rostov State University Press, Rostov-on-Don, **1**, (2001), **2** (2002) (in Russian)
4. Parinov, I.A. (ed.): Ferroelectrics and Superconductors: Properties and Applications. Nova Science Publishers, New York (2012)
5. Parinov, I.A. (ed.): Nano- and Piezoelectric Technologies, Materials and Devices. Nova Science Publishers, New York (2013)
6. Parinov, I.A.: Microstructure and Properties of High-Temperature Superconductors, 2nd edn. Springer, Heidelberg, New York, Dordrecht, London (2013)
7. Parinov, I.A., Zubkov, S.V., Skaliukh, A.S., et al.: Advanced Ferroelectric and Piezoelectric Materials with Improved Properties and their Applications. World Scientific Publishers, Singapore (2024)
8. Parinov, I.A. (ed.): Piezoceramic Materials and Devices. Nova Science Publishers, New York (2010)
9. Parinov, I.A. (ed.): Piezoelectric Materials and Devices. Nova Science Publishers, New York (2012)

[1] The study was carried out with the financial support of the Ministry of Science and Higher Education of the Russian Federation (State task in the field of scientific activity. Project No. FENW-2023-0010/GZ0110/23-11-IF).

10. Parinov, I.A. (ed.): Piezoelectrics and Nanomaterials: Fundamentals, Developments and Applications. Nova Science Publishers, New York (2015)
11. Parinov, I.A. (ed.): Piezoelectrics and Related Materials: Investigations and Applications. Nova Science Publishers, New York (2012)
12. Shevtsov, S.N., Soloviev, A.N., Parinov, I.A., et al.: Piezoelectric Actuators and Generators for Energy Harvesting—Research and Development. Springer Cham, Switzerland (2018)
13. Zakharov, Y.N., Sitalo, Y.I., Parinov, I.A., Boldyrev, N.A.: Pyroelectric, Flexoelectric and Related Effects in Ferroelectrics, Antiferroelectrics, Ferroelectric Relaxors and Multiferroics. Southern Federal University Press, Rostov-on-Don (2021) (in Russian)

Competing Interests The authors have no competing interests to declare that are relevant to the content of this manuscript.

Contents

Chapter 1
Current State of Research in the Field of Environmentally Friendly Smart Materials, Technologies and Devices

1.1 Development of Lead-Free Ferro-Piezomaterials Science at the Turn of the Millennium

The basis of most modern ferroelectric piezoceramic materials (FPCMs) used in production are currently solid solutions (SS) of lead-containing complex oxides ($PbTiO_3$–$PbZrO_3$) (PZT), $Pb(Nb_{2/3}Mg_{1/3})O_3$–$PbTiO_3$ (PMN–PT) and other. The most common technology for their manufacture involves solid-phase synthesis and sintering at high temperatures. In connection with the need to transition to environmentally friendly methods of production of FPCM and products made from them, the task of developing new, lead-free analog materials of PZT becomes urgent [1, 2]. For more than 40 years, the Research Institute of Physics of the Southern Federal University, Russia has been working on the search, study and creation of such materials based on lead-free solids, including those with the participation of alkali metal niobates (AMNs) [3]. The main directions and results of this activity are given in Table 1.1.

Comprehensive studies of more than 100 n-component systems (more than 4 thousand compositions) have been carried out. The specific morphology of the phase diagrams of these systems has been established, including a huge variety of polymorphic and morphotropic phase transformations of various natures; some features of composition–structure–properties–application have been identified; patterns of changes in physical properties during iso- and heterovalent ion substitutions have been established; the possibility of controlling the electrophysical parameters of solid solution by varying the conditions of their structure formation is shown.

All this made it possible to develop several new lead-free FPCMs and technologies for their production, protected by more than 80 security documents for methods of synthesis, hot pressing, optimization of filling compositions, identification of the microstructure of niobate ferroelectric ceramics; for piezoelectric ceramic materials for high-frequency transducers for various purposes; for highly sensitive materials;

I. A. Verbenko et al., *Environmentally Friendly Smart Materials with Special Electrical and Magnetic Properties*, Engineering Materials,
https://doi.org/10.1007/978-3-031-90387-8_1

Table 1.1 Promising areas of work to create lead-free, environmentally friendly ferropiezoelectric ceramic materials

Functional groups of industrially produced FPCM (according to Russian OST 110444-87)		Industrial materials	Basic characteristics of the foundations of the developed FPCM	Directions of work	Expected results
No p/p	Name	Abbre-viation			
I	Highly sensitive materials operating in receiving and/or emitting modes	TB-1 NBPZT-1 PZT-19	$\varepsilon^{T}{}_{33}/\varepsilon_0 = 1040–3550$; $\tan\delta \cdot 10^2 = 1.6$; $K_p > 0.3$; $\lvert d_{31}\rvert > 50$ pC/N; $\lvert g_{31}\rvert > 5$ mV m/N	Increasing piezo parameters, decreasing Q_M	Approximation in parameters to industrial materials at higher V_R, T_C, at lower ρ
I-1	Materials for ultrasonic shear vibrations	PZT-36	$\varepsilon^{T}{}_{33}/\varepsilon_0 < 800$; $K_p > 0.3–0.5$		
II	Materials operating in reception and/or emission mode under conditions of strong electric fields and/or mechanical stress	TBK-3 PZT-23 PZT-24 PZTB-3	$\varepsilon^{T}{}_{33}/\varepsilon_0 = 380–1240$; $K_p > 0.2$; $\lvert d_{31}\rvert = 30–60$ pC/N; $V_R > 5$ km/s; $Q_M = 250–3000$; $\rho < 4.5$ g/cm^3	Decrease $\tan\delta$, increase Q_M, d_{jg}, K_{jg}	Getting closer to industrial materials at lower ρ
III-1 III-2 III-3	Materials with increased stability of frequency characteristics for frequency-selective devices on volumetric (planar mode (3–1), tension–compression along the thickness (3–2)) and surface acoustic waves (3–3) in the temperature range from –60 to + 85 °C	PZT-22 PZT-35 PZT-35 PZT-35U PZT-33	$\varepsilon^{T}{}_{33}/\varepsilon_0 = 120–490$; $\tan\delta\ 10^2 = 1.6$; $K_p \sim 0.3$; $\lvert d_{31}\rvert = 12–30$ pC/N; $V_R = 4.8–5.9$ km/s; $Q_M = 1000–4000$; $\rho < 4.5$ g/cm^3 $TK3 \cdot 10^6 \cdot K^{-1} = 130$; $\delta f_0/f_r < 1\%$	Expanding the ranges of values ε_{jg}, K_{jg}	Obtaining materials with better temperature stability of the resonant frequency compared to industrial materials

(continued)

Table 1.1 (continued)

Functional groups of industrially produced FPCM (according to Russian OST 110444-87)		Industrial materials	Basic characteristics of the foundations of the developed FPCM	Directions of work	Expected results
No p/p	Name	Abbre-viation			
IV	Materials operating at temperatures of 250 °C and above, with increased stability of piezoelectric characteristics in each temperature range and/or mechanical stress	PZT-21 PZT-26 TNB-1	$\varepsilon^{T}{}_{33}/\varepsilon_0 = 50–1070$; $\tan\delta \cdot 10^2 = 0.1–2.0$; $K_p = 0–0.42$; $\lvert d_{31} \rvert = 0–60$ pC/N; $d_{33}/\lvert d_{31} \rvert \to \infty$; $T_K > 1470$ K; $Q_M = 4–3000$	Increase d_{jg}, K_{jg}	Obtaining materials with a wider range of operating temperatures at higher V_R and lower ρ
V	Addendum to Russian OST 110,444–87. Materials with low dielectric constant and high pyroelectric coefficient for high-frequency converters and pyroelectric receivers	PCR-3 PCR-10 PCR-11 PCR-20	$\varepsilon^{T}{}_{33}/\varepsilon_0 < 100–300$; $K_p = 0.15–0.30$; $\lvert d_{31} \rvert = 6–15$ pC/N; $V_R > 5.5$ km/s; $Q_M = 30–1000$; $\gamma \times 10^4$ C/m^{2}K	Increasing K_{jg}, pyrocoefficient γ	Approximation to industrial materials, but with higher V_R and lower ρ

for glass–ceramic materials; for posistor materials; for high-temperature ferropiezoelectric materials; for semiconductor material; for capacitor materials; for material for ferroelectric substrates; for material for high voltage sources; for electret materials; for pyroelectric materials; for materials for low and medium frequency applications.

The overwhelming majority of the developed materials belong to SS systems based on sodium, lithium and potassium niobates, which is explained by the unique properties of them, which are fundamentally different from lead-containing materials.

The main "attraction" of such objects is that many of them have a unique set of parameters: extremely low dielectric constant, high speed of sound with a wide range of mechanical quality factors (from units to thousands) and low specific gravity (less than 4.5 g/cm^3). This made it possible to develop, based on such materials, a new direction in physical instrumentation: acoustoelectronics, using both ultrasonic waves in the high-frequency range and hypersonic (from 10 MHz to 1.5 GHz) volumetric and surface waves. The fact that the mentioned materials have a specific gravity 1.5 times less than all known Pb-containing FPCMs that has made it possible to use them in devices where weight characteristics are decisive, including making it possible to produce from them lightweight miniature sensors used in various types of control devices for dynamic structures (aircraft). The ultra-high temperatures of the Curie point and the infinite anisotropy of the piezoelectric properties of some of them are associated with the development of flaw detectors and thickness gauges for non-destructive quality control of hot rolled products, as well as expert diagnostic systems for assessing the condition of equipment in various energy complexes (including nuclear reactors) [4]. Despite the obvious advantages, this group of lead-free FPCM (Table 1.2) occupied only a very narrow segment of piezotechnical applications [5] and could not replace the known industrial Pb-containing analogues, since their piezoelectric parameters did not reach the level of the corresponding properties of PZT materials.

In addition, to realize sufficiently high parameters of such lead-free materials, the use of the hot pressing (HP) method (sintering under conditions of externally applied pressure [6]) is most often required. Despite the advantages of this method (expressiveness, versatility, microstructure control), the use of HP involves the use of expensive alundum equipment, which provides a short operating time. Thus, the use of HP technology is advisable only in conditions of piece and small-scale production. In past years, the demand for such materials was determined by the development of high-tech industries and, above all, the defense industry, which made it possible to use almost any expensive methods to achieve the desired parameters. At the present stage of development, the sphere of implementation of electrical innovations is increasingly shifting to the field of development of household appliances.

This process requires new approaches in materials science [7]. Instead of requirements that determine the optimum properties only in the high-frequency range, the need to increase piezoelectric properties in general, including in the middle and low-frequency regions, comes to the fore. At the same time, the requirements for environmental safety of production and production costs are being sharply tightened. Note that, although during operation, PZT ceramics are practically harmless,

Table 1.2 Basic properties of some lead-free FPCM developed at the Scientific Research Institute of Physics of the Southern Federal University

Materials	Parameters								Areas of use
	Conventional							Specific	
	T_C,°C	$\varepsilon_{33}{}^T/\varepsilon_0$	K_p	$\|d_{31}\|$, pC/N	$\|g_{31}\|$, mV m/N	Q_M	V_R, km/s		
PCR-34	420	460	0.42	45	10.0	150	5.4	$K_{15} \sim 0.6$; $\tan\delta \cdot 10^2$ ($E = 50$ V/cm) $= 2.5$	High-frequency acoustoelectric transducers
PCR-35	370	120	0.22	12	10.5	1000	5.9	$V_s = 3.45$ km/s; $K_s = 0.16$; $\alpha_p = 1.3$ dB/cm; (f, MHz); TKZ·10^6, K^{-1}	Surface acoustic wave devices
PCR-90	335	100	0.30	15	16.9	245	6.0	$\|d_{31}\|/\sqrt{\varepsilon_{33}^T/\varepsilon_0} =$ 1.5 pC/N	Highly sensitive accelerometers, flaw detectors, medical diagnostic devices
PCR-61	More than 1200	48	0	0	0	400	5.2	$T_{\text{working limit}} > 950$°C; $K_t = 0.29$; $d_{33} = 14$ pC/N; $K_t/K_p \to \infty$; $d_{33}/\|d_{31}\| \to \infty$	High temperature pressure sensors
PCR-92	320	2700	0.2	189	7.3	360	5.5		Low and medium frequency converters
PCR-100	–	2020	–	–	–	1060	4.7		Low frequency converters
PCR-101	–	1090	0.43	84	203	50	4.3	$\|d_{31}\|/\sqrt{\varepsilon_{33}^T/\varepsilon_0} =$ 6.1 pC/N; $\|g_{31}\| =$ 20.9 mV m/N; $Z = 19.4$ mrayl	Mid-frequency radio-electronic devices operating in receive mode
PCR-102	–	120	0.2	10	45	550	5.8	$g_{33} = 42.6$ mV m/N; $Z = 25.7$ mrayl	High frequency converters
PCR-103	–	150	0.2	10	47	750	5.6	$\|d_{31}\|/\sqrt{\varepsilon_{33}^T/\varepsilon_0} =$ 6.1 pC/N; $\|g_{31}\| = 20.9$ mV m/N; $Z = 19.4$ mrayl	High-frequency converters operating in power modes

their production and disposal require large financial investments to comply with work safety regulations and careful monitoring.

Based on the need to reorient the FPCMs industry in the current economic situation, we previously formulated the task of developing lead-free materials that meet the needs of the main directions of modern industry. The most important stage of work in this direction is the development of relatively inexpensive (Fig. 1.1) lead-free FPCMs that can compete with the SS PZT system in classical applications of ferroelectrics [8].

To implement these plans, the most suitable system is $NaNbO_3$–$KNbO_3$, in which the highest piezoelectric parameters are potentially possible [9–11]. SS of this system are used primarily in microwave technology, which requires the highest possible density, fine grain structure, and low porosity of ceramics. For this reason, the HP method was applied to the $NaNbO_3$–$KNbO_3$ solid solution for the first time in the piezotechnical industry. However, for the widespread use of system materials, a transition to conventional ceramic technology (CCT) is necessary. However, in the case of AMNs, this is hampered by a number of technological difficulties associated with their specificity: hydrolysis of the starting substances during the synthesis process, a narrow range of optimal temperatures for synthesis and sintering, increased reactivity of the reagents, high volatility of AM at relatively low temperatures and in the result

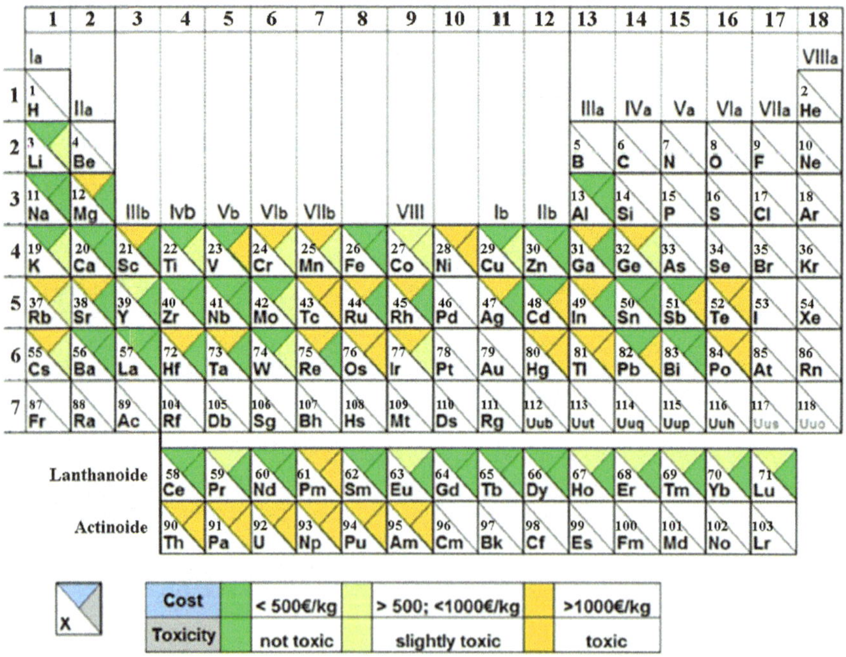

Fig. 1.1 Distribution of chemical elements by degree of toxicity and cost [8]

of the above phenomena, low experimental density and high electrical conductivity of ceramics at elevated temperatures.

In recent years, works aimed at solving this problem have appeared in several international journals. Analysis of statistical data obtained from processing foreign scientific, technical and other literary sources using specialized (www.scirus.com, www.scopus.com, http://ru.espacenet.com) and non-specialized (www.yandex.ru, www.google.ru) electronic search engines, allowed us to identify several patterns illustrated in Figs. 1.2 and 1.3. As shown in Figs. 1.2 and 1.3, the growing interest in lead-free FPCMs is demonstrated according to various sources.

The total number of publications increased most sharply in the period from 2007 to 2009, and with a sharp saturation in the following years. It should be noted that the number of mentions in the central press and patent literature at the current stage continues to grow steadily. Such dynamics are typical for new technologies at the stage of introduction into industrial production, which suggests the prospects of work aimed at expanding the range of lead-free FPCMs. Existing strategies for the synthesis of ferroelectric piezoceramic materials are present in Table 1.3.

Figure 1.4 shows the diagrams illustrating the proportion of the world countries participation in the development and research of lead-free FPCMs by year. If in 1999–2003 the leaders in this direction were the countries of the European Union, which traditionally invest heavily in the development of fundamental science, then since

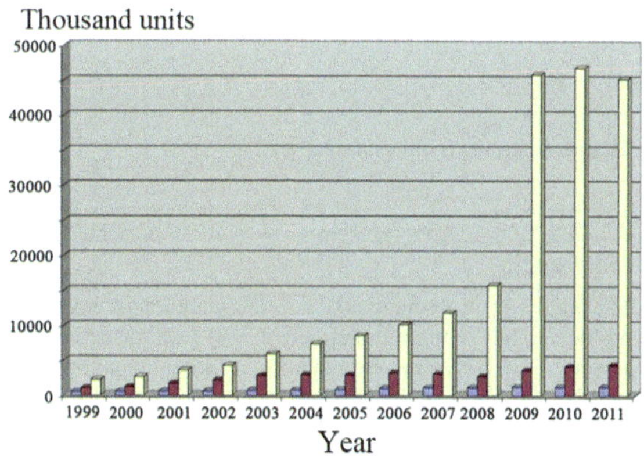

Fig. 1.2 Growth dynamics of references to lead-free ceramics in printed sources: blue—articles in leading world magazines; red—technical and patent documentation; yellow—total number of mentions in printed sources

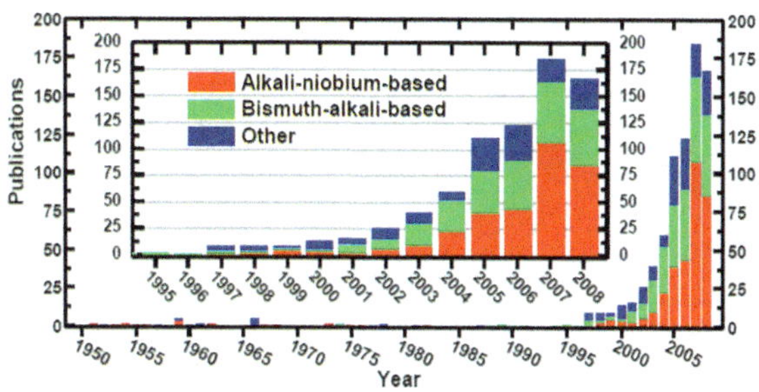

Fig. 1.3 Dynamics of growth in mentions of lead-free ceramics in the central press according to data from [8]

Table 1.3 Strategies for the synthesis of ferroelectric piezoceramic materials

Characteristics	Synthesis strategy		
	Physico-chemical	Technological	Compositional
Period dominance	1950–1970	1970–1990	1990–2010
The main problem to be solved	Achieving maximum piezoactivity	Variation of properties within one chemical composition, control of properties due to manufacturing technology, acceleration of industrial production methods [12]	Creation of multifunctional materials with a combination of properties not realized in monomaterials [13]
Objects	Ferroelectric compounds ($BaTiO_3$, etc.). Binary and three-component SS based on PZT	Multicomponent systems of solid solutions with multidimensional morphotropic regions [14]	Composite materials, porous ceramics, multilayer structures, statistical mixtures
Breakthrough technologies	CCT	Hot pressing [15], wet synthesis methods	Layer-by-layer sintering, slip casting, "coral" technology [16]
Properties optimization method	Physico-chemical analysis	Application of methods for controlling the reactivity of solids [17]	The use of structures with different types of connectivity, control of phenomena at phase boundaries [18]
Characteristics	Synthesis strategy		
	Nanoscale	Meta strategy	Self-organizing
Period of dominance[1]	2004–present	2010–present	Future
The main problem to be solved	Reducing the minimum effective volume of material (miniaturization of piezotechnics) [19]	Creation of environments with electromagnetic properties not found in nature [20]	Creation of adaptive materials with self-regulating properties
Objects	Low-dimensional structures, nanocrystals, thin films [21]	Metamaterials	Complexly ordered materials at the nanoscale level

(continued)

[1] A period of dominance is defined as a specific time frame during which certain research direction exhibited a dominant position.

Table 1.3 (continued)

Characteristics	Synthesis strategy		
	Nanoscale	Meta strategy	Self-organizing
Breakthrough technologies	Topochemical texturing, chemical synthesis methods [22], vacuum thermal spraying, CVD process, PVD process [23]	Creation of artificial composite structures with a complex, controlled, regular structure [24], as well as structures with long-range order, but without periodic properties (icosahedral phases, quasicrystals) [17]	Atomic design, the concept of self-organization of chemical systems [25, 26]
Control	Control of material structure at the nanoscale level	Search for the exact relationship between the resonant properties and the structure of the piezocomposite [27, 28]	–

2004 the most active have been shown by technologically developed Asian countries (Japan, and then China), paying attention, firstly turn, the practical application of scientific developments. This pattern of changes in the diagrams indicates a growing interest in lead-free FPCMs, primarily from the viewpoint of their industrial use. The Russian segment in the diagrams, which was relatively large in 1999–2002, decreased significantly against the backdrop of a rapid increase in the number of publications in 2004–2009, which, on the one hand, indicates the existence of a significant scientific reserve, and on the other hand, about well-known Russian problems, in including at the stage of implementation of developments.

A common problem with the "lead-free direction" is the desire to greatly complicate the chemical composition, which is understandable, since, in essence, it is necessary to bring together the parameters of fundamentally different groups of materials (the situation is reminiscent of a recipe for "axe soup"). In [28–30], for example,

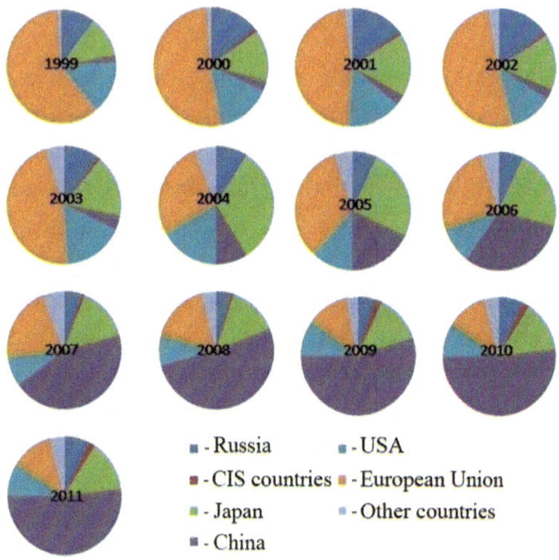

Fig. 1.4 Dynamics of share participation of different countries of the world in the creation and research of lead-free FPCMs

attempts are described to obtain the necessary properties by constructing multicomponent systems. In [31, 32], the use of rather exotic methods for producing AMNs is considered: topochemical texturing and synthesis in a plasma discharge. The data from the above works are quite contradictory and, in addition, the best of the results presented were obtained using labor-intensive, complex techniques that can only be considered as laboratory ones, which also quite naturally indicates the search for ways to solve the problem of lead-free FPCMs within the framework of more modern strategies for the synthesis of FPCMs (Table 1.3).

It is obvious that, despite the rapid development of modern technology to produce lead-free ceramics, the improvement of electrical devices requires the search for fundamentally new approaches to their creation. In this case, the most promising may be:

(i) nanoferroelectrics obtained through the artificial formation of a structure by layer-by-layer assembly;
(ii) metamaterials built on ordered structures of quantum threads and dots [18, 27];
(iii) compositional structures built on the base of the concept of self-organization of chemical systems [17, 25].

This book presents the results of more than 40 years of activity of the Active Materials Department aimed at studying the properties of solid solutions based on alkali metal niobates and creating, on their basis, environmentally friendly ferropiezoceramic materials for a wide range of applications. An analysis of the dynamics of the retrospective developments in the field of lead-free FPCMs in Russia and around the world was carried out. Possible future directions of development are highlighted.

1.2 Lead-Free Ferroelectric Ceramics Based on Alkali Metal Niobates: History, Technology, and Prospects

For more than half a century, piezoelectrics have been under the close attention of researchers. Initially, with the discovery of ferroelectricity in polycrystalline ceramics, due to their high dielectric constant, they were used as materials for capacitors. However, with the expansion of the range of piezoelectric ceramics, particularly the discovery of the PZT system, the range of their applications also grew (sonars, transductors, micropositioning devices, actuators and many others [33]). Nowadays, almost no complex product of modern technology, from an airplane to a cell phone, can do without electronic devices consisting of piezoelectric ceramics. However, the world does not stand still, and a change in the vector of technological development contributes to the emergence of new requirements for the technologies and materials used. Today, among the most stringent requirements, environmental safety requirements take on a special role.

Thus, recently there has been an increase in the efforts of international organizations aimed at displacing toxic substances from areas of the national economy. In particular, the European Union adopted the Waste Electrical and Electronic Equipment (WEEE) and Restriction of Hazardous Substances in Electrical and Electronic Equipment (RoHS) directives in 2003 [2]. While WEEE regulates the disposal, reuse and recycling of such equipment, RoHS defines the necessary conditions to ensure the safety of the environment and human health when operating electronic components containing toxic substances. Mercury, cadmium, hexavalent chromium, several organic compounds and, especially, lead are identified as the main carriers of a threat to environmental safety during recycling, disposal and simply improper use of electronic equipment. Thus, the EU countries point, since June 2006, that any product containing more than 0.1 wt% of the listed substances, except for cadmium, the maximum permissible concentration of which is even lower, namely 0.01 wt%.

Currently, an exception is made for piezoelectrics. This is due to the lack of lead-free ceramics that can replace lead-containing ones in the most important areas of their application.

Even though almost all manufacturers of complex electronic equipment are interested in extending the exclusion of piezoelectrics from objects falling under the jurisdiction of RoHS, this issue is reviewed annually by the European Union Commission to analyze the possibility of switching to the use of lead-free piezoelements. The introduction of lead-containing materials can only take place if it is scientifically proven and technically confirmed that the replacement or elimination of lead-containing components is impossible. However, the use of lead has already been eliminated in almost all new consumer electronics. Thus, the products based on the use of alternative lead-containing materials will dominate the European market in the future.

Despite significant efforts in developing alternative lead-containing materials over the past 10 years and in just over 10 years in hundreds of laboratories around the world, this problem remains completely unresolved. This review is devoted to one of the main directions in the search for highly effective lead-free materials, the objects of which are solid solutions based on sodium–potassium niobates.

1.2.1 Sodium–Potassium Niobate

Sodium–potassium niobates (KNNs) are one of the most studied lead-free ferroelectrics (FEs) systems, second in popularity, perhaps, only to the sodium–bismuth titanate–barium titanate system. KNN is the only lead-free system with a perovskite structure that has a high Curie temperature, higher than in PZT, and piezoelectric properties comparable to it. High chemical inertness and compatibility with human tissues make KNN SS promising for medical applications. Their high electromechanical coupling coefficients and low dielectric constant open the possibility of application in ultrasonic technology.

Since 2005, after the works [1, 32, 34], the number of publications devoted to KNN exceeded the number of works on all other lead-free piezoactive compounds.

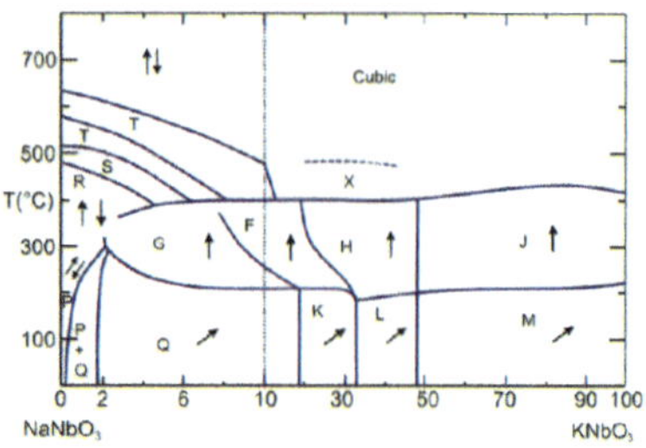

Fig. 1.5 KNN phase diagram. The regions designated Q, K and L are monoclinic (or, as is often indicated in the literature, orthorhombic) ferroelectric phases, M, G are orthorhombic ferroelectric phases, F, H, J are tetragonal ferroelectric phases

However, widespread industrial application is still hampered by several disadvantages. KNN are formed from volatile compounds of alkali metals (AMs), which lead to several technological difficulties in the synthesis and sintering of ceramics [35]. Moreover, the most effective ceramics based on KNN were obtained in the region of the polymorphic transition between the tetragonal and orthorhombic phases. This leads to the occurrence of phase transition fatigue even with slight heating of the ceramics, significantly limiting the range of possible practical applications [36, 37].

KNN is a solid-state system comprising two principal components: orthorhombic potassium niobate and antiferroelectric sodium niobate. The potassium niobate component undergoes a sequence of phase transitions analogous to those observed in barium titanate, albeit at higher temperatures. The sodium niobate component exhibits a multitude of phase transitions of diverse natures. The KNN phase diagram is a complex system comprising multiple thermally induced phase transitions and morphotropic phase boundaries (Fig. 1.5).

At room temperature, the morphotropic regions correspond to 17.5, 32.5 and 47.5% sodium niobate. It is noteworthy that, over a wide range, the transition temperatures between the FE and FE-PE phases do not depend on the ratio of the components. At the same time, even a small addition of potassium niobate to sodium niobate can cause a transition from the AFE to the FE state.

1.2.2 Synthesis of Sodium–Potassium Niobates

As already noted, the main difficulty in obtaining alkali metal niobates (AMNs) lies in achieving high density ceramics. Hydrolysis of alkali metal compounds during the synthesis of solid solution leads to the formation of low-melting, highly volatile, very aggressive media, in particular, alkali metal hydroxides with melting points "lying" significantly lower than the sintering temperatures of ceramics [T_m (KOH) = 400 °C, T_m (NaOH) = 320 °C] and forming a liquid phase (LP) during sintering. Harmful consequences of this: porosity of ceramics due to the volatility of fluxes, low density, interaction with ceramic substrates (alundum, zirconate and others), heterogeneity (stratification) of ceramics along the height of the sample, high conductivity, violations of stoichiometry cause low experimental density of objects and irreproducibility of properties. In addition, the destructive consequence of the increased volatility of compounds is the rapid degradation of technological equipment (substrates, crucibles, boats) [31, 35, 38–41].

To solve the problem of volatility, the introduction of an excess amount of alkali metal oxides is usually used [42], as well as firing and sintering of powders in a controlled atmosphere [43] in closed crucibles, when the mass loss becomes negligible [31], using bedding [35].

The conventional raw materials for the synthesis of alkaline metals are the alkali metal carbonates. They are cheap, readily available, have a higher degree of purity than the corresponding oxides, and upon calcination they completely decompose into the corresponding oxides and carbon dioxide gas. However, one of the main disadvantages of these compounds is their hydroscopic nature [44], which makes them difficult to store and difficult determination of the exact ratios during mixing of raw materials [45].

In recent years, several works have described the achievement of better results using nitrates [46] or bicarbonates [35] as starting compounds, which are characterized by higher resistance to hydrolysis processes, which helps maintain the stoichiometry of a given composition.

Stirring of the starting mixtures for the synthesis of KNNs is usually carried out in ethanol or acetone. Ethanol is the most used as it combines reasonable price and low solubility of ionic (polar) materials such as oxides, carbonates, nitrates. Higher alcohols can also be used to increase the viscosity of the medium, and, consequently, reduce wear on containers and grinding balls, which will also lead to a decrease in product contamination by impurities.

KNNs are very sensitive to deviations from stoichiometry, and even small deviations from a given composition can lead to the formation of impurity phases. They are hygroscopic, which can lead to destruction of samples during firing because of intense evaporation of absorbed moisture [44].

Some reduction in the number of impurities in already synthesized products can be achieved using "wet" chemistry methods. Thus, even in [47], the successful use of washing with a 2% solution of potassium carbonate after firing, which led to the dissolution of non-perovskite impurity phases, was described.

Several works in recent years have been devoted to the use of the mechanical activation method for the synthesis of KNNs. However, this can result in the formation of aggressive alkaline environments that interact with the equipment elements and, as a result, contaminate the synthesis products [48]. Thus, in [49], the synthesis of sodium niobate using mechanical activation in a Retch PM 400 planetary ball mill using yttrium-stabilized zirconium oxide balls with a mass ratio of powder and grinders of 25/1 was reported. After 5 h of stirring, signs of a reaction between the starting components appeared. After grinding for 20 h, the reaction was completed, but reflexes appeared on the X-ray diffraction patterns, indicating contamination of the product by the material of the balls and container. The formation of potassium niobate and KNN SS using the same technique turned out to be completely impossible.

Several authors have reported that attrition grinding is more effective in increasing the density of KNNs than planetary milling. Using abrasive grinding, it is possible to achieve a density of 98.5% of the theoretical, while at the same time, when using a planetary mill, it is possible to achieve a density of no more than 94% of the theoretical [50]. The physical and chemical state of niobium oxide is also of great importance for the technology of niobate ceramics [see Ref. [51] and references therein].

1.2.3 Sintering of Sodium–Potassium Niobates

Due to its simplicity and industrial readiness for its mass use, conventional ceramic technology (CCT) is the most common method for manufacturing KNN system ceramics [52]. This is especially noticeable when it comes to the manufacture of elements specialized for specific devices (for example, multilayer actuators). In this case, CCT has no alternatives at all, since the use of hot pressing [39, 41, 45, 53] or sintering in a spark plasma discharge [54, 55] will inevitably lead to damage of the electrodes, located between the ferroelectric layers in these products.

KNN SS have a melting point (T_m) of about 1120 °C, which limits the maximum sintering temperature of ceramics. In the pre-melting region, anomalous grain growth and non-uniform melting are observed due to the appearance of regions enriched in potassium, which melt at an even lower temperature [43]. It has been established that the heating rate during sintering has a great influence on the density and electrophysical characteristics (coefficient of electromechanical coupling of the planar vibration mode, K_p) of ceramics (see Fig. 1.6). Under these conditions, optimal properties, particularly, the maximum density of conditionally pure KNN ceramics (97.6%) are achieved at a heating rate of 5° per minute. However, technological parameters such as heating rate depend on the particle size and state of the initial reagents and are therefore individual in each specific case. Typically, the maximum achievable density for KNN is between 94 and 96% of the theoretical one [56].

HP is commonly used to further densify KNN ceramics when the density is less than 96% of theoretical density. In [39, 41], it was reported that by hot isostatic pressing of KNN, it was possible to achieve a density higher than 99% of the theoretical one. It was also shown that hot-pressed ceramics have a finer microstructure compared to sintered ceramics according to CCT. At the same time, the electrical characteristics are significantly improved [53, 55].

Sintering in a cold spark plasma discharge is a very effective method of increasing the sinter ability of materials, that is, better compaction of ceramics with minimal grain growth. It was shown that this statement is also true for ceramics based on KNN. According to the results described in [54], samples sintered using this method had a density of 99% of the theoretical one and grain sizes between 200 and 500 nm.

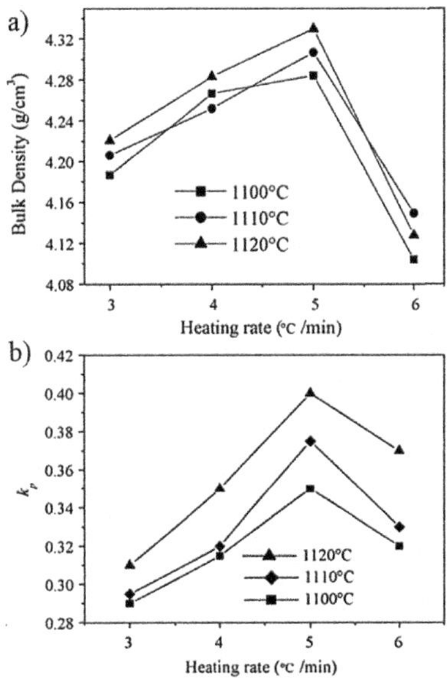

Fig. 1.6 Density (**a**) and K_p (**b**) of KNN ceramics as a function of the heating rate and synthesis temperature

Moreover, the obtained piezoelectric moduli (148 pC/N) were comparable to those achieved in hot-pressed ceramics. However, a significant disadvantage is that firing at 900 °C for several hours requires through conductivity, reduced to an acceptable level, to ensure complete polarization of the sample.

Single crystals can be obtained by liquid phase sintering (LF sintering) [57], if potassium copper niobates are used as LF. Cadmium oxide can also act as a source of LF, but this method contradicts the provisions of RoHS that have already entered into force and similar rules adopted outside the EU [56].

1.2.4 Effect of Modifiers on the Properties of Niobates Sodium–Potassium Modification Method

One of the most effective methods for changing the physical properties of ferroelectrics is the method of introducing small (no more than 10 at%) additives of monoxides of various elements (modification) [58]. The great possibilities of this method are due to the existence of a significant amount of metal monoxides and their combinations that can dissolve in complex oxides, in particular, in FE oxides with a perovskite-type structure (PTS), over a wide range of concentrations, as well as oxides of elements that form when introduced into the original objects "fluxes" (B_2O_3, SiO_2, V_2O_5), and, finally, additives of combined action that influence the properties of the original systems due to liquid phase and cation-exchange interaction with the base (modifying glasses). The undoubted advantage of the method is the ability to: while maintaining a set of specific properties of the modified objects, by insignificantly varying their composition, selectively change the values of some parameters in the desired direction.

In the case of a significant change (improvement) in the ceramic and technological characteristics of the modified materials, as well as such changes in the dielectric and piezoelectric parameters that can be considered as "tuning" them to target values, the modification is represented as a method of optimizing the parameters. Changes in these characteristics to such an extent that leads to the formation of a new set of properties allow us to consider modification as a method of creating qualitatively new materials with adjustable parameters.

From a theoretical viewpoint, the establishment of general patterns of changes in physical properties when modifying PTS provides information about the composition–structure–property relationships, which allows us to consider it as a method for studying the ferroelectric state. Thus, the results of modification have a wide range of applications, namely from fundamental research to technological processes at the atomic-molecular level. In this regard, the relevance of the problem of modifying FE PTS becomes clear.

As bibliographic data shows, the modification of SS based on the PZT system has been most fully studied. At the same time, the possibilities for modifying the system itself are currently practically exhausted. Multicomponent lead-containing solids have also been examined in sufficient detail. Analysis and generalization of enormous experimental material (more than 300 variants of combinations of modifying oxides are known for the PZT system alone) made it possible to classify all known modifiers of these systems and, on this basis, to establish general patterns of changes in their physical properties [59]. This information was successfully used to predict the properties of FPCMs and search for optimal ways to create them [12].

The possibilities of modifying SS based on AMNs have been studied in more detail in [58, 60–62]. A number of works describe the influence of various modifiers, including ions Pb^{2+} [63–66], Cd^{2+} [64–66], Mg^{2+} [67–70], Ca^{2+} [67–70], Sr^{2+} [67–70], Ba^{2+} [67, 68, 70], Bi^{3+} [63], La^{3+} [63], Ti^{4+} [64–66], Ge^{4+} [64–66] on the properties of the binary system $(Na_{1-x}K_x)NbO_3$. The effect of some of them (Mg^{2+}, Sr^{2+}, La^{3+}, Sb^{3+}) and modifying glasses on the ceramic and electrical properties of the solid solution of the binary system $(Na_{1-x}Li_x)$ NbO_3 is considered in [61, 71, 72].

KNN modifiers can be divided into two groups. The first group includes modifiers that affect the behavior of the material during sintering and do not change the phase diagram of the system. They are used to improve specific properties such as piezoelectric moduli and coupling coefficients. The second group includes modifiers, the introduction of which leads to a shift in the transition temperature between the tetragonal and orthorhombic phases (200 °C in pure KNN [73, 74]) closer or further from room temperature. In [75], based on the results of the influence of various impurities on KNN, it is concluded that changing the processing conditions has a comparable effect on the piezoelectric and mechanical properties of KNN as the introduction of modifiers. This conclusion is confirmed in [38, 76].

1.2.5 Lithium Modification

Modification of KNN by Li is usually aimed at shifting the O–T phase transition. Thus, in [32], the authors, using Li, Ta and Sb as additives, obtained ceramics with a piezoelectric modulus d_{33} > 200 pC/N (without texturing) and d_{33} > 300 pC/N (with texturing) with a Curie temperature of 253 °C. Subsequently, these data were supplemented in [30, 77–79]. Before work [32], there were only two reports in the press devoted to the modification of KNN with lithium-containing modifiers [31, 80].

Later work was aimed at studying the dependence of the phase filling of the final product on the chemical composition and optimizing the processing conditions of ceramics [81]. Typically, d_{33} values of about (200–300) pC/N are achieved at a Curie temperature of 350 °C. An increase of 100 pC/N in the piezoelectric module usually corresponds to a decrease in the Curie temperature by 100 K.

Similar systems with different compositions of Li-containing components ($LiTaO_3$, $LiNbO_3$, $LiSbO_3$ or Ta alone) were studied in detail to optimize the microstructure by varying sintering conditions. The effect of pure Sb_2O_3 on the properties of KNN has not yet been described. The effects first used in [32] generally boil down to the following: the introduction of tantalum prevents the anomalous growth of grains and reduces the Curie temperatures and T–O transitions. Lithium, as already noted, reduces the temperature of the T–O transition to room temperature when introducing 6 at% into the *A*-sublattice and increases the Curie temperature due to compaction. At the same time, it is possible to achieve an increase in the piezoelectric modulus to 300 pC/N at a Curie temperature of 350 °C only by introducing lithium [82]. The effect of $LiSbO_3$ is like $LiNbO_3$ and allows, with the introduction of 1 mol% shift the temperature of the T–O transition to room temperature [83, 84]. Even though the results [32] are encouraging, there are several controversial issues, in particular, the discrepancy between the position of the morphotropic region: according to the original work, the morphotropic boundary is located near 5 mol% lithium tantalates, according to later data—at 6–7 mol% [31, 85].

Subsequently, it became obvious that other aspects, such as low temperature stability of polarization and insufficiently high piezoelectric characteristics [36, 37], the complexity of the production process and the high cost of tantalum oxide would hinder the development of large-scale industrial production. It was shown in [37] that thermal cycling along the O–T phase transition leads to a decrease in piezoelectric properties by approximately 30%. At the same time, many applications require frequent heating (sterilization of medical instruments at 140 °C). The introduction of 6% lithium reduces the phase transition temperature [86]. The introduction of an even larger amount of lithium, for example, 7%, leads to a further decrease in the temperature of the O–T transition to a region below room temperature, which eliminates the negative influence of the O–T transition during thermal cycling. However, the introduction of additional lithium leads to a decrease in the piezoelectric properties to a level close to the maximum achievable during thermal cycling of materials of optimal composition. Considering that the Curie temperature of such a material increases to 480 °C, we can conclude that there are no undesirable phase transitions in it over a wider region than, for example, in the SS of the PZT system [87]. The authors of [81] studied the behavior of KNN modified with lithium using high-temperature impedance spectroscopy and force electron piezomicroscopy and found an increase in piezocontrast in the modification region of 6.5% lithium, which apparently corresponds to the maximum piezoelectric properties. The work [88] describes unipolar deformations in the KNN SS, the values of which are close to those known for PZT. The authors also tried to reproduce the extreme values of d_{33} described in [32], however, they failed.

In [89], transparent electro-optical lead-free ceramics $(K_{0.5}Na_{0.5})_{1-x}Li_xNb_{1-x}Bi_xO_3$ were obtained by sintering using the HP method. Due to the grain growth suppression effect of joint modification with lithium and bismuth, a dense, fine-grained microstructure is usually formed. Simultaneous modification also contributes to the implementation of a structure close to cubic with minimal optical anisotropy. This can help reduce light scattering on grains, their boundaries, and domain walls and, consequently, increase the optical transparency of ceramics. For ceramics modified with 5 mol% of these ions, the optical transparency was 60% in the region close to the infrared. Ceramics also had a high linear electro-optical coefficient.

The simultaneous influence of Li and La on the structure and electrical properties of ceramics was studied in [90]. X-ray diffraction analysis showed that the samples have a perovskite-type structure with orthorhombic distortion. The existence of only a small impurity of a foreign phase (3% of $K_6Li_4Nb_{10}O_{30}$) was discovered. The temperature of the FE-PE transition decreases with increasing content of modifiers. Lanthanum doped ceramics had a relatively low dielectric loss tangent in the range from room temperature to 350 °C. Detailed studies of electrophysical and dielectric characteristics have shown that the Curie temperature shifts to lower temperatures with increasing concentrations of lithium and lanthanum. Moreover, an even sharper decrease in the maximum dielectric constant, blurring of the phase transition and weakening of the dielectric properties are observed. For ceramics with $x = 0.065$, it can be explained by the formation of pseudocubic structures with a high degree of lithium and lanthanum incorporation. Conductivity in a constant field increases with increasing temperature, which indicates a negative temperature coefficient of electrical conductivity. The low activation energy obtained for ceramic samples is associated with the electronic contribution to the electrical conductivity of the samples. Temporary aging of ceramics was studied in [91].

It was shown in [92] that $(K_{0.5}Na_{0.5})_{0.97}Bi_{0.01}(Nb_{1-x}Zr_x)O_3$ can also be considered as a promising material. This ceramic is produced using conventional solid-phase technology. The structure of ceramics was studied in the context of the influence of zirconium content on it: it was found that with an increase in its concentration, ceramics undergo a phase transition from the pseudocubic to the orthorhombic phase. The best piezoelectric properties were observed in ceramics polarized at 100 °C, since the orthorhombic and tetragonal phases coexist in this region. It has been shown that when doped with zirconium, the piezoelectric properties improve and reach the optimal value at a Zr content of 0.02 mol% ($d_{33} = 161$ pC/N, $K_p = 0.41$, $Q_m = 81$, $T_C = 370$°C, $T_{T\text{-}O} = 130$°C).

1.2.6 Compounds of Copper

Copper compounds have proven to be effective additives when introduced in the form, for example, $K_4Cu(Nb/Ta)_8O_{23}$ [93]. It has been shown that the number of additives should remain within 0.5%, otherwise degradation of the piezoelectric properties occurs [94]. The introduction of CuO also makes it possible to reduce the sintering temperature to less than 950 °C [95, 96]. Copper has also been used to modify more complex KNN-based systems with similar effects [97, 98]. The influence of CuO on the microstructure and electrical properties of ceramics was also studied by us in [35]. Within the studied concentration ranges, the introduction of copper leads to a simultaneous decrease in the temperatures of the FE and T–O transitions.

1.2.7 Double Copper Niobate $CuNb_2O_6$

One of the most promising materials that have been actively studied in recent years are multicomponent solid solutions based on AMNs containing double copper niobate [99, 100]. In this case, quite promising results were obtained: mechanical quality factor $Q_m > 1000$ and electromechanical coupling coefficient $K_p > 0.42$, allowing the use of these materials in ultrasound generators commonly used in medicine (ultrasound therapeutic systems, plastic surgery, prosthetics [100]) and household appliances [99–102]. The authors of the works associate the observed effects with the formation of liquid phases based on copper oxide compounds. However, such high Q_m values were observed only when using $CuNb_2O_6$, which probably indicates the importance of this compound in the formation of the mechanical properties of the structure. During the production of the SS system $(1 - x)NaNbO_3$–$xCu_{0.5}NbO_3$, even at relatively small x, complete dissolution is not observed: impurity phases are formed [103], which indicates an extremely narrow range of existence of the SS.

1.2.8 Alkaline Earth Metal Titanates and Zirconates

It was shown in [104] that the introduction of calcium titanate almost eliminates electrical fatigue in SS based on the KNN system and leads to a decrease in the temperature dependence of piezoelectric characteristics.

The introduction of strontium titanate into KNN initiates the relaxor behavior of SS [31, 105] with high dielectric constants (over 3000 for titanate and over 5000 for strontium zirconate). At the same time, such ceramics have greater stability of properties during thermal cycling than lanthanum-modified ceramics of the PZT system. Similar results were obtained when modifying KNN $BiScO_3$ [106]. This relaxor effect is also observed in solid solutions modified with $CaTiO_3$ [107], but it is much less pronounced, and the dielectric constant in such ceramics does not reach values greater than 1000.

In [108], the morphotropic phase boundary between the orthorhombic and tetragonal phases in solid solutions modified with $SrTiO_3$ is in the region with 4% modifier content. The authors also found that modified SS have a piezoelectric modulus d_{33} of about 100 and a coupling coefficient K_p from 0.266 to 0.325, depending on the Curie temperature, which sharply decreases with increasing strontium titanate content. These observations were further confirmed in [109, 110], where KNN with 6% lithium was modified with strontium titanate. A significant reduction in piezoelectric modulus and Curie temperature with the addition of 0.5% strontium titanate to KNN was reported in [107]. Despite the high relative density (98.4%), the values of residual polarization, K_p, T_C and d_{33} are reduced by (5–15) %.

In [111], lead-free ceramics with various additions of calcium titanate (CT) were prepared by the conventional solid-phase method. Experimental results indicate the formation of SS at additive contents of up to 7 mol%. An analysis of the dependencies of dielectric constant on temperature showed that with increasing calcium titanate content, the Curie temperatures and T–O transition temperatures monotonically decrease. It is shown that the morphotropic phase boundary in the studied system (KNN-CT) is located at $x = 0.03$ and the corresponding compositions have optimal properties: high relative density (98% of theoretical) with a grain size of 1–2 μm, excellent piezoelectric properties: $d_{33} = 117$ pC/N, $K_p = 0.39$, $P_r = 21$ μC/cm^2, $T_C = 333$ °C.

In [112], lead-free $Na_{1/2}K_{1/2}NbO_3$ ceramics, modified with alkaline earth metals Ca^{2+} and Ba^{2+}, were prepared by the conventional solid-phase reaction method. Disks doped with 0–2 mol% barium and calcium were sintered at 1125 °C for 2 h. The authors found that the introduction of more than 0.5 mol% of alkaline earth elements as additives led to a sharp change in grain size and microstructure, in general. Introduction of alkaline earth additives in the range of up to 0.5 mol% led to an increase in the ferroelectric softness of the ceramics (an increase in the piezoelectric modulus and a decrease in the coercive field), but with deeper modification, the characteristics deteriorated.

Purely electrostrictive behavior has been described in SS modified with 15% or more strontium titanate [113]. The electro-optical properties of KNN with 20% strontium titanate were studied in [114]. These SS exhibit a pseudocubic structure and relaxor behavior [105]. Mixing with 33% strontium zirconate leads to the formation of a single-phase FE relaxor [115].

It was shown in [116] that the design of ceramics based on compositions from the region of coexistence of rhombohedral and orthorhombic phases is an effective way to increase the piezoelectric activity of KNN-based ceramics. Additions of barium and tantalum zirconate were used to improve the piezoelectric properties of lead-free ceramics with the formation of the composition $(1-x)K_{0.48}Na_{0.52}(Nb_{0.95}Ta_{0.05})O_3$–$xBaZrO_3$ [$(1-x)$KNNT–xBZ] by changing the position of the phase borders. It has been established that the rhombohedral and orthorhombic phases coexist in the x range from 0.05 to 0.07. With an increase in the content of strontium zirconate, the Curie temperature and the temperature of the O–T polymorphic transition sharply decrease, and the dielectric constant increases almost linearly. Ceramics with $x = 0.06$ have maximum electrical characteristics ($d_{33} = 193$ pC/N and $K_p = 32.6\%$), since they are in the two-phase region and have a dense microstructure.

It was found in [117] that the introduction of $Ba_{0.8}Ca_{0.2}ZrO_3$ can improve the properties of KNN ceramics. Lead-free ceramics $(1 - x)K_{0.5}Na_{0.5}NbO_3$–$xBa_{0.8}Ca_{0.2}ZrO_3$ was obtained by the conventional solid-phase method, and the effect of $Ba_{0.8}Ca_{0.2}ZrO_3$ on its structure and piezoelectric properties was studied. The coexistence of the rhombohedral and orthorhombic phases was established in the x range from 0.04 to 0.08. With an increase in the $Ba_{0.8}Ca_{0.2}ZrO_3$ content, the grain size of the ceramic decreases and the Curie temperature sharply decreases. Optimal piezoelectric properties ($d_{33} = 197$ pC/N, $K_p = 40.6\%$) were obtained in compositions lying in the two-phase region.

1.2.9 Solid Solutions with BNT and Other Titanates

KNN has a morphotropic phase boundary between the orthorhombic and tetragonal phases in ternary systems with lead, strontium and barium titanates at a content of approximately 5%. In the system with barium titanate, the morphotropic region is in the vicinity of 6% barium titanate. The solubility limit of barium titanate was found to be in the region of 20%. Piezoelectric properties decrease with the introduction of barium titanate. An increase in density and a decrease in grain size in these compositions is described in [118] and confirmed in [119], but at the same time contradicts the results of [120]. The KNN ternary system SS with sodium bismuth titanate (BNT) are orthorhombic up to 2% BNT, then tetragonal up to 9% BNT and cubic at higher concentrations. In morphotropic compositions with 2% BNT, d_{33} is 195 pC/N, at a Curie temperature of 395 °C [121]. BNT promotes the manifestation of relaxor properties with low temperature stability of dielectric constant.

KNN SS with barium titanate, as well as double sodium-bismuth and potassium-bismuth titanates are characterized by electromechanical coupling coefficients not exceeding 0.15 and depolarization temperatures not exceeding 100 °C [122–124]. The additional introduction of bismuth and scandium oxides leads to the formation of a pseudo-ternary system, but at the same time, with an increase in the coercive field, the depolarization temperatures remain below 200 °C [125]. The potassium niobate–barium titanate system is characterized by relaxor behavior [126]. It is reported [127] that 5 mol% barium titanates when introduced into KNN causes the appearance of a morphotropic phase boundary. The sintering of these materials is improved due to the formation of liquid phases at temperatures above 1060 °C, but this, unfortunately, is accompanied by anomalous grain growth [119]. Mixing with other ferroelectrics at a relatively low content, for example, $BiFeO_3$, $Bi_{1/2}K_{1/2}TiO_3$, BNT, or BT, also leads to the manifestation of properties characteristic of compositions at the O–T transition boundary but does not lead to the formation of new phase boundaries.

1.2.10 Other AMN Modifiers

Silver oxide, as well as silver in the form of Ag(Nb/Ta)O_3 solution, leads to an increase in the Curie temperature of KNN modified with Li, Ta and Sb, moreover, more strongly than with the introduction of $LiNbO_3$, which is associated with a sharp decrease in the temperature of O–T transition [128–130]. Substitution of 30 at% niobium with tantalum enhances electrostriction and provides 0.11% unipolar strain at 4 kV/mm. Manganese helps reduce electrical conductivity losses, reduces dielectric constant, but helps increase Q_m and K_p [131].

Doping 0.5 mol% of alkaline earth metals shows that strontium and calcium improve the sintering properties of ceramics, while magnesium oxide, on the contrary, prevents it. All of them contribute to a decrease in the temperature of the O–T transition; it is noted that barium oxide leads to the formation of impurity phases, and magnesium reduces the relative density more than all other modifiers.

Zinc oxide and tin dioxide promote sintering, while tungsten, tin, scandium and cerium oxides inhibit grain growth. All the listed compounds, except tin dioxide, prevent compaction of ceramics [56]. However, paper [132] states that zinc oxide promotes compaction and prevents the common cracking of KNN ceramics. As stated in [133], zirconium inhibits grain growth. Boron oxide helps to increase the phase stability of KNN at high temperatures [134], although it leads to an increase in the sintering temperature. Iron oxide and lanthanum ferrite were used to reduce the sintering temperature of pure potassium niobate [135–137].

In [138], $(K_{0.5}Na_{0.5})_{0.95}Li_{0.05}Sb_{0.05}Nb_{0.95}O_3$ ceramics were obtained using CCT. The solid solutions of the indicated compositions had a perovskite-type structure with tetragonal distortion at $x < 0.0075$ and pseudocubic distortion at $x = 0.010$. The Curie temperature decreased with increasing $SmAlO_3$ content. Moreover, the introduction of samarium aluminate led to a significant expansion of the sintering temperature range. Ceramics with $x = 0.005$ had optimal electrical properties (d_{33} = 226 pC/N, K_p = 38%, dielectric loss tangent 3%, Q_m = 60, Curie temperature 327 °C) and represented a promising candidate for piezoelectric applications.

In [139], the effect of firing temperature on the piezoelectric properties of (Na, K)NbO_3–$LiTaO_3$ ceramics modified with Ag was studied. The samples were obtained using solid-phase synthesis and CCT sintering. The crystal structure and microstructure of these lead-free piezoelectric ceramics fired in the range from 750 °C to 950 °C were studied using X-ray diffraction and scanning electron microscopy. Electrophysical and dielectric properties were studied in the range of 1 kHz–1 MHz.

In [140], 0.992 $(0.95K_{0.5}Na_{0.5}NbO_3$–$0.05LiSbO_3)$–$0.008BiScO_3$ (KNN–LS–BS) ceramics were successfully prepared by the thermal decomposition of organometallic compounds. The highest content of a phase with a perovskite structure was observed in samples fired at 700 °C for 1 h. The samples had an orthorhombic structure.

However, samples sintered at temperatures from 1025 °C to 1100 °C had tetragonal distortion of the unit cell. The microstructure of sintered ceramics was represented by square or rectangular grains, and the grain size increased with increasing sintering temperature. The density of the ceramics increased with increasing sintering temperature to 1075 °C and amounted to 97.5% of the theoretical value at this point, and then sharply decreased with a further increase in temperature. In the densest ceramics, excellent electrical characteristics were obtained: dielectric constant after polarization = 6600 (at the Curie temperature), tanδ = 0.04 (at the Curie temperature), residual polarization = 19.4 μC/cm^2 and coercive field = 24.1 kV/cm (when a voltage of 40 kV/cm is applied).

In [141], using $Li_{0.08}(Na_{0.52-x}K_{0.48})_{0.92}NbO_3$ ceramics as an example, the possibility of controlling the polymorphic phase transition between the tetragonal and orthorhombic phases by changing the sodium content was studied. Grain growth accelerated with the introduction of sodium additives, leading to an increase in grain size from 2–3 μm to 7–8 μm and the appearance of a bimodal structure for samples with $x = 0.015$–0.04. The optimal piezoelectric properties were $d_{33} = 226$ pC/N and $K_p = 36.8\%$ in samples with $x = 0.04$. In this case, the Curie temperature, residual polarization and coercive field were 468 °C, 25.5 μC/cm^2 and 1.2 kV/mm, respectively.

The fact that there are still no methods for increasing the temperature of the O–T transition: the introduction of all known modifiers leads to the opposite effect. A few compounds, for example, $LiNbO_3$ and $AgNbO_3$, only contribute to an increase in the Curie temperature of KNN, forming a purely tetragonal material with obviously lower piezoelectric characteristics. Figure 1.7 illustrates the influence of the position of the T–O transition on the value of d_{33} [142, 143]. Maximum piezoelectric properties appear when the temperature of the T–O transition of an optimally synthesized material is 30–50° higher than its operating temperature. Mixtures with other ferroelectrics and solid solutions based on them also do not lead to a change in the nature of the dependence of the piezoelectric properties on the position of the T–O transition. Texturing, at present, is the only way to increase piezoelectric properties, without complicating the composition and state of objects in the low-temperature region.

The highest values of the piezoelectric modulus at room temperature are observed in materials with a T–O transition temperature between 55 °C and 70 °C. It is clearly seen that on the tetragonal side the piezoelectric modulus decreases faster than on

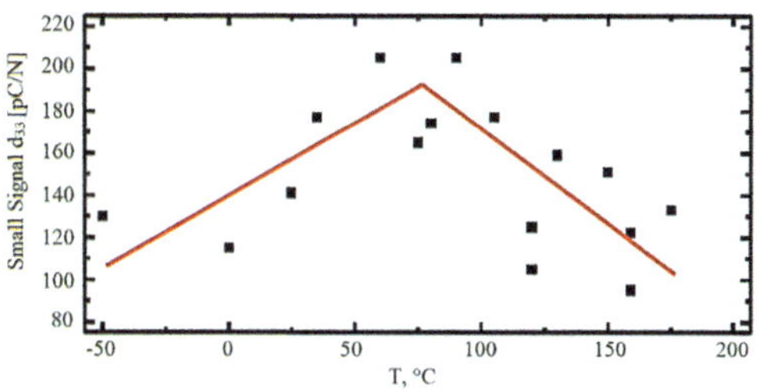

Fig. 1.7 Dependence of the piezoelectric modulus d_{33} on the temperature of the T–O transition in KNN modified with lithium

the orthorhombic side. The wide range of values is apparently due to the loose interpretation of the term "room temperature" in various laboratories around the world, which usually means temperatures from 17 to 30 °C [7].

So, an analysis of the technological aspects of creating highly efficient piezoelectric ceramic materials based on alkali metal niobates has been carried out. The experience of using basic technological techniques aimed at optimizing the properties of ceramics is summarized. The most promising directions for modifying this group of materials have been identified. Ceramic materials based on KNN are promising for the creation of new lead-free piezoceramic materials, while the selection of production conditions and modifying additives can significantly expand the range of possible practical applications, covering almost their entire range.

1.3 Mesoscopically Inhomogeneous Media with Special Electrical and Magnetic Properties: Background, Main Problems and Application Prospects

The basis of the vast majority of modern micro- and nanoelectronics devices are solid-state media that are complexly organized at the macroscopic, microscopic and mesoscopic levels with special electrical and magnetic properties. Moreover, the formation of their parameters occurs because of a complex sequence of physicochemical processes (phase transitions, diffusion and percolation effects, adsorption and desorption phenomena, hydrolytic and electrolytic reactions). Each of the listed phenomena leads to qualitative and quantitative changes in the entire set of characteristics. This determines the need considering thermodynamic prehistory as a series of physical phenomena and conditions for their occurrence, forming the qualitative originality of macro-responses of material environments, construction and detailed study of phase diagrams of states.

Among the latter, multi-element compositions are the most promising, especially for applications in critical conditions (nuclear energy, telecommunications technologies, aircraft and rocketry, medical technology), that have a wide variety of electrophysical parameters and provide greater opportunities for "design" of the properties of solid solutions. In addition, in recent years, environmentally friendly lead-free ceramics, multicomponent solid solutions with a special character of manifestation of ferroelectric properties, for example, ferroelectric relaxors and media, combining ferroelectric and magnetic properties, have become of particular importance.

Over more than half a century of history, the evolution of ferroactive media has proceeded in two complementary directions (Fig. 1.8): (i) complication of the "phase composition" (1) and (ii) complication of the "chemical composition" (2).

Within each of them, methods for constructing objects with the necessary combination of physical characteristics were developed. In recent years, due to the active implementation concepts of nanoscale structures, metamaterials and additive technologies, significant progress has been achieved within the framework of direction

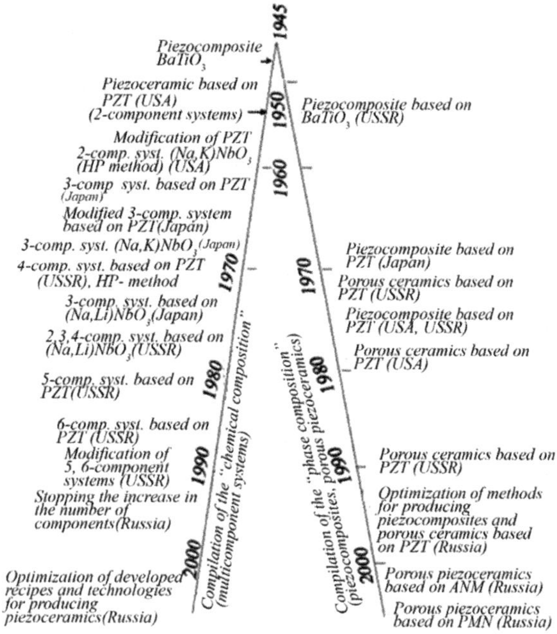

Fig. 1.8 Development of materials science in the physics of ferroelectricity

(1). However, the harmonic development of the scientific and technical basis requires additional efforts in the second direction, within the framework of which the main vector of the presented work is built.

The most promising in this regard are: (i) alkali metal niobates; (ii) multiferroics based on bismuth ferrite; (iii) ferroelectric relaxors. The diagram (Fig. 1.9) shows the problems-goals of creating each of these groups and the main disadvantages, to overcome which, in many respects, our work is devoted. However, all the listed objects are not without drawbacks. Let us consider the origin of some of them using the example of alkali metal niobates, considering their thermodynamic history.

Since the main structure-forming component in the case of alkali metal niobates is niobium oxide, it has a significant effect on their structure and properties. The effects of niobium oxide can be described by the following:

1. *Physico-chemical state of* Nb_2O_5*:*

 (i) effect of cation–anion composition on formation of low-melting impurities leads to changes in the kinetics of sintering of solid state (from solid-phase to liquid-phase), the nature of recrystallization (from continuous to intermittent secondary), grain growth conditions (from isotropic solid phase to anisotropic, when the resulting liquid phase, acting as an active solvent, favors the crystallization of large grains from the melt), their final shape (from equiaxed to ideomorphic) and packaging (from dense to loose);
 (ii) effect of polymorphic composition on a high difference in the activation energies of various modifications of Nb_2O_5 leads to the formation of impurities of

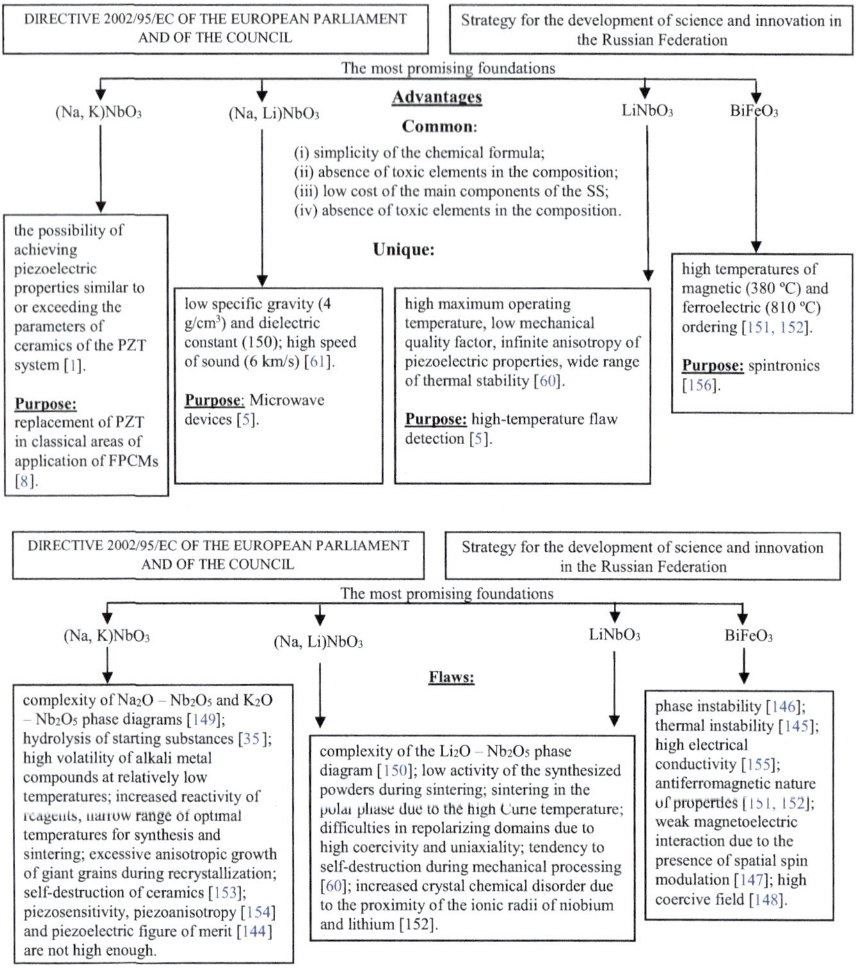

Fig. 1.9 Promising functional materials of modern piezotechnics, micro- and nanoelectronics, spintronics

foreign phases, significant differences in the crystal structure and properties of the compounds;

(iii) effect of granulometric composition on a significant increase or decrease in the specific surface area of Nb_2O_5 from the standard (950–1000 m^2/kg) leads to a decrease in the strength of ceramic cakes;

(iv) effect of rheological composition on needle-shaped grains, which determines a highly developed hydration surface and degradation as a raw material component.

2. *Variable oxidation state* on high electrical conductivity, thermal instability, the occurrence of additional anomalies in dielectric characteristics when exposed to an electric field and temperature.
3. *Tendency to form defective block structures* on quasi-composite nature at the mesostructural level, Maxwell–Wagner type polarization.
4. *Coordinatively unsaturated nature* on the formation of implantation type defects, the absorption of impurity elements from the external environment, the formation of irregular solid solutions.

There are currently no standards to consider all these phenomena. The next series of factors influencing the formation of the structure and properties of niobate ceramics arises at the level of binary oxide systems (Fig. 1.10a, b). Here we can highlight: (i) rich phase content of the diagrams of the corresponding binary systems; (ii) low melting temperatures of several formed compounds and their increased chemical activity.

The following series of phenomena affecting the processes of phase formation in these materials can be considered when analyzing the characteristics of compounds with the general formula $A\mathrm{NbO}_3$. In this case, it is necessary consider the effects:

(i) discrepancy between the crystallographic characteristics of alkali metal ions and NbO_6 octahedra on increased defectiveness of the solution (Fig. 1.11);
(ii) presence of vacant octahedra on absorption of impurity ions;
(iii) the proximity of the ionic radii of Nb^{5+} and Li^+, which determines the crystal chemical disorder;
(iv) sintering in the polar phase of $LiNbO_3$;
(v) hydrolysis of alkali metal niobates on increased electrical conductivity.
(vi) polarization electrolysis;
(vii) high coefficients of thermal expansion and their anisotropy, leading to self-destruction of ceramics;
(viii) intermittent secondary recrystallization, leading to self-destruction and the formation of single-crystal-ceramic composites;
(ix) reduced reactivity in some cases on increased duration of synthesis;
(x) high volatility of alkali metal compounds on violation of stoichiometry.
(xi) narrow sintering interval.

In addition, when analyzing phase formation processes in these objects, one should consider the features of multicomponent systems, based on them:

(i) complex morphology of phase diagrams (Fig. 1.12);
(ii) in many cases, the range of existence of solid solutions is limited;
(iii) smaller (compared to PZT) possibilities for varying properties, limiting the possible range of practical applications due to the relatively small dipole moments of unit cells.

Thus, it is relevant to establish the patterns of formation of macroclines of multicomponent mesoscopically inhomogeneous media with special electrical and

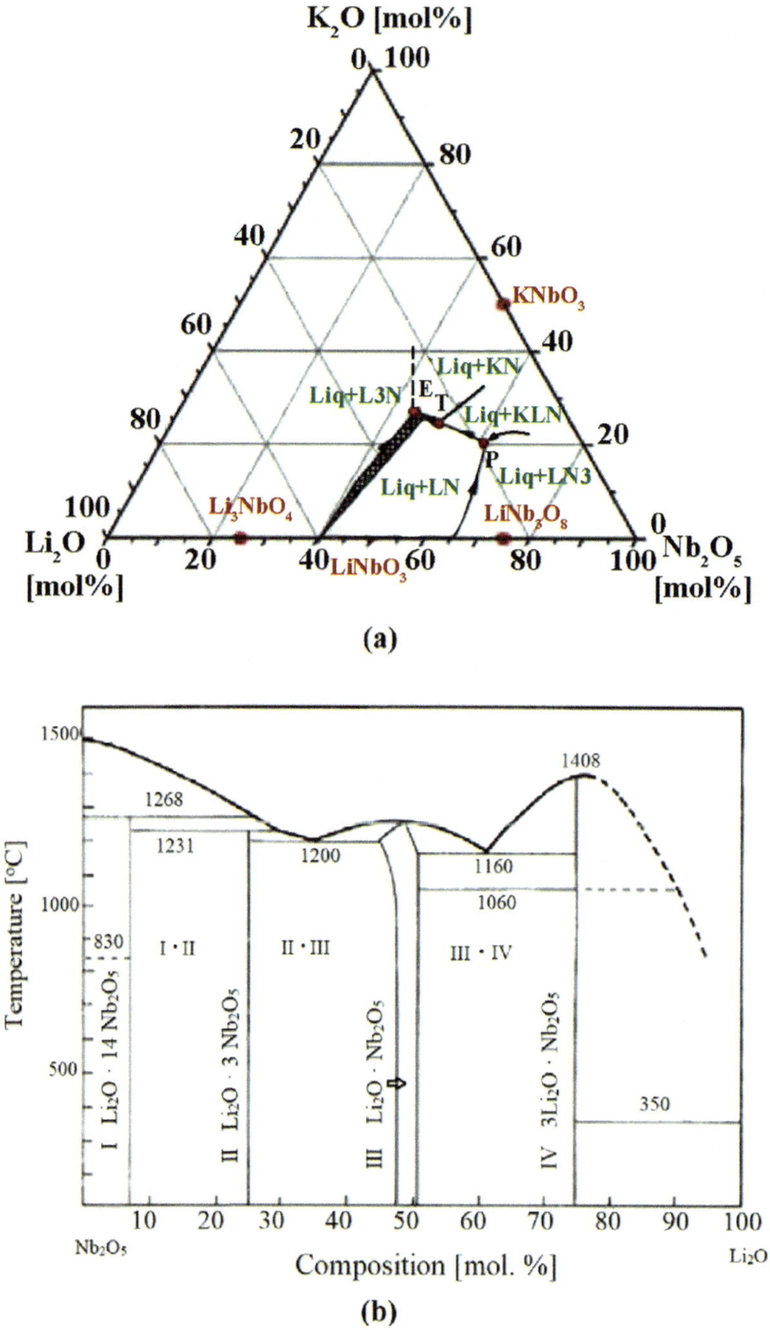

Fig. 1.10 **a** Phase diagram of the ternary system K_2O–Li_2O–Nb_2O_5; **b** fusibility diagram of the binary system Nb_2O_5–Li_2O

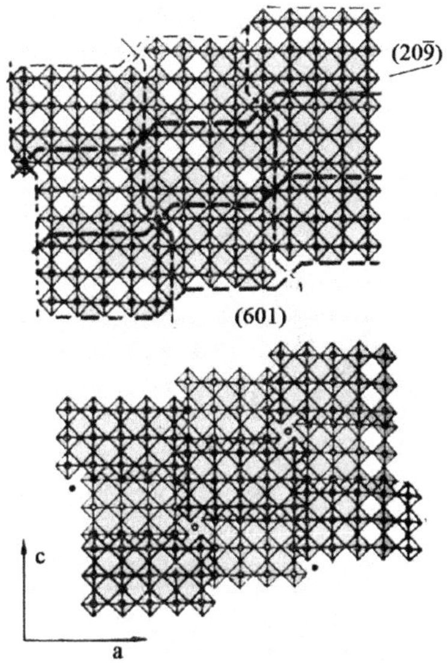

Fig. 1.11 Scheme of the formation of extended defects based on crystallographic shift in oxygen-deficient structures

magnetic properties based on the analysis of their equilibrium and nonequilibrium phase diagrams and the thermodynamic path along which the specified state parameters are achieved.

1.4 Lead-Free Ferropiezoelectric, Multiferroic Materials and Metamaterials Based on Spatially Inhomogeneous Media: Evolution of Synthesis Strategies

In spatially inhomogeneous media (ceramics, composites, porous heterogeneous systems, etc.) with a periodic crystal structure, the thermodynamic parameters of the state change several critical and subcritical phenomena (reconstructive, phase and percolation transitions) [157–159]. Moreover, in the result of the entire set of preceding transitions, successively emerging phases are formed in a crystalline medium, structured not only at the microscopic, but also at the meso- and macroscopic levels. This causes the appearance of structures formed by coexisting phases with a complex mesoscopic pattern, the characteristics of the elements of which largely determine the extrema of the properties of crystalline substances in areas of structural instability. For real objects, this picture is even more complicated due to violations of the stoichiometry of the composition, the effects of the interaction of intrinsic, biographical and deformation defects, as well as phase transition "fatigue", several nonequilibrium processes accompanying the synthesis of ordered crystalline phases. With a broader approach, we can talk about the problem of the formation of phase states, considering the crystal physical features of substances, as well as the

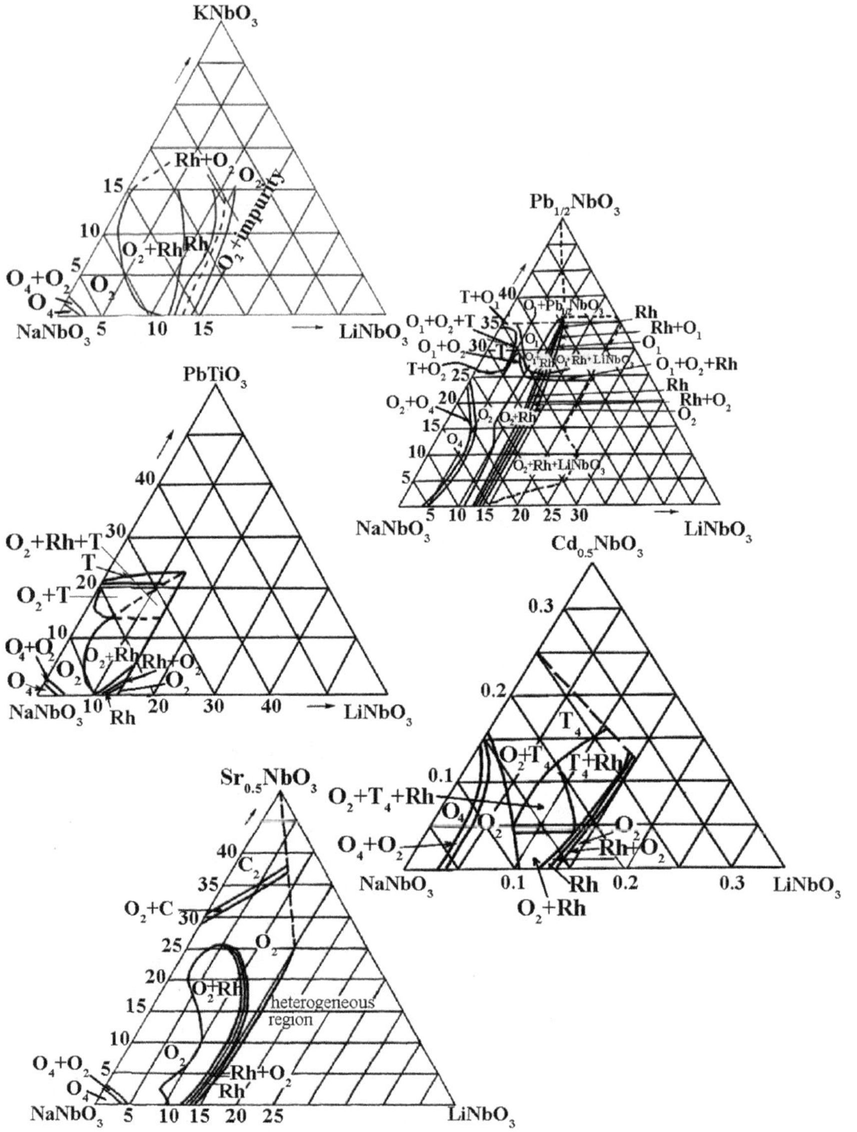

Fig. 1.12 Three-component systems based on alkali metal niobate

dependence of their macro- and mesoscopic structure on the “thermodynamic path” along which the specified values of state parameters are achieved.

Despite the extensive bibliographic material available, the nature of structural instabilities, features of phase states and many details of transformations in these objects are not fully understood. This situation is largely due to the extreme complexity of their preparation. Since it is spatially inhomogeneous media with structural instabilities that are the basis of almost all active materials used in practice, the study of critical and subcritical phenomena occurring in them is relevant and scientifically significant.

In recent years, in the practice of using active materials, there has been a transition to new technologies, which determines the need to create fundamentally new means of monitoring, diagnostics and management of processes occurring in systems and devices, which initiates several trends in the development of modern piezoelectric materials science, aimed at creating materials with:

(i) periodic structure of acoustic inhomogeneities for microwave applications;
(ii) reduced mass capacity, weight and dimensions for their use in devices where these characteristics are decisive, as well as an extended high frequency operating range, ensuring the possibility of their use in digital information processing devices;
(iii) extended operating temperature range (up to 2000 K) for creating contact sensors;
(iv) ferroelectric and magnetic properties combined in one object to create a four-bit magnetoelectric memory, which has an order of magnitude greater information capacity and energy efficiency in comparison with all currently known methods of storing information, as well as to implement the direct transformation of magnetic and electric fields necessary for integration objects of spintronics and conventional electronic technology, as well as with high dielectric memory necessary for creating sources of external electrostatic fields and batteries of electrical energy;
(v) high emission of polarization energy for the creation of high voltage pulsed electromechanical converters in rocket launch systems, in emergency signaling systems and blocking of critical processes (including on the International Space Station), when separating the booster blocks of launch vehicles, and launching hydrometeorological probes.

Work on the search, creation and research of highly efficient materials with special electrical and magnetic properties has been carried out in Russia and abroad for more than 50 years (see Fig. 1.8). Thus, when creating the Scientific Research Institute of Physics of the Southern Federal University, this direction was identified by the Government of the country as a priority. During this time, the institute synthesized about 30 thousands new compositions of various structural types; crystal physical conditions for their existence have been established; natural relationships between composition-structure-properties were determined, a methodology for a targeted search for promising substances was developed, and more than 200 types of materials and methods for their production were created, protected by security documents and "overlapping" almost all known piezotechnical industries [5].

Materials of the 1st group, resistant to electrical and mechanical stress, are intended for devices operating in power modes, namely piezotransformers, piezoelectric motors, ultrasonic emitters, high-voltage generators, multiple-action ignition systems. Materials of the 2nd group with high dielectric constant are used in low frequency receiving devices, namely hydrophones, microphones, seismic receivers.

Materials of the 3rd group with high piezosensitivity are used in ultrasonic flaw detectors, devices for non-destructive testing using the acoustic emission method, and ultrasonic medical diagnostic devices. The 4th group materials are designed to work in the mid-frequency range. The 5th group materials with high anisotropy of piezoelectric parameters are used for accelerometry and thickness measurement. The 6th group materials with high stability of the resonant frequency are used in frequency-selective devices as filters with different bandwidths. The 7th group materials with low dielectric constant and high speed of sound, can be used in high-frequency electroacoustic transducers on volume and surface waves. The 8th group materials with a high pyroelectric coefficient are intended for use as working elements of sensors for pyroelectric receivers of radiant (thermal) energy. High-temperature materials of the 9th group materials are intended for non-destructive testing of heated bodies and processes occurring at high and ultra-high temperatures. Electrostrictive materials of the 10th group record changes in the linear dimensions of objects. Materials with low specific gravity belonging to the 11th group materials can be used in devices in which weight characteristics are decisive [5].

Among such materials, the most knowledge-intensive, in demand, and, recently, relevant, are materials that combine ferromagnetic and piezo (ferroelectric) activity, high piezosensitivities, extreme (up to infinite) piezoanisotropy, because these materials "cover almost all the main piezotechnical, micro- and nanoelectronic areas (medical diagnostics, hydroacoustics, ultrasonic flaw detection, information technology and telecommunications, spintronics). The current transition to environmentally friendly products, due to the formation of a new Russian [160] and European [2] legislative framework, forces us to look for alternative materials science solutions, to exclude toxic elements, primarily lead, from the composition of all known world analogues. However, the values of the main parameters of lead-free materials do not reach the level of lead-containing prototypes (solid solutions of the $PbTiO_3$–$PbZrO_3$ (PZT) system), required for practical applications.

There are no examples of instant implementation of piezo (ferroelectric) and magnetic properties, which are necessary for many practical applications (magnetoelectric memory, filters, modulators, switches, phase shifters, etc.) in bulk materials. Note that so far only ferromagnetic (for example, $BiFeO_3$) [151, 152], magnetodielectric (for example, $TbMnO_3$) [161] and piezomagnetic (for example, CoF_2) [162] materials are known. In this regard, the development of materials with these properties also seems very relevant.

At the moment, it has become obvious that when creating a number of materials: bi-negative media with a negative refractive index of electromagnetic waves for microwave technical applications [20, 163], nanolasers for biomedical diagnostics and information technology [164–166], nonlinear optical media for high-speed optical switches [167], devices based on surface acoustic waves [168] and many other applications require a combination of properties that are not found in natural materials or materials formed on the basis of classical technologies [167]. Recently, a new direction has begun to develop rapidly, namely the creation of "macrocrystals" (phonon [169], photonic [170], magnon [171]), as well as metamaterials [172], the effective material parameters of which can change periodically and take zero or negative values in some frequency ranges.

However, the creation of such media seems extremely difficult due to technological difficulties at each stage of the process of obtaining them. As noted in [173], the problem is that the period of a photonic or phonon crystal must be of the same order of magnitude as the wavelength of radiation propagating in the crystal, and for light and high-frequency acoustic waves the wavelengths are extremely short. This forces the production of two-dimensional microstructural media to move to the technology of drawing microstructural optical waveguides, which are optical fibers with air channels oriented in a certain way [174]. While solving technological problems of creating these materials, the main one remains aside, namely attempts to move from expensive, complex single-crystal technology to a simpler, more accessible and cheaper one, which is conventional ceramic production. However, there are no examples of creating such polycrystalline (ceramic or composite) media with a periodic structure of acoustic inhomogeneities based on materials with different (sharply different) speeds of propagation of sound waves. The technologies for "conjugating" them into one "metamaterial" and the final dimensions of its heterogeneous (in terms of electrical and acoustic parameters) parts are not obvious. One of the main modern trends in the development of sensor equipment is to improve its metrological (expanding the measurement range, increasing reliability and resistance to influencing factors), weight and size characteristics, as well as expanding functionality through the development of combined sensors. Thus, the creation of high-temperature (HT) pressure sensors and corresponding technical diagnostic devices seems to be one of the primary tasks of the corresponding piezotechnical direction. The expansion of the operating temperature range is limited by the lack of materials with ultra-high operating temperature limits (> 2000 K) [14], which also have sufficient piezosensitivity, vibration reliability, and resistance to external influences, allowing their use under combined external factors. "Suppliers" of high-temperature ceramic materials are most often compounds and solid solutions of four groups, differing in chemical bases, type of crystal structure and, therefore, operating temperature limits ($T_{work.lim.}$): titanates–lead zirconates with $T_{work.lim.}$ ~ 500 K, lead titanates with $T_{work.lim.}$ ~ 600–650 K (perovskites), lead metaniobates (potassium–tungsten bronzes) with $T_{work.lim.}$ ~ 550–600 K, Bi-containing with a layered perovskite-like structure with $T_{work.lim.}$ ~ 750–800 K [175]. Higher $T_{work.lim.}$ can be realized based on pseudo-ilmenite lithium niobate ($T_{work.lim.}$ ~ 1200 K), including its SS with other alkali metal niobates and alkaline earth metal pyroniobates (with $T_{work.lim.}$ ~ 1500 K). Already the creation of the last two groups of HT materials is a very difficult task due to the crystal chemical features of the elements included in the compositions [61], as well as the structure of the compounds [176], which provoke their anion-deficient non-stoichiometry, increased sorption capacity, critical dependence on the thermodynamic prehistory (conditions preparation), including the physicochemical state of precursors, a high degree of compositional disorder, multi-stage processes for the formation of final products, hydrolysis of the initial and final products, a narrow range of optimal temperatures for synthesis and sintering, high volatility of *A*-cations, excessive anisotropic growth of giant grains due to the developing process of intermittent secondary recrystallization, self-destruction of ceramics, etc., which in many cases leads to low experimental density and high electrical conductivity of materials [177]. However, nevertheless, we have solved such problems (materials with $T_{work.}$ up to ~ 1200 K have been created). Shift of $T_{work.}$ moving towards even higher temperatures is a task that has not yet been solved. We see the path to success in the creation of composite materials, including porous ones with different types of connectivity, the design of multicomponent systems with subsequent nanostructuring of selected promising bases.

It is obvious that one of the main characteristics of the materials being developed is their specific gravity, the reduction of which makes it possible to significantly reduce the mass intensity and dimensions of the products made from them. Almost all known industrially produced piezomaterials are based on lead-containing compositions having a specific gravity of about 8.0 g/cm^3. Reducing this characteristic by 2–3 times will significantly "lighten" diagnostic systems.

An indisputable alternative to global materials science "brands" are materials based on AMNs, including composite and porous ones with a specific gravity of 3–4 g/cm^3 and lower. However, such materials have not yet found wide application in practice. This is largely due to the impossibility of obtaining such materials with a sufficient level of piezoresponse by traditional methods, which, in turn, is associated with the complexity of their crystal structure, the presence of a large number of phase transitions of various natures that destabilize the structure of materials, and the ease of violating the stoichiometry of a given composition due to increased volatility of alkaline cations, polarization electrolysis, strong dependence of properties on the conditions of structure formation [61].

Figure 1.9 in Sect. 1.3 presents promising functional materials of modern piezotechniques, micro- and nanoelectronics, spintronics, indicating their advantages and disadvantages.

Overcoming these disadvantages and eliminating the negative phenomena associated with them, while maintaining the advantages of materials, requires the use of a synthesis strategy based on the use of relatively cheap industrial methods based on conventional ceramic technology. At the same time, to date, several methods have been developed to optimize the properties of active ceramic materials, which are presented in Fig. 1.13.

Thus, the synthesis of each of the materials must include both the stage of sequential optimization of the technology and modification of the chemical composition of the ceramics. Let us consider some results of the practical application of the chosen strategy to the development of materials based on AMNs. One of the significant factors determining the properties of these ceramic materials is the physicochemical state of the starting substances: alkali metal carbonates and niobium oxide.

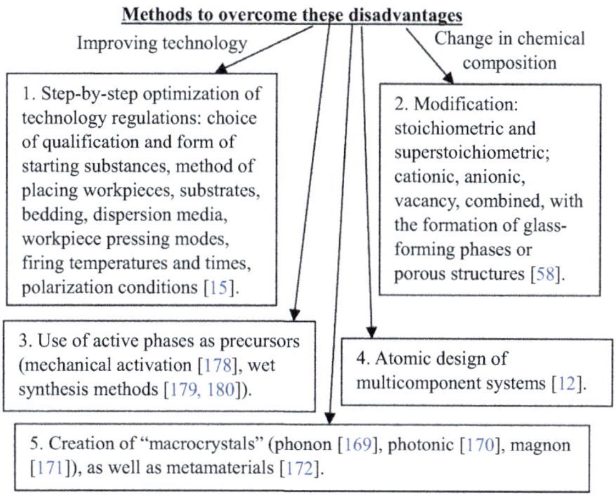

Fig. 1.13 Strategy for the synthesis of a new generation of active electrical materials

1.4.1 *Selection of the Optimal Form of Starting Substances*

In [181], it is shown that the use of stable raw materials in the synthesis, namely alkali metal bicarbonates, is less prone to hydrolysis; allows one to achieve a 30% increase in the electromechanical coupling coefficient of the planar vibration mode (K_p) and improve other performance characteristics of ceramics (Fig. 1.14).

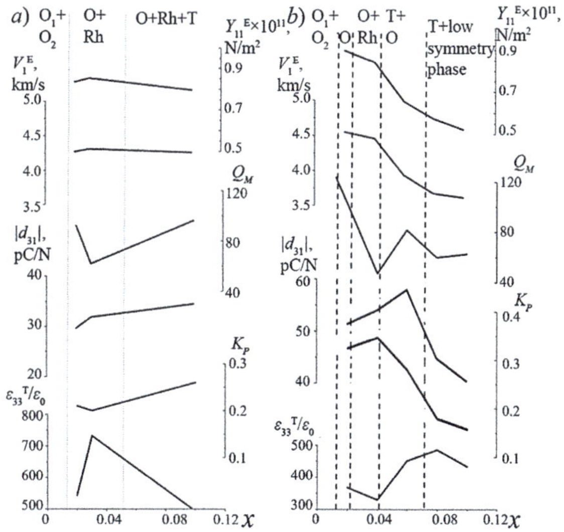

Fig. 1.14 Dependencies of the phase composition and electrical properties of solid solution obtained from carbonates (**a**) and hydrocarbonates (**b**) of alkali metals, and on the concentrations of the components

1.4.2 Selection of the Optimal Physicochemical State of Nb_2O_5

Niobium pentoxide Nb_2O_5 is the most thermally stable reagent among those participating in the synthesis of SS, "covered" with reaction products, contained in quantities that are significantly greater than other components in SS charges. In this regard, its physicochemical state (defective, polymorphic, granulometric, cation–anion, rheological) has, as shown in [56], a decisive influence on the formation of the properties of niobate ceramics. Considering that these parameters (except of the impurity composition) are not regulated by current Russian GOSTs, it seemed relevant to establish the degree of their influence on the final properties of AMNs, which was done in [61, 182–204].

It was shown in [61] that the defect structure of niobate ceramics has a block structure "inherited" from Nb_2O_5. This structure leads to the appearance of tetrahedral vacant positions and a deficiency of positions in the *A*-sublattice, leading to the emergence of structural nonstoichiometry in the result of crystallographic shift and, therefore, the precipitation of impurity phases. It has been shown that with an increase in the synthesis and sintering temperature and hot-pressing pressure, it is possible to achieve the dissolution of impurities in the solid solution, an increase in the experimental density and an improvement in the electrical parameters of objects [198, 203].

Polymorphic state of Nb_2O_5. Industrially produced Nb_2O_5 is almost always multiphase, with a predominance of the high-temperature modification α. It has been established that when using thermal firing of Nb_2O_5, the optimum properties of niobate ceramics are observed in the region of transition to the α-modification. This resembles the corresponding behavior of parameters near morphotropic regions with the only difference that the response of ceramics to structural rearrangement was realized essentially in the third "generation", since the transition took place in the initial reagent.

Thus, it has been shown that optimization of the properties of niobate ceramics is facilitated by preliminary firing of niobium oxide at the transition temperature between a mixture of modifications and the most stable α-modification.

The influence of *granulometric composition* was studied in detail in [182, 190–194, 204]. It has been established that the optimal characteristics, including the highest mechanical strength, are possessed by solids obtained from niobium oxide with specific surface area S_{sp} for various modifications equal to 550–650 m^2/kg (for Nb_2O_5 (I), "NbO–Pt"), 950–1000 m^2/kg (for Nb_2O_5 (II), "special purity"), 2500–3500 m^2/kg (for Nb_2O_5 (III), "special purity").

The complex *rheology of* Nb_2O_5 and niobate FPCMs is characterized by high caking, tendency to clump and low flowability. This is due, on the one hand, to the dendritic structure of niobium pentoxide crystallites, and on the other, to its high hygroscopicity. It is possible to achieve a significant improvement in the rheological characteristics and, therefore, the parameters of ceramics by using various methods of activating the corresponding oxides.

In [189, 195, 197], the influence of the *Nb_2O_5 impurity composition* on the properties of niobate ceramics is considered. It has been established that a harmful admixture of fluorine (up to 0.2 wt%, according to the current Russia GOST), which forms aggressive compounds to Nb-containing oxides. Due to the incompleteness of the reaction, the powder of the synthesized material always contains a certain amount of carbonate compounds of alkali metals, which, when interacting with fluorides, form fusible compounds. The presence of the latter and the associated composition fluctuations determine the corresponding changes in the kinetics of solute-phase sintering (from solid-phase to liquid-phase), the nature of recrystallization (from continuous to intermittent secondary), and the conditions of grain growth (from isotropic from the solid phase to anisotropic, when the resulting liquid phase acts as active solvent, favors the crystallization of large grains from the melt), their final shape (from equiaxed to ideomorphic) and packing (from dense to loose). Crystallizing in the form of small single crystals, ideomorphic grains form numerous layers in the bulk of the samples, contributing to the formation of a composite-like state: single crystal-ceramics. Insufficient intergranular cleavage of such particles softens the samples, favoring the spontaneous destruction of ceramics during the sintering process. In [189, 195, 197], it is shown that a significant improvement in the properties of the studied solid solutions can be achieved by reducing the fluorine concentration to 0.01–0.05%.

A detailed description of the influence of the phase, granulometric and cation–anion state of niobium pentoxide on the properties of materials obtained from charges with its participation is given in [183–189, 195–197, 199–202].

1.4.3 Determination of Rational Regulations for Synthesis and Sintering

Table 1.4 shows the technological methods used, the effects they lead to, and the reasons for what was observed. In [60] it is noted that achieving a sufficiently high “activity” of the synthesized powders for subsequent sintering requires carrying out the synthesis in the temperature range of transition from a fine-grained multiphase mixture of $LiNbO_3$ and initial components to coarse-grained $LiNbO_3$. Sintering of ceramics based on lithium methaniobate is also complicated by the fact that, due to the extremely high temperature of the phase transition (1200 °C), it occurs in the polar phase, which contributes to the development of mechanical stresses and increased fragility of the sintered materials [60].

It has been established [205] that increased attention should be paid to the selection of technological regulations when synthesizing materials are based on bismuth ferrite. Thus, in [146] calculations of thermodynamic potentials for the formation reaction of the latter are given, from which it follows that it is not the most stable of the possible compounds in the Bi_2O_3–Fe_2O_3 system.

Table 1.4 Technological methods used, common to AMNs, the effects they lead to, and the reasons for what was observed

Technological method	Effect	Causes
1. Use of pre-synthesized SS bases	Increasing the density, piezoelectric activity and electrical resistance of ceramics	More precise control of the stoichiometry of products, reducing the degree of hydrolysis at the final stage of synthesis
2. Sintering of small workpieces placed in one layer	Preservation of the integrity of the sinter and the stoichiometry of the solid solution	Reduction in the degree of delamination in the result of the accumulation of low-melting hydrolysis products in the lower part of the workpiece
3. Use of special substrates (corundum)	Preserving the integrity of the cakes and the technological equipment used	Reduced adhesion of the sample to the substrate due to the inertness of corundum to hydrolysis products
4. Use of special bedding (AM and Ca zirconates)	Increasing the density, piezoelectric activity and electrical resistance of ceramics	Reducing the sublimation of volatile compounds of AM due to the creation of an atmosphere enriched with the latter
5. Selection based on a series ($\Delta T_{synth.}$ = 100 °C, in steps of 20 °C; $\Delta T_{sint.}$ = 30 °C, in steps of 5 °C) of samples of optimal temperature–time regimes of synthesis and sintering and, therefore, their strict regulation	The maximum possible density and homogeneity of ceramics within the framework of conventional technology and raw materials have been achieved	Removal of residual hydrolysis products that interfere with sintering

In [145], the results of a study of its temperature stability are presented, from which it follows that at elevated temperatures (> 700 °C) bismuth ferrite slowly decomposes. Attempts to obtain pure material based on $BiFeO_3$ using exotic, more energy-consuming and labor-intensive methods than the solid-phase reaction method were made in [180, 206–209].

Within the framework of the chosen strategy, aimed at the production of multiferroic ceramics by careful selection of technological regulations within the framework of CCT, it is possible to obtain $BiFeO_3$ with an impurity content (5–7%) [205].

1.4.4 Selection of Polarization Modes

It was established in [210] that the polarization of ceramics based on AMNs is complicated, especially at the initial stage, by the process of electrolysis of impurity inclusions, formed in the result of hydrolysis during synthesis and sintering. Considering this phenomenon, as well as the process of evaporation of moisture, adsorbed by hygroscopic AMNs, required optimization of the polarization scheme (Fig. 1.15).

Optimization of the chemical composition of compounds is carried out through the construction of multicomponent systems and modification.

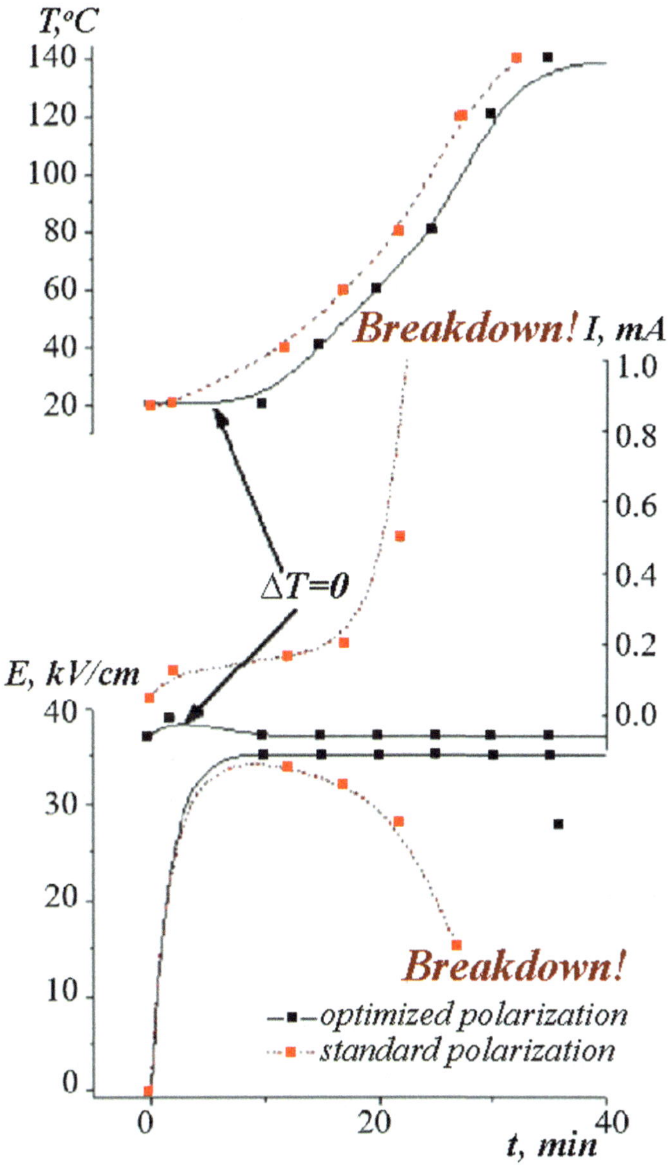

Fig. 1.15 Optimization of the polarization scheme for ceramics based on AMNs

(1) *Construction of multicomponent systems.* Multicomponent systems have a number of undeniable advantages over their binary components [12]: with an increase in the number of components, the dimensionality of the morphotropic region increases, which significantly expands the possibilities for choosing a SS with a given combination of parameters; the introduction of new components expands the possibilities of changing the composition, and, consequently, the parameters of the SS; with an increase in the number of components, the efficiency of the SS increases; as the number of components increases to 4–5, as a rule, the manufacturability of the systems improves. This is confirmed by an increase in K_p by almost one and a half times in multicomponent systems based on (Na, K)NbO_3 compared to the basic binary one [32].

(2) *Design of modifiers (determining the optimal qualitative and quantitative composition and methods of introducing additives).* One of the most effective methods for changing the physical properties of ferroelectrics is the method of introducing small (no more than 10 at%) additives of monoxides of various elements (modification). The great potential of this method is due to the existence of a significant amount of metal monoxides and their combinations that can dissolve in complex oxides, in particular, in solar cell oxides with a perovskite-type structure, over a wide range of concentrations, as well as oxides of elements that form "fluxes" when introduced into the original objects (B_2O_3, SiO_2, V_2O_5), and, finally, additives of combined action that influence the properties of the original systems due to the formation of a liquid phase and cation-exchange interaction with the base (modifying glasses). The undoubted advantage of the method is the ability, while maintaining a set of specific properties of the modified objects, to selectively change the values of some parameters in the desired direction by slightly varying their composition.

(3) *The work undertaken within the framework of the chosen strategy to modify AMNs.* Due to the introduction of NiO, it allows one to increase the coefficient of electromechanical coupling in ceramics based on (Na, K)NbO_3 by 15% [211]; by introducing MnO_2, it increases the piezoanisotropy and piezosensitivity [154] of ceramics based on (Na, Li)NbO_3 > 20%. Due to the modification of $LiNbO_3$-based ceramics with pyroniobates and the use of nanotexturing, it is possible to significantly (up to 1400 °C) increase the upper limit of operating temperatures and piezoanisotropy of high-temperature FPCMs [212].

(4) *The use of rare earth elements as modifiers for $BiFeO_3$.* It allows one to obtain pure ceramics, enhance their electrical resistance, eliminate the influence of spatial spin modulation [97], and in some cases, initiate the formation of ferromagnetic phases [213].

For each of the materials being developed, the technological problems described have not been fully resolved. Essentially, the strategy is aimed at creating materials not with improved, but with fundamentally new properties. In this sense, its technological vector reflects the focus on creating "breakthrough" technologies that ensure competitiveness in the global market.

Research carried out within the framework of the described strategy was also aimed at the development of composite and porous ceramics. Standard powders of sodium lithium (potassium) niobates, bismuth ferrite, lead zirconate titanate, lead titanate and methaniobate, bismuth titanate, lead magnoniobate and lead potassium niobate were used as the starting piezoceramic materials. The largest volume of experimental results was obtained for materials of the lead zirconate-titanate system [5, 214]. Also interesting are the results obtained for the first time for magnoniobate/ lead titanate ceramics, as well as lead methaniobate and sodium potassium niobate [13, 215, 216].

Within the framework of the holistic concept of microstructural design of polycrystalline materials [214], a universal technology to produce porous piezoceramics, as well as polycrystalline composites based on it, has been developed, suitable for mass production, providing reproducible results for a wide class of piezoceramic materials. The developed technology ensures the production of porous piezoceramic materials with specified properties and porosity up to 70%. A technology has also been developed for the manufacture of ultrasonic transducers and devices based on porous piezoceramics and polycrystalline composites in the frequency range 50 kHz–20 MHz for a wide class of applications, including medical diagnostic equipment, non-destructive testing and diagnostics, level and flow metering, hydroacoustics, as well as power ultrasonic systems medical and technological purposes. The developed technology for producing porous ceramics is based on a modified method of burning out a porogen, which uses complexes of organic compounds. The main difference between the method used and those usually used in the production of porous ceramics [217–220] is the use of sets of organic powders with a given dispersion (1–150 μm), passivators/activators of compaction and special processing modes at all stages of manufacturing powders suitable for sintering homogeneous ceramics with controlled parameters (porosity up to 70% and average pore size from 5 to 150 μm). The developed technology also allows the use of the slip casting method to obtain thin, porous films and makes it possible to obtain ceramics with anisotropic porous structures and a specified porosity distribution over the thickness of the sample. Theoretical models have been developed to calculate the parameters of porous and composite materials [214]. By varying the conditions for producing sodium niobate ceramics, a directed change in closed porosity (3–0 connectivity) was carried out from 1.0 to 13.6%. It is shown that as it increases, changes occur in the phase state, nature of the grain structure and electrical properties of $NaNbO_3$. This is a consequence of the simultaneous occurrence of several processes: the disappearance of the ferroelectric *Q*-phase, the formation and interaction of point defects (vacancies) in the alkali component, the appearance of extended defects (crystallographic shear planes), the formation and ordering of "pore–multilayer grain shell" clusters, a sharp increase in the total surface area pores and its significant excess of the area of the external boundaries of the sample. The latter is probably associated with the giant piezoanisotropy of porous sodium niobate ceramics. The possibility of practical use of porous sodium niobate ceramics in piezotechnical industries has been shown [220]. Porous ceramics based on the following materials were obtained and studied:

(i) PCR-34 and PCR-35 are lead-free piezoceramics, based on sodium–potassium and sodium–lithium niobates, respectively (preliminary studies);
(ii) high-temperature piezoceramics—PCR-40, sodium-bismuth titanate PCR-50, TNB-1, based on lead titanate, and lead-potassium niobate PCR-43;
(iii) soft ferroelectric PZT—TsTSNV-1, PCR-7 M, PZ-29 (Ferroperm), APC-855 (APC), TsTSSt-2;
(iv) ferro-rigid PZT—PCR-78, PCR-8, TsTSSt-3, APC 841;
(v) highly sensitive PZT—PCR-1, PCR-37, TsTS-19;
(vi) anisotropic piezoceramics, based on lead titanate and lead methaniobate;
(vii) piezoceramics based on relaxor SS (magnoniobate-lead titanate).

Special attention is paid to the accurate measurement of the parameters of porous piezoceramics (d_{33}, k_t, Q_M), which is a significant difficulty for low-quality materials with a heterogeneous structure. The measurements were carried out in accordance with the Russian Standards on samples of porous piezoceramics obtained using the same technology in compliance with all technological regulations. The experimental results were processed using standard statistical methods. The nature of the obtained dependencies of dielectric, elastic and piezoelectric parameters on porosity for all materials studied (except for materials with initial piezoelectric anisotropy) was the same within the limits of measurement accuracy. This made it possible to summarize extensive experimental data and present them in the form of values normalized to the corresponding values of the parameters of dense ceramics [214].

Additional possibilities for controlling the special electrical and magnetic properties of functional materials, aimed both at improving practically required parameters and at creating their new combinations, are associated with the synthesis of nanostructures and nanostructured materials [22]. The Scientific Research Institute of Physics of the Southern Federal University, together with the Institute of Solid-State Chemistry and Mechanochemistry of the Siberian Branch of the Russian Academy of Sciences (ISSCM SB RAS) (Novosibirsk) and the Department of General and Inorganic Chemistry (DGIC) of the South Russian State Technical University (SRSTU-NPI) (Novocherkassk) developed physicochemical methods for producing nanoobjects (by dispersing macroscopic materials and using other methods).

In [221–224], the influence of mechanical activation of precursors on the phase formation and evolution of dielectric spectra of solid solutions of various groups of functionally active materials (multiferroics and ferroelectric relaxors (FERs)) was studied. In FERs, due to mechanical activation, it was possible to reduce the synthesis temperature (T_{synth}) by 150 °C, reduce the firing time by half, increase the relative density of sintered ceramics to 97%, and obtain samples that do not contain a pyrochlore phase. The use of mechanical activation in relation to SS based on bismuth ferrite makes it possible to reduce the total time of isothermal exposure of $BiFeO_3$ by almost an order of magnitude and, in combination with "hardening" of samples during synthesis, to get rid of impurities.

Preliminary results of the search for multiferroic materials based on a different structural type of substances, namely crystals with a spinel structure were obtained.

The objects of experimental and theoretical research are multicomponent oxide systems, including chromites and ferrites of transition elements. This attempt is justified by the fact that some simple chrome spinels ACr_2O_4 (A = Mg, Zn, Cd, Hg, Fe, Co, Cu, Ni, Mn) had previously been discovered at low temperatures with unusual magnetic and electrical properties that appear in the result of phase transitions. Therefore, it is advisable to study chromites and ferrites with a spinel structure that are more complex in composition.

The ternary spinel-like systems $CuCr_2O_4$–$FeCr_2O_4$–$NiCr_2O_4$ and $CuCr_2O_4$–$NiCr_2O_4$–$NiFe_2O_4$ were synthesized for the first time. Phase diagrams of systems, characterized by complex phase equilibria have been constructed [5]. The presence in them of many tetragonal and orthorhombic phases, as well as morphotropic regions with coexisting phases of different symmetries, determined interest in the study of not only their structural features, but also, which is especially valuable from the viewpoint of practical application and physicochemical properties. Currently, the electrophysical and magnetic properties of these solid solutions are being studied in detail.

Numerous methods have been developed for creating ceramic and composite lead-free materials for practical applications: methods of synthesis, hot pressing and molding, including large-size blocks, optimal backfill compositions, methods for controlling the microstructure of objects, creating states with different manifestations of ferroelectric properties (relaxor, electret, posistor, pyroelectric, etc.) [225]; and also ceramic and composite materials have been developed for high frequency converters for various purposes [226], highly sensitive materials [227], high-temperature materials [228], materials for high voltage sources [229], glass–ceramic materials [230], posistor materials [231], semiconductor materials [232], capacitor materials [233], pyroelectric materials [234], materials for low and medium frequency applications [235], material for ferroelectric substrates [236], materials for frequency selective devices [237], electret materials [238, 239].

Methods have been developed aimed at creating a methodology for constructing multicomponent SS systems with morphotropic regions and accompanying extreme parameters of objects [240], a methodology for predicting and searching for zones of optimal characteristics [14], a concept for the modular "design" of spinel-like phases with a combination of electrical and magnetic properties [194], technologies for the synthesis of nanostructures and nanostructured materials [194], studies of heterogeneous systems, including composites and statistical mixtures [240–243], multicomponent multiphase systems such as natural silicates, the basis of which is volcanic glass, formed by frame-type aluminosilicates and microlites (small crystals of prismatic cut, glued together by a glassy phase having a felt-like structure) [244].

Superconducting materials [245–247], as well as superionic conductors [248], have been created and studied.

According to the generally accepted theory of the outstanding economist N.D. Kondratiev, the foundation of economic development is a change in technological structures. The core and at the same time the result of any technology created by man is the material. It is with materials that one of the most widespread periodicals of human development is associated: Stone Age–Bronze–Iron–.... Of course, at the present stage it seems very difficult to single out any one material that determines the vector of technology improvement: there are not even dozens of them now, but hundreds and even thousands. Thus, the core of the sixth technological order, which is associated with the development of science and technology at the beginning of the twenty-first century [249], is nanoelectronics, nanophotonics, nanomaterials and nanostructured coatings, nanobiotechnology and nanosystem technology. When comparing facts within the development of one branch, in this case, piezoelectric materials science, it is possible to determine a clear pattern of changes in the dominant synthesis strategies and the corresponding methodological apparatus (see Table 1.3).

It is obvious that, despite the rapid development of modern technology for the manufacture of special ceramics, the improvement of electrical devices requires the search for fundamentally new approaches to the creation of electroactive materials. In this case, the most promising may be:

(i) nanoferroelectrics obtained through the artificial formation of a structure by layer-by-layer assembly;
(ii) metamaterials built on ordered structures of quantum threads and dots [27, 28];
(iii) compositional structures built on the base of the concept of self-organization of chemical systems [25, 26].

1.5 Lead-Free Piezoceramics: The End or the Beginning?

More than 10 years have passed since the birth of the "lead-free" direction in the world ferropiezoelectric science, the beginning of which, according to many authors [250–257], is dated to 2004 and the famous work of Saitoet al. [115]. In this article, it was shown that using technological methods based on the achievements of wet or soft chemistry, it is possible to achieve in ceramics based on alkali metal niobates, the implementation of values of the main piezoelectric characteristics close to those in industrially developed materials based on the PZT system ($PbZr_{1-x}Ti_xO_3$). Even though in the next 10 years neither materials with similar parameters nor devices based on them appeared on the market, thanks to this publication, as well as the simultaneously published review by Cross [1], a discussion was initiated on the issue of complete, or partial replacement of lead-containing piezoceramics in electronic equipment.... As a result, many scientific groups around the world received funding and conducted research aimed at creating new, highly effective lead-free materials.

Over the course of several years, there has been a sharp increase in interest in lead-free ceramics [7, 258–260]. Publication activity was assessed by hundreds of papers in leading journals, and related areas were actively developing. A logical continuation of such efforts was the tendency to generalize the results obtained and draw intermediate results, which was manifested in several recent reviews and monographs [250, 251, 255, 261–263].

Almost all presented publications include a fundamental research part (1), based on a discussion of phase transition phenomena; technological part (2), which discusses the effects of modification; comparative and applied part (3), considering the relationship between various technically important electrophysical parameters and prospects for the use of ceramics in sectors of the national economy. At the same time, it is traditional, and much to the regret of the authors of this book, to suppress the results obtained in Russia in foreign publications. This omission is probably due to the insufficiently complete representation of the works of domestic authors in the foreign press. This gap certainly needs to be filled, and this is partly what this book aims to do.

The domestic history of lead-free materials science is significantly longer than that, which dates to the works of Saito and Cross, as well as the adoption of corresponding environmental programs in several countries [264–267]. At the same time, it has a solid basis associated with the introduction of several special materials into production [35].

At the Scientific Research Institute of Physics of the Southern Federal University, in the scientific school created by Professor E.G. Fesenko, during almost 50 years, the design, study and creation of materials, based on lead-free solid solutions (SS), including with the participation of AMNs have been carried out [61]. However, the tasks facing physics, chemistry and technology of functional materials during the formation of this direction were dictated by no ecology but the need to create piezoceramics with a unique combination of properties for special applications. The main "attraction" of such objects is that many of them have a unique set of parameters: extremely low dielectric constant, high sound speed with a wide range of mechanical quality factors (from units to thousands) and low specific gravity (less than 4.5 g/cm^3). This made it possible to develop, based on such materials, a new direction in physical instrumentation: acoustoelectronics, using both ultrasonic waves in the high-frequency range and hypersonic (from 10 MHz to 1.5 GHz) volumetric and surface waves.

Despite the obvious advantages, this group of lead-free ceramics occupied only a very narrow segment of piezotechnical applications [5] and could not replace the well-known industrial Pb-containing analogues, since their piezoelectric parameters did not reach the level of the corresponding properties of PZT materials. Moreover, to realize sufficiently high parameters of the first lead-free materials based on the $NaNbO_3$–$LiNbO_3$ binary system, the use of the hot pressing (HP) method (sintering under conditions of externally applied pressure [6]) is required. Despite the advantages of this method (expressiveness, versatility, microstructure control), the use of HP involves the use of expensive alundum equipment, which provides a short (1–2 cycles) operating time. Thus, the use of HP technology is advisable only in conditions of piece and small-scale production. Thus, it was urgent to develop lead-free ceramics that could compete with the PZT SS system in classical areas of application of ferroelectrics.

To implement these plans, the $NaNbO_3$–$KNbO_3$ system and solid solutions based on it are more suitable, in which, due to the high polarizability of the potassium cation and the rich filling of the corresponding phase diagram, it is possible to realize higher piezoelectric parameters than in the $NaNbO_3$–$LiNbO_3$ system [9–11].

1.5.1 Alkali Metal Niobates

Since 2000, sodium potassium niobates have been the most widely studied representative of lead-free ferroelectric materials. Currently, there are several reviews devoted to their study [253–256, 258–261, 263, 268, 269]. The first fundamental studies of pseudobinary phase equilibria in this system were undertaken in the late 1960s and early 1970s [270, 271]. At the same time, unlimited mutual solubility of these components was established (Fig. 1.16) [18].

Sodium niobate crystallizes at room temperature as an orthorhombic phase, space group Pbma. $NaNbO_3$ is a well-known model antiferroelectric [272, 273], with a huge number of phase transitions (Fig. 1.17) [10, 270, 274, 275]. Potassium niobate, on the contrary, is a ferroelectric with orthorhombic symmetry (Amm2). The same can be said about SS with sodium niobate with a $KNbO_3$ content of more than 2 mol%. The most studied composition is $Na_{0.5}K_{0.5}NbO_3$. It undergoes successive phase transitions from the rhombohedral phase (R3c) at low temperatures to the orthorhombic phase (exists in the range of 160–200 °C), then to the tetragonal phase (P4mm) and at 400 °C (Curie temperature) to cubic phase (Pm3m) [276]. Moreover, several researchers attribute monoclinic structure to the phase at room temperature [277]. The solidus temperature of $Na_{0.5}K_{0.5}NbO_3$ is 1140 °C. In addition, the existence of a morphotropic phase boundary between two orthorhombic phases at $x =$

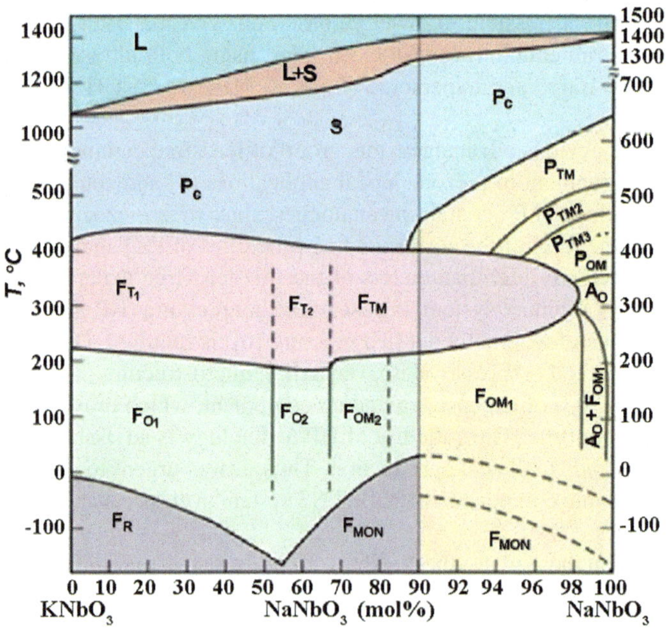

Fig. 1.16 Refined fragments of the phase diagram of the system $(1-x)NaNbO_3$–$xKNbO_3$

0.475 was also previously indicated [271], however, its existence was subsequently not confirmed [277, 278].

The piezoelectric characteristics of this $Na_{0.5}K_{0.5}NbO_3$ ceramics, obtained by air sintering are not too great. Thus, the piezoelectric modulus d_{33} is 80 pC/N [270, 279]. Nevertheless, by changing the ratio of components, it can be increased by 10–15 pC/N and the electromechanical coupling coefficient (k_{33}) can be increased to 0.38 [41, 280–282].

The highest values of these parameters d_{33} (160 pC/N) and k_{33} (50.45%) were obtained using HP and sintering in a spark plasma discharge [41, 53, 54]. Thus, the properties of pure $Na_{0.5}K_{0.5}NbO_3$, which are most in demand for practical applications, cannot compete with the parameters of materials of the PZT system [270]. However, there are also no industrial materials based on the unmodified PZT system [5].

A new stage of active research into piezoelectrics based on $Na_{1-x}K_xNbO_3$, initiated, as already noted, by growing interest in environmental problems [264–267] and several optimistic publications [1, 32], led to extensive research, focused on studying the influence of modifiers and the construction of multicomponent systems.

Thus, in [32], it was discussed a possibility to achieve a value of $d_{33} \approx 300$ pC/N in a modified composition, and even $d_{33} \approx 400$ pC/N using texturing, although at the expense of reducing the Curie temperature to 253 °C. The work indicated that the increase in the piezoelectric response was explained by the proximity of the morphotropic phase boundary between the orthorhombic (O) and tetragonal (T) phases. Later, it was recognized that modifiers reduced the temperature of the O–T transition, and, therefore, the increase in d_{33} could be associated, rather, with the proximity of the polymorphic transformation [256]. The proximity of both phase boundaries contributes to enhanced properties [35, 254, 256, 283]. The temperature instability of materials with polymorphic transformation temperatures close to room temperature is also obvious [37, 284]. The situation is even more complicated due to the significant dependence of the properties of AMNs on their thermodynamic prehistory [153, 177].

An increase in thermal stability can be ensured by a shift of the T–O transition to the temperature range below room temperature [285–287], however, in some cases

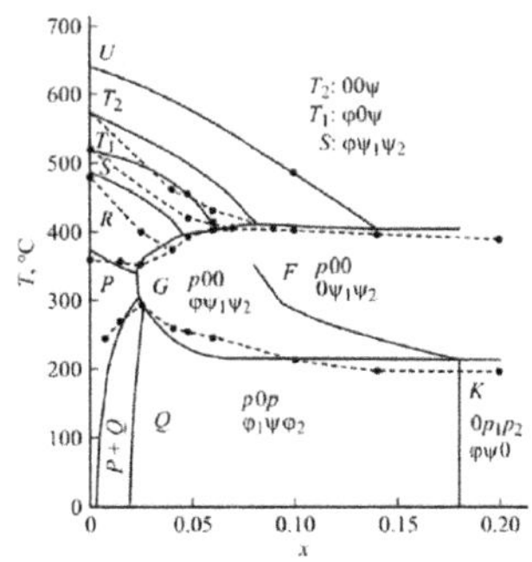

Fig. 1.17 Refined fragment of the phase diagram of the system $(1-x)NaNbO_3$–$xKNbO_3$

this leads to a significant decrease in properties [104, 288]. Thus, the introduction of Li (6 mol%) helps to reduce the temperature of the T–O transition to room temperature. It is also interesting that lithium modification simultaneously leads to an increase in the Curie temperature and, therefore, expands the working range [35–37, 85, 87, 254, 289, 290]. Modification with tantalum helps to reduce both the Curie temperature and the temperature of the T–O transition, while improving the sinter ability of ceramics and allowing control of its microstructure. However, the high cost of tantalum does not allow one to classify it as an easily accessible element of materials science [260].

There are many works that describe the influence of various components and modifiers on the T–O transition and Curie temperatures in KNN-based SS, and, therefore, changes in the temperature dependence of piezoelectric properties [254, 289, 291–293]. Thus, modification of $LiSbO_3$, just like $LiTaO_3$, leads to a decrease in both the Curie temperature and the T–O transition, which is accompanied by an increase in piezoelectric activity with a decrease in the temperature stability of properties [35, 258–260, 291, 294, 295]. In addition, according to several works [24, 67–70], a significant reduction in the temperature of the T–O transition in modified ceramics was achieved through the introduction of zirconates and titanates of alkaline earth metals.

Another way to form a morphotropic phase boundary in solid solutions close in composition to $Na_{0.5}K_{0.5}NbO_3$, and, consequently, to enhance piezoelectric activity, is to stabilize the low-temperature rhombohedral phase at room temperature [110, 296–298]. According to [298], by constructing a six-component system, it was possible to obtain ceramics with $d_{33} = 490$ pC/N. At the same time, we note that, considering the multicomponent nature of the system, as well as the presence of three volatile components (Na, K, Bi), such values in ceramics will have problems with technological achievability and reproducibility [6].

To increase the density of ceramics based on $Na_{0.5}K_{0.5}NbO_3$, classical glass-forming additives were used [51, 132, 299–302], but significantly higher results were achieved using copper compounds such as CuO [292, 303], $K_4CuNb_8O_{23}$ [94, 300], $K_{5.4}Cu_{1.3}Ta_{10}O_{29}$ [304] and $CuNb_2O_6$ [99–103]. When using them, as a rule, the sintering ability of ceramics not only improves, but also the mechanical quality factor increases significantly. Thus, it was shown [305] that the addition of 0.38 mol% $K_{5.4}Cu_{1.3}Ta_{10}O_{29}$ makes it possible to increase the mechanical quality factor $Na_{0.5}K_{0.5}NbO_3$ from 90 to 1300. Even more significant success was achieved using $CuNb_2O_6$ ($Q_m \approx 1700$) [258]. Further modification of CuO can lead to Q_m values from 2600 to 3000 [306, 307], which the authors associated with an increased concentration of defect dipoles in the material [308]. Apparently, the transition from conventional solid-phase sintering to sintering with the participation of a liquid phase is also significant [309].

1.5.2 The Problem of Efficiency

One of the conditions for the widespread replacement of lead-containing piezoceramics with environmentally friendly ones is the achievement of both high piezoelectric activity and mechanical quality factor [144, 255, 310]. The corresponding group of lead-containing ceramics is characterized by values of $Q_m = 1000–2000$ with an average value of $K_p \sim 0.6$ [5].

Assessments of the possibility of implementing the above characteristics in lead-free ceramics [144, 310, 311], carried out by various authors (Figs. 1.18 and 1.19), show that, firstly, lead-free ceramics are significantly inferior in this ratio to lead-containing ones, and secondly, the most promising among them are solid solutions based on (Na, K)NbO_3, modified with copper compounds.

There is a reason for the decrease in Q_m with increasing K_p. There may be an increase in internal friction with an increasing amplitude of mechanical vibrations [144]. For example, for vibrator No. 1 from [312] in the form of a rod, performing longitudinal oscillations in the direction X_2 in an electric field, perpendicular to its length, with amplitude $E_{3m} = U_m/b$, the displacement of the rod element is:

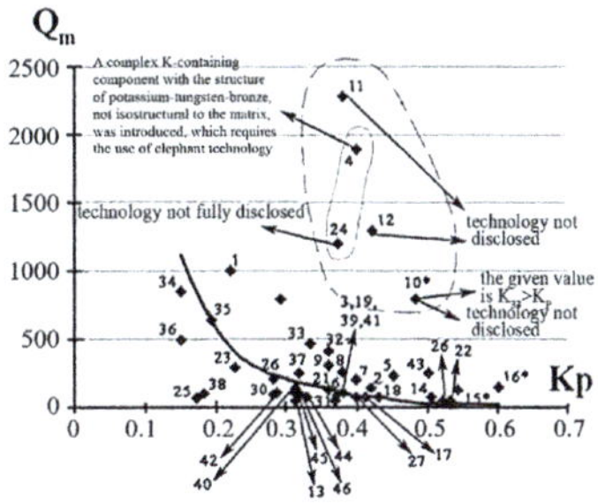

Fig. 1.18 Q_m versus K_p in lead-free ceramics

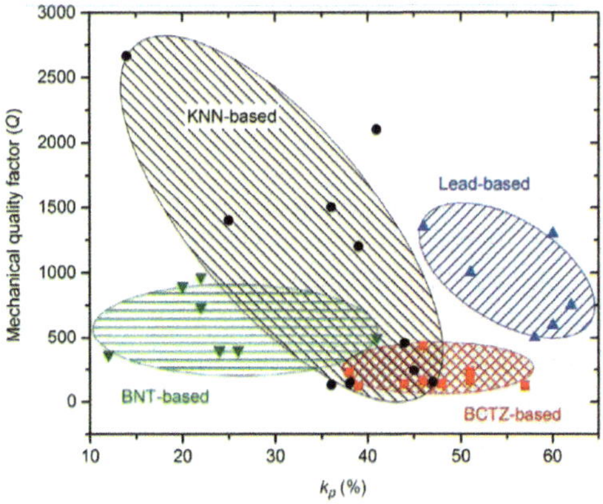

Fig. 1.19 Comparison of dependencies Q_m–K_p in different groups of piezomaterials

$$u_2(X,t) = \frac{d_{31}}{kb}(\sin kX - \tan\frac{kl}{2}\cos kX)U_m e^{i\omega t};\ \frac{1}{V_1^E} = \sqrt{\rho s_{11}^E}$$

where k is the wave number, V_1^E is the speed of propagation of longitudinal elastic waves in the rod at a constant electric field, l is the length, b is the thickness of the rod, $|d_{31}|$ is the piezoelectric modulus, U_m and ω are the amplitude and frequency of the applied electrical voltage.

Mechanical quality factor $Q_m = (2\pi f_s L)/R = 1/(2\pi f_s CR)$, where f_s is the series resonance frequency, L, C and R are the parameters of the equivalent circuit [313]), corresponding to mechanical inertia, elasticity and losses (for example, internal friction). It is logical to assume that the value of internal friction proportional to R increases with increasing amplitude of mechanical vibrations, which, according to (1), is proportional to $|d_{31}|$ and increases with increasing K_{31} ($K_{31}^2 = d_{33}^2/\varepsilon_{33}^T s_{11}^E$) and K_p.

Unfortunately, the theoretical description of piezoelectric oscillations is carried out in the absence of friction ($R = 0$), that is, with an infinite quality factor Q_m, after which losses are formally introduced, therefore, from the theory of Q_m and K_p. Nevertheless, considering the definitions of these characteristics and the corresponding formulae, we will try to identify such a connection.

As is known [270], the electromechanical coupling coefficient is the square root of the fraction of mechanical energy that is converted into electrical energy and vice versa. The remaining energy during each cycle is lost to elastic or dielectric hysteresis or dissipated. The reciprocal of the mechanical quality factor, Q_m, that is, internal friction ($1/Q_m$) (or mechanical losses), characterizes the ability of a solid body to irreversibly dissipate the energy of mechanical vibrations [314]. In ferroelectrics, internal friction is largely determined by the reorientation of domains [315], and, as was shown in [316], the part of the internal friction, caused by the inelastic displacement of domain boundaries, is one of the most sensitive parameters to changes in the structure.

In accordance with [270, 317, 318], we have:

$$Q_m = \frac{10^{12}}{2\pi R_k C_0 f_r\left[\left(f_a^2 - f_r^2\right)/f_a^2\right]}.$$

The expression in the denominator in square brackets is equal to K_{eff}^2, where $K_{eff.}$ is the effective coefficient of electromechanical coupling of a piezoelectric element of any size. If the latter has the form of a ring, $K_{eff.} = K_{31}$ for the radial vibration mode [319]. In the formula for Q_m, the parameters f_r and f_a are the resonant and anti-resonant frequencies of the main overtone (disk, rod), respectively, Hz; C_0 is the electrical capacitance of a mechanically clamped piezoelectric element, pF; R_k is the resistance of the piezoelectric element to the resonant frequency of the first overtone (main), Ω.

Considering that $\varepsilon_{33}^T/\varepsilon_0 = 11.3C_0 b/S$ (where b is the thickness, cm; S is the area, cm^2, equal to al, where a is the width, cm; l is the length of the bar, cm), V_1^E (sound speed) = $(1/\rho S_{11})^{1/2}$ (ρ is the sample density, g/cm^3, S_{11} is the modulus of elasticity, $1/S_{11} = Y_{11}^E$ (Young's modulus, N/m^2), we obtain the following expression for Q_m:

$$Q_m = \frac{10^{12}}{11.3ab}\sqrt{\rho}/R_k\sqrt{Y_{11}^E\frac{\varepsilon_{33}^T}{\varepsilon_0}K_{31}^2}.$$

An inverse dependence of Q_m on K_p was also observed in [12]. Thus, it is noted in [12] that in a strong field (at a large dynamic stress $\sigma_m = 13$ MPa); $Q_m^\sigma < Q_m$ (where Q_m is the mechanical quality factor at weak excitation) that is, at large amplitudes of mechanical stress σ_m, domain boundaries become larger mobile, leading to a decrease in mechanical quality factor (Fig. 1.20a, b) [12]. Moreover, Fig. 1.20b shows that the structural characteristic δ is a homogeneous deformation parameter, characterizing the mobility of domain boundaries, and η is the degree of residual domain reorientation other than 180°, from the concentration of components in one of the multicomponent systems with a morphotropic region (MR). It is obvious that Q_m should increase, firstly, with a decrease in the mobility of domain boundaries, that is, with an increase in δ (in the tetragonal, T-phase), and, secondly, with a decrease in the number of domain boundaries other than 180° ones, that is with increasing η, since an increase in η corresponds to partial monodomainization (in the rhombohedral, Rh, phase).

Since the indicated structural parameters δ and η are approximately inversely related one to another [12], they are factors acting on Q_m in opposite directions in the phases under consideration. In the T-phase, the decisive role is played by the decrease in the mobility of domain boundaries due to an increase in δ (here η practically does not change), and in the Re-phase, a decrease in the number of domain boundaries due to an increase in η.

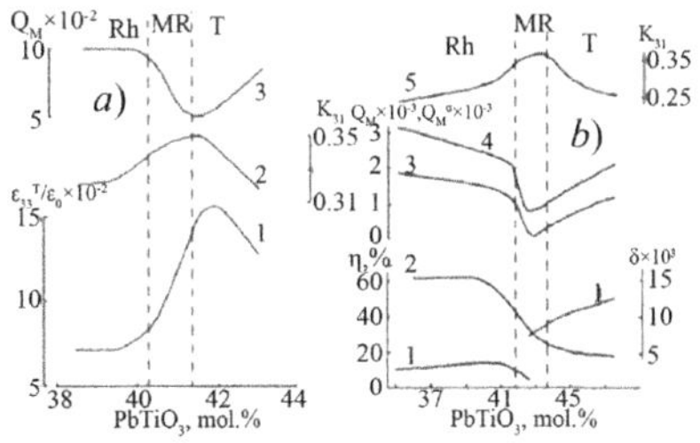

Fig. 1.20 **a** Dependencies of $\varepsilon_{33}{}^{\mathrm{T}}/\varepsilon_0$ (1); K_{31} (2); Q_m (3) on the composition in the system $PbTiO_3$–$PbZrO_3$–$PbNb_{2/3}Zn_{1/3}O_3$–$PbNb_{2/3}Mn_{1/3}O_3$; **b** dependencies of δ (1); η (2); Q_M^σ (3); Q_M (4); K_{31} (5) on the composition in the $PbTiO_3$–$PbZrO_3$–$PbNb_{2/3}Mn_{1/3}O_3$–$PbW_{1/2}Mg_{1/2}O_3$ system

1.5.3 Prospects for Application

Thus, it becomes clear that the values of Q_m and K_p cannot simultaneously reach large values, which extremely limits the prospects for creating universal materials, based on lead-free ceramics, like the most widely used ceramics, based on PZT–TsTS-19. However, the importance of Q_m and K_p for different applications is quite different, and if K_p is the most important parameter when operating a piezoelectric element in sensor mode, the role of Q_m is more significant for a resonator [5, 250]. Moving away from simple and universal solutions is a natural tendency by creating ever more advanced and safer technology.

Let us consider some realities and prospects for the use of more complex, but at the same time safer, lead-free ceramics [250]. The total market volume of ferroelectric materials was $38.4 billion in 2017, its structure is presented in Fig. 1.21. Analysis of the consumption of these products indicates the continuously increasing role of active elements in the most high-tech industries, such as biomedicine, information technology and robotics, precision engineering, non-destructive testing, special measuring equipment, military and civil sonar, and ultrasonic power technologies.

Despite the huge volume and growth of the market, experts also note growing competition from alternative technologies (semiconductor composites, ferromagnets, compounds with giant magnetoresistance), which can lead to a rapid redistribution of the market with a relatively weak impact, for example, with the introduction of more stringent environmental requirements in relation to manufactured products [251].

At the same time, there are several applications for which the disadvantages of lead-free ceramics are not such, and in some cases, they provide some advantages. One of the promising areas for using lead-free materials is bioprosthetics, for example, to restore hearing. In the human ear, there are two complementary mechanisms for the perception of sound: direct perception of sound waves and perception of vibrations of bone tissue. The idea of the using a piezo actuator is to directly

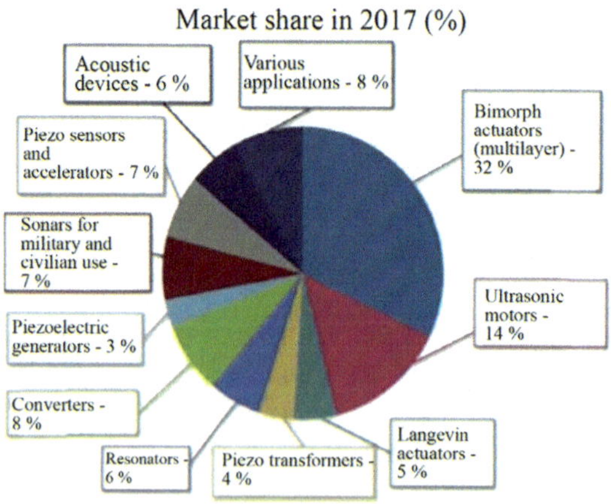

Fig. 1.21 Market structure of ferroelectric piezomaterials in 2017

influence the bone responsible for the second mechanism. This will make it possible, if necessary (telephone conversation in the subway), to transmit the necessary sound information, despite the level of ambient noise, and to develop a new generation of hearing aids that are significantly less demanding on the preservation of external auditory receptors [320] (see Fig. 1.22a). The high reliability and speed of this technology makes it promising for use in drives of propulsion systems (from ship diesel engines to printer heads). We also note that this technology does not require the implementation of record piezoelectric parameters. For its implementation, biocompatibility is much more important.

A promising area for the use of piezoceramic multilayer structures is highly efficient, fast-acting simulators of tactile sensations with a large signal bandwidth (Fig. 1.22b). The main requirements in this case are the largest possible value of $|d_{31}|$

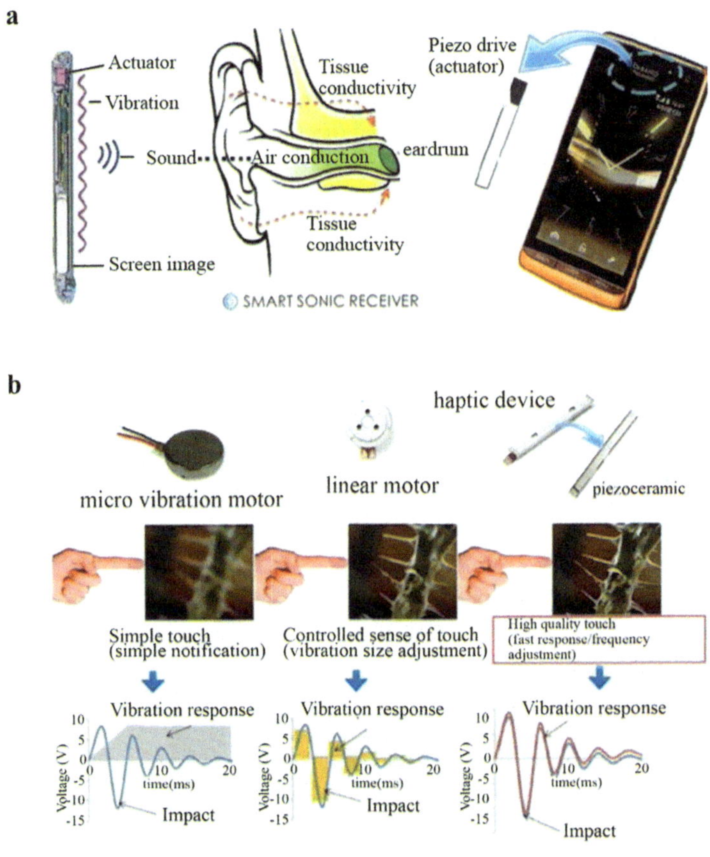

Fig. 1.22 **a** Using a piezo sound transducer to create a directional channel for transmitting audio information (Retrieved from http://www.kyoceramobile.com/smart-sonic-receiver/). **b** Comparison of various technologies for transmitting tactile sensations

with a minimum dielectric constant, which is quite achievable in lead-free ceramics [321].

Another example for which lead-free ceramics have undeniable competitive advantages is audio speakers. Speaker is a sensor that converts an electrical signal into sound waves. The advantages of piezoceramics in this direction are associated with the size factor. The use of piezotechnology makes it possible to significantly reduce the size of speakers and give them an almost arbitrary shape. Placing such sensors in direct contact with biological tissues places significant demands on their biocompatibility.

Another promising technology for using piezomaterials is micro-air intakes. Miniaturization of electronic householders and special equipment, increasing energy efficiency and wear resistance of instrument systems, is limited. In the previous decade, there were numerous attempts to introduce lead-free elements into production activities as an alternative to lead-containing ones. Unfortunately, almost all of them concerned the replacement of lead-containing piezoelements in already finished, long-used devices. However, it is obvious that the design of new instruments and devices based on lead-free piezo technologies is more promising. The most famous example of the development of lead-free material is the PIC700 material created in 2014 by PI Ceramic GmbH [322, 323], which has a consistently high K_p and is proposed for use in ultrasonic technology, transducers and sensors. The company announced high production volumes of the material in the amount of 50 kg per one production cycle (minimum batch). At the same time, Russian developments that are not inferior in most parameters [321] are still in demand only for scientific and development activities. Moreover, there is the lack of effective small-sized air-cooling systems. The most promising and closest to practical application are micro-air intakes (Fig. 1.23), based on piezoelectric elements, acting as an ultrasonic vibration pump that discharges air under high pressure [324].

Other examples include:

(i) lead-free camera autofocus module introduced by HanaUtech Co.; it was only slightly inferior in parameters to a similar device based on PZT [323];

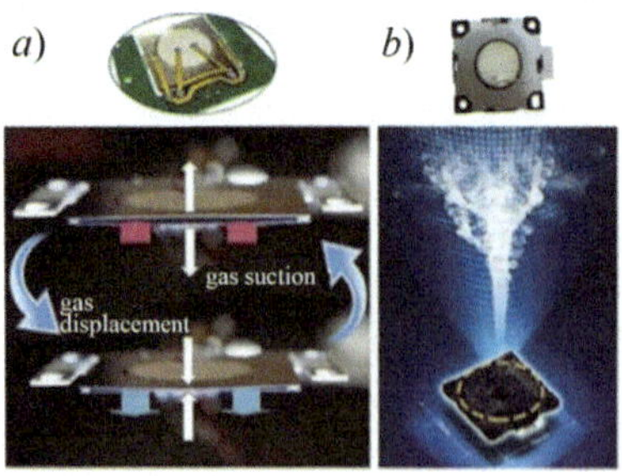

Fig. 1.23 Micro-air intake technologies, based on piezoelectric elements: **a** proposed by General ElectricCo. (Retrieved from http://www.geglobalresearch.com/innovation/gesdualpiezoelectriccoolingjets-dcj-cool-quiet); and **b** offered by Murata Co., Ltd. (Retrieved from http://www.murata.com/enus/products/mechatronics/fluid/feature)

(ii) an ultrasonic sensor based on lead-free material, presented by Honda Electronics Co., which, although inferior in basic parameters to lead-containing ones, could still compete in a number of them [322];
(iii) a material, based on alkali metal niobate tantalate for a non-resonant fuel knock sensor, has been patented by NGK Co. The sensor has a knock bandwidth of (5–15) kHz, demonstrates flexibility in relation to the type of system used and high temperature stability.

Note that a similar development exists in Russia [325].

1.6 Conclusions

This initial chapter presented a comprehensive literature review of environmentally friendly ferroelectric piezoelectric ceramic materials, and the methodologies employed for their preparation. The results were discussed for more than 40 years of work of the Active Materials Department of the Southern Federal University (former Rostov State University, Russia) aimed at studying the properties of solid solutions, based on alkali metal niobates, and creating environmentally friendly ferro-piezoceramic materials (FPCMs) on their basis for a wide range of applications.

An analysis of the dynamics of the retrospective developments in the field of lead-free FPCMs in Russia and around the world was carried out. Possible future directions of development are highlighted.

A discussion of the technological aspects of creating highly efficient piezoelectric ceramic materials, based on alkali metal niobates, has been performed. The experience of using basic technological methods aimed at optimizing the properties of ceramics is summarized. The most promising directions for modifying this group of materials have been identified.

Ceramic materials based on KNN are promising for the creation of new lead-free piezoceramic materials, while the selection of production conditions and modifying additives can significantly expand the range of possible practical applications, covering almost their entire range.

Mesoscopically inhomogeneous media with special electrical and magnetic properties were considered, including background, statement of main problems and application prospects. It was concluded on relevance to establish the patterns of formation of macroclines of multicomponent mesoscopically inhomogeneous media with special electrical and magnetic properties based on the analysis of their equilibrium and nonequilibrium phase diagrams and the thermodynamic path along which the specified state parameters are achieved.

Evolution of synthesis strategies of lead-free ferro-piezoelectric, multiferroic materials and metamaterials, based on spatially inhomogeneous media, was discussed in detail. It is obvious that, despite the rapid development of modern technology for the manufacture of special ceramics, the improvement of electrical devices requires the search for fundamentally new approaches to the creation of electroactive materials.

Today, one cannot but agree with the authors of most review works on lead-free ceramics that the motivation of electronic equipment manufacturers to introduce lead-free piezomaterials is still insufficient for a sharp increase in funding for design work and the next, final step on the ladder technology readiness levels. The technology itself to produce lead-free smart materials has not been fully developed; the parameters are specific, which requires significant costs for adapting production, and in some cases, the search for new engineering solutions.

References

1. Cross, E.: Lead-free at last. Nature **432**(4), 24–25 (2004)
2. Directive 2002/95/EC of the European Parliament and of the Council of 27 January 2003 on the restriction of the use of certain hazardous substances in electronic equipment. Official Journal of the European Union **3**, 19–23 (2003)
3. Reznichenko, L.A.: Phase states and properties of spatially inhomogeneous ferroactive media with different thermodynamic prehistory. DrSc Thesis (Physical and Mathematical Sciences). Rostov State University Press, Rostov-on-Don (2002) (in Russian)
4. Trofimov, A.I.: Ultrasonic Systems for Monitoring Curvatures of Technological Channels of Nuclear Reactors. Energoatomizdat, Moscow (1994). (in Russian)
5. Danciger, A. Ya., et al.: Highly efficient piezoceramic materials. Directory, Rostov-on-Don, **30** (1994) (in Russian)
6. Fesenko, E.G.: The Perovskite Family and Ferroelectricity. Atomizdat, Moscow (1972). (in Russian)
7. Seifert, K.: Lead-Free Piezoelectric Ceramics. Doctor of Engineering Thesis. Dermstadt (2010)
8. Verbenko, I.A., et al.: Lead-free piezoelectric ceramics and environmentally friendly technology for its production. Ecol. Indus. Prod. **4**, 45–47 (2007). (in Russian)
9. Haertling, G.E., et al.: Ferroelectric ceramics: History and technology. J. Am. Ceram. Soc. **82**, 797–818 (1999)
10. Ivliev, M.P., et al.: Phase states and features of dielectric properties of solid solutions of sodium-potassium niobates. Phys. **45**(10), 1886–1891 (2003)
11. Raevsky, I.P., et al.: Crystal chemical aspects of the influence of thermodynamic prehistory on the type of phase diagrams temperature-composition of solid solutions of sodium-lithium niobate and sodium-potassium niobate. J. Tech. Phys. **72**(6), 120–124 (2002)
12. Fesenko, E.G., et al.: New Piezoceramic Materials. Rostov State University Press, Rostov-on-Don (1983). (in Russian)
13. Rybyanets, A.N., et al.: Polycrystalline composite materials: design concept and implementation examples. Mater. Int. Sci. Pract. Conf. "Fundamental Problems of Piezoelectric Instrument Making", Mirea, Moscow, pp. 112–114 (2003) (in Russian)
14. Danciger, A. Ya., et al.: Highly Efficient Piezoceramic Materials. Search Optimization. Pike Publishing House, Rostov-on-Don (1995) (in Russian)

15. Okazaki, K.: Ceramic Engineering for Dielectrics. Tokyo (1969)
16. Reed, J.S.: Principles of Ceramic Processing. Wiley, New York, NY (1995)
17. Boldyrev, V.V.: Control of chemical reactions in the solid phase. Soros Educ. J. **5**, 49–55 (1996). (in Russian)
18. Lewis, J.A.: Colloidal processing of ceramics. J. Am. Ceram. Soc. **83**(10), 2341–2359 (2000)
19. Vorotilov, K.A., et al.: Ferroelectric storage devices: promising technologies and materials. Nano- Microsyst. Technol. **10**, 30–42 (2008). (in Russian)
20. Slyusar, V.: Metamaterials in antenna technology. History and basic principles of creation. Electron. Sci. Technol. Bus. **7**, 70–79 (2009) (in Russian)
21. Sigov, A.S., et al.: Thin ferroelectric films: production and integration prospects. Phys. Technol. **52**(4), 709–717 (2010) (in Russian)
22. Talanov, V.M., Ereyskaya, G.P.: Methods for the Synthesis of Nanostructures and Nanostructured Materials. South-Russian State Technical University Press, Novocherkassk (2011). (in Russian)
23. Williams, L., et al.: Nanotechnology has No Secrets. EKSMO, Moscow (2009). (in Russian)
24. Denz, C., et al.: Nonlinearities in Periodic Structures and Metamaterials. Springer Series in Optical Sciences **150** (2009)
25. Prigozhin, I.: Introduction to the Thermodynamics of Irreversible Processes. Foreign Literature, Moscow (1960). (in Russian)
26. Zhdanov, Yu.A.: Historical method in chemistry. Quest. Philos. **2**, 125–141 (1980). (In Russian)
27. Demikhovsky, V. Ya., et al.: Physics of Quantum Low-dimensional Structures. Logos, Moscow (2000) (in Russian)
28. Kobayashi, N.: Introduction to Nanotechnology. BINOM, Knowledge Laboratory, Moscow (2007)
29. Lin, D., et al.: Structure, piezoelectric and ferroelectric properties of Li- and Sb-modified $K_{0.5}Na_{0.5}NbO_3$ lead-free ceramics. J. Phys. D Appl. Phys. **40**, 3500–3505 (2007)
30. Yang, Z., et al.: Phase transitional behavior and electrical properties of lead-free $(K_{0.44}Na_{0.52}Li_{0.04})(Nb_{0.96-x}Ta_xSb_{0.04})O_3$ piezoelectric ceramics. Appl. Phys. Lett. **90**, 042911 (2007)
31. Guo, Y.P., et al.: $(Na_{0.5}K_{0.5})NbO_3$–$LiTaO_3$ lead-free piezoelectric ceramics. Mater. Lett. **59**(2–3), 241–244 (2005)
32. Saito, Y., et al.: Lead-free piezoceramics. Nature **432**, 84–87 (2004)
33. Burfoot, J.C., et al.: Polar Dielectrics and Their Applications. University of California Press, Berkeley (1979)
34. Fuyuno, I.: Toyota's production line leads from lab to road. Nature **435**(7045), 1026–1027 (2005)
35. Verbenko, I.A., et al.: Preparation and dielectric properties of lead-free ceramics of the composition $[(Na_{0.5}K_{0.5})_{1-x}Li_x](Nb_{1-y--z}Ta_ySb_z)O_3$. Inorgan. Mater. **45**(6), 762–768 (2009)
36. Higashide, K., et al.: Temperature dependence on the piezoelectric property of $(1-x)(Na_{0.5}K_{0.5})NbO_{3}$–$xLiNbO_3$ ceramics. J. Eur. Ceram. Soc. **27**(13–15), 4107–4110 (2007)
37. Hollenstein, E., et al.: Temperature stability of the piezoelectric properties of Li-modified KNN ceramics. J. Eur. Ceram. Soc. **27**(13–15), 4093–4097 (2007)
38. Du, H.L., et al.: The microstructure and ferroelectric properties of $(K_{0.5}Na_{0.5})NbO_3$–$LiNbO_3$ lead-free piezoelectric ceramics. Mater. Sci. Eng. B-Solid State Mater. Adv. Technol. **136**(2–3), 165–169 (2007)

39. Egerton, L., et al.: Isostatically hot-pressed sodium-potassium niobate transducer material for ultrasonic devices. Am. Ceram. Soc. Bull. **47**(12), 151–156 (1968)
40. Guo, Y.P., et al.: Ferroelectric-relaxor behavior of $(Na_{0.5}K_{0.5})NbO_3$-based ceramics. J. Phys. Chem. Solids **65**(11), 1831–1835 (2004)
41. Jaeger, R.E., Egerton, L.: Hot pressing of potassium-sodium niobates. J. Am. Ceram. Soc. **45**(5), 209–213 (1962)
42. Zhao, P., et al.: Enhanced dielectric and piezoelectric properties in $LiTaO_3$-doped lead-free (K, Na)NbO_3 ceramics by optimizing sintering temperature. Scripta Mater. **58**(6), 429–432 (2008)
43. Zhen, Y.H., et al.: Normal sintering of (K, Na)NbO_3-based ceramics: influence of sintering temperature on densification, microstructure, and electrical properties. J. Am. Ceram. Soc. **89**(12), 3669–3675 (2006)
44. Reisman, A., et al.: Metastability in niobate systems. J. Am. Chem. Soc. **81**(6), 1292–1295 (1959)
45. Ringgaard, E., et al.: Properties of lead-free piezoceramics based on alkali niobates. Ferroelectrics **319**, 323–333 (2005)
46. Lanfredi, S., et al.: Dense ceramics of $NaNbO_3$ produced from powders prepared by a new chemical route. J. Eur. Ceram. Soc. **20**(7), 983–990 (2000)
47. Reisman, A., et al.: Preparation of pure potassium metaniobate. J. Am. Chem. Soc. **78**(4), 719–720 (1956)
48. Miller, A.I., et al.: Preparation, electron microscopic studies of mechanically activated compositions of various compositions. In: Phase Transitions, Ordered States and New Materials (11), 1 (2012) (in Russian)
49. Rojac, T., et al.: Mechanochemical synthesis of $NaNbO_3$, $KNbO_3$ and $K_{0.5}Na_{0.5}NbO_3$. Sci. Sinter. **37**(1), 61–67 (2005)
50. Chen, J., Feng, D.: *In situ* TEM studies of para-ferroelectric phase transitions in $NaNbO_3$. Phys. Stat. Sol (A) **109**, 427–434 (1988)
51. Chen, K.P., et al.: Effect of borax addition on sintering and electrical properties of $(K_{0.5}Na_{0.5})NbO_3$ lead-free piezoceramics. Ceram. Int. **41**(8), 10232–10236 (2015)
52. Abrahams, S. C., et al.: Ferroelectric lithium niobate. 4. Single crystal neutron diffraction study at 24 °C. J. Phys. Chem. Solids **27**(6–7), 1013–1018 (2012)
53. Haertling, G.H.: Properties of hot-pressed ferroelectric alkali niobate ceramics. J. Am. Ceram. Soc. **50**(6), 329–330 (1967)
54. Li, J. F., et al.: Ferroelectric and piezoelectric properties of fine-grained $Na_{0.5}K_{0.5}NbO_3$ lead-free piezoelectric ceramics prepared by spark plasma sintering. J. Am. Ceram. Soc. **89**(2), 706–709 (2006).
55. Wang, R.P., et al.: Fabrication and characterization of potassium-sodium niobate piezoelectric ceramics by spark-plasma-sintering method. Mater. Res. Bull. **39**(11), 1709–1715 (2004)
56. Zuo, R.Z., et al.: Sintering and electrical properties of lead-free $Na_{0.5}K_{0.5}NbO_3$ piezoelectric ceramics. J. Am. Ceram. Soc. **89**(6), 2010–2015 (2006)
57. Fisher, J.G., et al.: Growth of (Na, K, Li)(Nb, Ta)O_3 single crystals by solid state crystal growth. J. Eur. Ceram. Soc. **27**(13–15), 4103–4106 (2007)
58. Titov, S.V.: Effects of modification in alkali metal niobates, lead titanate, lead zirconate and their solid solutions. Ph.D. Thesis (Physical and Mathematical Sciences). Rostov State University Press, Rostov-on-Don (2001) (in Russian)
59. Grineva, L. D.: Study of the patterns of changes in physical properties when modifying ferroelectric solid solutions of complex oxides with a perovskite-type structure. Ph.D. Thesis (Physical and Mathematical Sciences). Rostov State University Press, Rostov-on-Don (1975) (in Russian).

60. Chernyshkov, V.A.: Physical properties of piezoceramics of lithium metaniobate and solid solutions based on it. Ph.D. Thesis (Physical and Mathematical Sciences). Rostov State University Press, Rostov-on-Don (1990) (in Russian)
61. Danciger, A.Ya., et al: Multicomponent systems of ferroelectric complex oxides: physics, crystal chemistry, technology. Design Aspects of ferro-piezoelectric materials. Rostov State University Press, Rostov-on-Don, **1** (2001), **2** (2002) (in Russian)
62. Gavrilyatchenko, S.V., et al.: The stabilization of the properties of ferro-piezoelectric ceramics for filters. Ferroelectrics **214**, 273–274 (1998)
63. Ananyeva, A.A., et al.: Piezo- and dielectric properties of modified potassium-sodium niobate ceramics. In: Methods for Obtaining and Studying Piezo- and Ferroelectric Materials and Feedstock for Them, pp. 89–95. MDNTP, Moscow (1966) (in Russian)
64. Ananyeva, A.A., Ugryumova, M.A.: Development and research of ferroelectric ceramics for electroacoustic transducers, pp. 201–207. In: Barium Titanate. Science, Moscow (1973) (in Russian)
65. Ugryumova, M.A.: Research and development of new piezoceramics based on solid solutions of barium and lead niobate, potassium and sodium niobate. Ph.D. Thesis (Technical Sciences). Moscow (1968) (in Russian)
66. Ugryumova, M.A.: Piezoceramics based on (K, Na)NbO_3 solid solutions. Electron. Technol. Ser. 14. Mater. (Inorganic Dielectrics) **14**(20), 130–138 (1969) (in Russian)
67. Dambekalne, M.Ya., et al.: The influence of modification on the piezoelectric properties and texture of ceramics of solid solutions of potassium and sodium niobate. Electron. Technol. Ser. 14. Mater. (Inorganic Dielectrics) **8**, 33–35 (1969) (in Russian)
68. Freidenfeld, E.Zh.: Physico-chemical principles of synthesis and electrical properties of ceramic and glass-crystalline piezoelectrics. DrSc Thesis (Technical Sciences), Moscow (1972) (in Russian)
69. Yanson, G.D., et al.: Ceramic solid solutions of potassium and sodium metaniobates modified with alkaline earth metal ions. In: Piezo- and Ferroelectric Materials and Their Application, pp. 36–38. MDNTP, Moscow (1972) (in Russian)
70. Yanson, G.D., et al.: Kinetics of sintering of ferroelectric ceramics based on alkali niobates modified with oxides of alkaline earth elements, pp. 40–45. In: Proceedings on IV Intersectoral Meeting, Moscow (1972) (in Russian)
71. Grineva, L.D., et al.: Study of the patterns of changes in physical properties when modifying solid solutions based on alkali metal niobates. In: Crystallization and Properties of Crystals, vol. (6), pp. 57–65. Novocherkassk (1979) (in Russian)
72. Nassan, K., et al.: Stacking-fault model for stoichiometry deviations in $LiNbO_3$ and $LiTaO_3$ and the effect on the curie temperature. J. Appl. Phys. **41**, 533–537 (1970)
73. Ahtee, M., et al.: Lattice parameters and tilted octahedra in sodium-potassium niobate solid solutions. Acta Crystallogr. A **32**, 434–445 (1976)
74. Ahtee, M., et al.: Structural phase-transitions in sodium-potassium niobate solid-solutions by neutron powder diffraction. Acta Crystallogr. A **34**, 309–317 (1978)
75. Malic, B., et al.: Alkaline-earth doping in (K, Na)NbO_3 based piezoceramics. J. Eur. Ceram. Soc. **25**(12), 2707–2711 (2005)
76. Tashiro, S., et al.: Influence of mixing condition and nonstoichiometry on piezoelectric properties of (K, Na, Pb)NbO_3 ceramics. Jpn. J. Appl. Phys. **43**(9B), 6711–6715 (2004)
77. Ming, B.Q., et al.: Piezoelectric properties of (Li, Sb, Ta) modified (Na, K)NbO_3 lead-free ceramics. J. Appl. Phys. **101**(5), 054103 (2007)
78. Rubio-Marcos, F., et al.: Sintering and properties of lead-free (K, Na, Li)(Nb, Ta, Sb)O_3 ceramics. J. Eur. Ceram. Soc. **27**(13–15), 4125–4129 (2007)
79. Yoo, J., et al.: Piezoelectric and dielectric properties of (Li–Na–K)(Nb–Ta–Sb)O_3 ceramics with variation in poling temperature. Jpn. J. Appl. Phys. **45**(9B), 7444–7448 (2006)

80. Chu, S.Y., et al.: Properties of (Na, K)NbO_3 and (Li, Na, K)NbO_3 ceramic mixed systems. Ferroelectrics **287**, 23–33 (2003)
81. Rai, R., et al.: Impedance spectroscopy and piezoresponse force microscopy analysis of lead-free (1-x)$K_{0.5}Na_{0.5}NbO_3$–$x$$LiNbO_3$ ceramics. Current Applied Physics **13**, 430–440 (2013).
82. Zhao, P., et al.: High piezoelectric d_{33} coefficient in Li-modified lead-free (Na, K)NbO_3 ceramics sintered at optimal temperature. Appl. Phys. Lett. **90**(24), 242909 (2007)
83. Wu, J.G., et al.: Piezoelectric properties of $LiSbO_3$-modified ($K_{0.48}Na_{0.52}$)NbO_3 lead-free ceramics. Jpn. J. Appl. Phys. **46**(11), 7375–7377 (2007)
84. Zhang, S., et al.: Piezoelectric properties in perovskite 0.948($K_{0.5}Na_{0.5}$])NbO_3–0.052$LiSbO_3$ lead-free ceramics. J. Appl. Phys. **100**(10), 104108 (2006)
85. Hollenstein, E., et al.: Piezoelectric properties of Li- and Ta-modified ($K_{0.5}Na_{0.5}$)NbO_3 ceramics. Appl. Phys. Lett. **87**(18), 182905 (2005)
86. Matsubara, M., et al.: Effect of Li substitution on the piezoelectric properties of potassium sodium niobate ceramics. Jpn. J. Appl. Phys. **44**(8), 6136–6142 (2005)
87. Guo, Y.P., et al.: Phase transitional behavior and piezoelectric properties of ($Na_{0.5}K_{0.5}$)NbO_3–$LiNbO_3$ ceramics. Appl. Phys. Lett. **85**(18), 4121–4123 (2004)
88. Chernyshkov, V.A., et al.: Influence of composition non-stoichiometry and conditions of production on the properties of lithium ferropiezoceramic metaniobate. In: Proceedings on VIII All-Union Conference State and Prospects of Development of Methods for Obtaining and Analyzing Ferrite, Ferropiezoelectric, Capacitor and Resistive Materials and Raw Materials for Them. Pt. 1. Donetsk, 102 (1983) (in Russian)
89. Li, F., et al.: Fabrication of transparent electro-optic $(K_{0.5}Na_{0.5})_{1-x}Li_xNb_{1-x}Bi_xO_3$ lead-free ceramics. J. Eur. Ceram. Soc. **33**, 123–130 (2013)
90. Rai, R., et al.: Influence of Li and La content on phase structures and electrical properties of $K_{0.5}Na_{0.5}NbO_3$ lead-free piezoelectric ceramics. J. Alloys Compds. **577**, 575–580 (2013)
91. Saito, Y., et al.: High performance lead-free piezoelectric ceramics in the (K, Na)NbO_3–$LiTaO_3$ solid solution system. Ferroelectrics **338**, 1433–1448 (2006)
92. Wang, X., et al.: Compositional dependence of phases structure and electrical properties in $(K_{0.50}Na_{0.50})_{0.97}Bi_{0.01}(Nb_{1-x}Zr_x)O_3$ lead-free ceramics. Ceram. Int. **139**, 8021–8024 (2013)
93. Matsubara, M., et al.: Piezoelectric properties of $(K_{0.5}Na_{0.5})(Nb_{1-x}Ta_x)O_3$–$K_{5.4}CuTa_{10}O_{29}$ ceramics. J. Appl. Phys. **97**(11), 114105 (2005)
94. Chen, Q., et al.: Piezoelectric properties of $K_4CuNb_8O_{23}$ modified ($Na_{0.5}K_{0.5}$)NbO_3 lead-free piezoceramics. J. Appl. Phys. **102**(10), 104109 (2007)
95. Park, H.-Y., et al.: Low-temperature sintering and piezoelectric properties of CuO-added 0.95($Na_{0.5}K_{0.5}$)NbO_3–0.05$BaTiO_3$ ceramics. J. Am. Ceram. Soc. **90**(12), 4066–9 (2007)
96. Park, S.-J., et al.: Effect of CuO on the sintering temperature and piezoelectric properties of lead-free 0.95($Na_{0.5}K_{0.5}$)NbO_3–0.05$CaTiO_3$ ceramics. Mater. Res. Bull. **43**(12), 3580–3586 (2008)
97. Li, E., et al.: Influence of CuO on the structure and piezoelectric properties of the alkaline niobate-based lead-free ceramics. J. Am. Ceram. Soc. **90**(6), 1787–1791 (2007)
98. Lin, D., et al.: Structure, dielectric, and piezoelectric properties of CuO-doped $K_{0.5}Na_{0.5}NbO_3$–$BaTiO_3$ lead-free ceramics. J. Appl. Phys. **102**(7), 074113 (2007)
99. Yang, M.-R., et al.: Disk-type piezoelectric transformer of a $Na_{0.5}K_{0.5}NbO_3$–$CuNb_2O_6$ lead-free ceramic for driving T5 fluorescent lamp. J. Alloys Compd. **522**, 3–8 (2012)
100. Yang, M.-R., et al.: An ultrasonic therapeutic transducer using lead-free $Na_{0.5}K_{0.5}NbO_3$–$CuNb_2O_6$ ceramics. J. Alloys Compd. **507**, 433–438 (2010)
101. Yang, M.-R., et al.: Fabrication and characterization of $Na_{0.5}K_{0.5}NbO_3$–$CuNb_2O_6$ lead-free step-down piezoelectric transformers. J. Appl. Phys. **110**, 044503 (2011)

102. Yang, M.-R., et al.: Piezoelectric and ferroelectric properties of CN-doped $K_{0.5}Na_{0.5}NbO_3$ lead-free ceramics. J. Appl. Phys. **108**, 094103 (2010)
103. Sadykov, Kh.A., et al.: Features of the synthesis and sintering of environmentally friendly materials with the participation of sodium and copper niobates. Ecol. Indus. Prod. **2**, 44–49 (2013). (in Russian)
104. Zhang, S., et al.: Modified $(K_{0.5}Na_{0.5})NbO_3$ based lead-free piezoelectrics with broad temperature usage range. Appl. Phys. Lett. **91**(13), 132913 (2007)
105. Kosec, M., et al.: New lead-free relaxors based on the $K_{0.5}Na_{0.5}NbO_3$–$SrTiO_3$ solid solution. J. Mater. Res. **19**(6), 1849–1854 (2004)
106. Du, H.L., et al.: Design and electrical properties' investigation of $(K_{0.5}Na_{0.5})NbO_3$–$BiMeO_3$ lead-free piezoelectric ceramics. J. Appl. Phys. **104**(3), 034104 (2008)
107. Chang, R.-C., et al.: An investigation of $(Na_{0.5}K_{0.5})NbO_3$–$CaTiO_3$ based lead-free ceramics and surface acoustic wave devices. J. Eur. Ceram. Soc. **27**(16), 4453–4460 (2007)
108. Guo, Y., et al.: Dielectric and piezoelectric properties of lead-free $(Na_{0.5}K_{0.5})NbO_3$–$SrTiO_3$ ceramics. Solid State Commun. **129**(5), 279–284 (2004)
109. Kusumoto, K.: Dielectric and piezoelectric properties of $KNbO_3$–$NaNbO_3$–$LiNbO_3$–$SrTiO_3$ ceramics. Jpn. J. Appl. Phys. **45**(9B), 7440–7443 (2006)
110. Wang, R.P., et al.: Phase diagram and enhanced piezoelectricity in the strontium titanate doped potassium–sodium niobate solid solution. Phys. Status Solidi A Appl. Mater. Sci. **202**(6), R57–R59 (2005)
111. Chen, Ch.-S., et al.: Effects of $CaTiO_3$ addition on microstructures and electrical properties of $Na_{0.52}K_{0.48}NbO_3$ lead-free piezoelectric ceramics. Ceram. Int. **39**, S125–S128 (2013)
112. Taub, J., et al.: Phase structure and piezoelectric properties of Ca- and Ba-doped $K_{1/2}Na_{1/2}NbO_3$ lead-free ceramics. Ceram. Int. **39**, 3555–3561 (2013)
113. Bobnar, V., et al.: Electrostrictive effect in lead-free relaxor $K_{0.5}Na_{0.5}NbO_3$–$SrTiO_3$ ceramic system. J. Appl. Phys. **98**(2), 024113 (2005)
114. Kroupa, J., et al.: Electro-optic properties of KNN-STO lead-free ceramics. J. Phys. D-Appl. Phys. **38**(5), 679–681 (2005)
115. Bobnar, V., et al.: Relaxor like dielectric dynamics in the lead-free $K_{0.5}Na_{0.5}NbO_3$–$SrZrO_3$ ceramic system. Journal of Applied Physics 101(7), 074103 (2007).
116. Zhang, B., et al.: Rhombohedral-orthorhombic phase coexistence and electrical properties of Ta and $BaZrO_3$ co-modified (K, Na)NbO_3 lead-free ceramics. Curr. Appl. Phys. **13**, 1647–1650 (2013)
117. Chen, T., et al.: Piezoelectric behavior of $(1-x)K_{0.50}Na_{0.50}NbO_3$–$xBa_{0.80}Ca_{0.20}ZrO_3$ lead-free ceramics. Ceram. Int. **39**, 6619–6622 (2013)
118. Pibosic, I., et al.: Electrical properties of $KNbO_3$–$BaTiO_3$ ceramics. Euro Ceram. VII, Pt. 1–3 **206**(2), 1401–1404 (2002)
119. Park, H.Y., et al.: Microstructure and piezoelectric properties of $0.95(Na_{0.5}K_{0.5})NbO_3$–$0.05BaTiO_3$ ceramics. Appl. Phys. Lett. **89**(6), 062906 (2006)
120. Ahn, C.W., et al.: Effect of MnO_2 on the piezoelectric properties of $(1-x)(Na_{0.5}K_{0.5})NbO_3$–$xBaTiO_3$ ceramics. Jpn. J. Appl. Phys. Lett. Exp. Lett. **44**(42–45), L1361–L1364 (2005)
121. Zuo, R., et al.: Phase structures and electrical properties of new lead-free $(Na_{0.5}K_{0.5})NbO_3$–$(Bi_{0.5}Na_{0.5})TiO_3$ ceramics. Appl. Phys. Lett. **90**(9), 092904 (2007)
122. Li, Y.M., et al.: Dielectric and piezoelectric properties of lead-free $(Na_{0.5}Bi_{0.5})TiO_3$–$NaNbO_3$ ceramics. Mater. Sci. Eng. B Solid-State Mater. Adv. Technol. **112**(1), 5–9 (2004)
123. Takenaka, T., et al.: Lead-free piezoelectric ceramics based on $(Bi_{1/2}Na_{1/2})TiO_3$–$NaNbO_3$. Ferroelectrics **196**(1–4), 175–178 (1997)

124. Wada, T., et al.: Dielectric and piezoelectric properties of $(A_{0.5}Bi_{0.5})TiO_3$–$ANbO_3$(A = Na, K) systems. Jpn. J. Appl. Phys. s**40**(9B), 5703–5705 (2001).
125. Nagata, H., et al.: Lead-free piezoelectric ceramics of $(Bi_{1/2}Na_{1/2})TiO_3$–$KNbO_3$–1/2(Bi_2O_3·Sc_2O_3) system. Jpn. J. Appl. Phys. **37**(9B), 5311–5314 (1998)
126. Ravez, J., et al.: Relaxor ferroelectricity in ceramics with composition $Ba_{1-x}K_x(Ti_{1-x}Nb_x)O_3$. Mater. Lett. **36**(1–4), 81–84 (1998)
127. Guo, Y. P., et al.: Structure and electrical properties of lead-free $(Na_{0.5}K_{0.5})NbO_3$–$BaTiO_3$ ceramics. Jpn. J. Appl. Phys. **43**(9B), 6662–6666 (2004)
128. Wang, Y., et al.: High Curie temperature of (Li, K, Ag)-modified $(K_{0.50}Na_{0.50})NbO_3$ lead-free piezoelectric ceramics. J. Alloys Compd. **472**, 6–8 (2009)
129. Wang, Y.W., et al.: Microstructure and electrical properties of $[(K_{0.50}Na_{0.50})_{0.95-x}Li_{0.05}Ag_x](Nb_{0.95}Ta_{0.05})O_3$ lead-free ceramics. J. Am. Ceram. Soc. **91**(8), 2772–2775 (2008)
130. Wu, J., et al.: Effects of Ag content on the phase structure and piezoelectric properties of $(K_{0.44-x}Na_{0.52}Li_{0.04}Ag_x)$ $(Nb_{0.91}Ta_{0.05}Sb_{0.04})O_3$ lead-free ceramics. Appl. Phys. Lett. **91**(13), 132914 (2007)
131. Lin, D., et al.: Piezoelectric and ferroelectric properties of $K_xNa_{1-x}NbO_3$ lead-free ceramics with MnO_2 and CuO doping. J. Alloy. Compd. **461**(1–2), 273–278 (2008)
132. Park, S.H., et al.: Microstructure and piezoelectric properties of ZnO-added $(Na_{0.5}K_{0.5})NbO_3$ ceramics. Japanese J. Appl. Phys. 2 Lett. Exp. Lett. **43**(8B), L1072–L1074 (2004)
133. Malic, B., et al.: Influence of zirconia addition on the microstructure of $K_{0.5}Na_{0.5}NbO_3$ ceramics. J. Eur. Ceram. Soc. **28**(6), 1191–1196 (2008)
134. Du, H., et al.: Microstructure, piezoelectric, and ferroelectric properties of Bi_2O_3-added $(K_{0.5}Na_{0.5})NbO_3$ lead-free ceramics. J. Am. Ceram. Soc. **90**(9), 2824–2829 (2007)
135. Kakimoto, K., et al.: Ferroelectric and piezoelectric properties of $KNbO_3$ ceramics containing small amounts of $LaFeO_3$. Jpn. J. Appl. Phys. **42**(9B), 6102–6105 (2003)
136. Kakimoto, K., et al.: Lead-free $KNbO_3$ piezoceramics synthesized by pressure-less sintering. J. Eur. Ceram. Soc. **25**(12), 2719–2722 (2005)
137. Masuda, I., et al.: Ferroelectric property and crystal structure of $KNbO_3$ based ceramics. J. Electroceram. **13**(1–3), 555–559 (2004)
138. Du, J., et al.: $SmAlO_3$-modified $(K_{0.5}Na_{0.5})_{0.95}Li_{0.05}Sb_{0.05}Nb_{0.95}O_3$ lead-free ceramics with a wide sintering temperature range. Mater. Sci. Eng. B **178**, 1027–1031 (2013)
139. Shin, D.-J., Koh, J.-H.: Effects of calcination temperature on the piezoelectric properties of lead-free Ag doped (Na, K)NbO_3–$LiTaO_3$ piezoelectric ceramics. J. Alloy. Compd. **555**, 390–394 (2013)
140. Wattanawikkam, C., et al.: Low temperature fabrication of lead-free KNN-LS-BS ceramics via the combustion method. Ceram. Int. **39**, S399–S403 (2013)
141. Zhao, X.K., et al.: Controlling of the polymorphic phase transition on lead-free piezoelectric $Li_{0.08}(Na_{0.52-x}K_{0.48})_{0.92}NbO_3$ ceramics. Ceram. Int. **39**, 4475–4480 (2013)
142. Lin, D., et al.: Microstructure, phase transition, and electrical properties of $(K_{0.5}Na_{0.5})_{1-x}Li_x(Nb_{1-y}Ta_y)O_3$ lead-free piezoelectric ceramics. J. Appl. Phys. **102**(3), 034102–034107 (2007)
143. Lv, Y.G., et al.: Tantalum influence on physical properties of $(K_{0.5}Na_{0.5})(Nb_{1-x}Ta_x)O_3$ ceramics. Mater. Res. Bull. **44**(2), 284–287 (2009)
144. Andryushin, K.P., et al.: Lead-free ferroelectric materials with a wide range of mechanical quality factors, dielectric and piezoelectric activity. Constr. Compos. Mater. **2**, 53–59 (2011). (in Russian)
145. Carvalho, T.T., Tavares, P.B.: Synthesis and thermodynamic stability of multiferroic $BiFeO_3$. Mater. Lett. **62**, 3984–3986 (2008)

146. Phapale, S., et al.: Standard enthalpy of formation and heat capacity of compounds in the pseudo-binary Bi_2O_3–Fe_2O_3 system. J. Nucl. Mater. **373**, 137–141 (2008)
147. Popov, Yu.I., et al.: Linear magnetoelectric effect and phase-transitions in ferrite bismuth $BiFeO_3$. JETP Lett. **57**(1), 65–68 (1993)
148. Pradhan, S.K., Roul, B.K.: Improvement of multiferroic and leakage property in monophasic $BiFeO_3$. Physica B **406**, 3313–3317 (2011)
149. Shapiro, Z.E., et al.: Methods of preparation and physicochemical properties of potassium methaniobate. In: Synthesis of Ferroelectric Ceramics and Comprehensive Study of the Results, pp. 3–33. Riga (1981) (in Russian)
150. Shapiro, Z. N., et al.: Methods for obtaining alkali metal niobates. Reagents and highly pure substances. In: Niitekhim, Moscow, 72 (1978) (in Russian)
151. Smolensky, G.A., Chupis, I.E.: Ferromagnets. UFN, 137–415 (1982) (in Russian)
152. Venevtsev, Yu.N., et al.: Ferromagnets. Science, Moscow (1982). (in Russian)
153. Verbenko, I.A., et al.: Features of environmentally friendly technology for manufacturing lead-free ferroelectric ceramics based on alkali metal niobates. Ecol. Indus. Prod. **3**, 22–30 (2008). (in Russian)
154. Verbenko, I. A., et al.: Modification as a method for creating non-lead-containing ferroelectric ceramics with high piezosensitivity and piezoanisotropy. Third Int. Environmental Congress (Fifth Int. Sci. Techn. Conf.) (ELPIT 2011), Tolyatti (2011) (In Russian).
155. Wang, Y.P., et al.: Electrical and magnetic properties of single-phased and highly resistive ferroelectromagnet $BiFeO_3$ ceramic. J. Phys. D. Appl. Phys. **39**, 2019–2023 (2006)
156. Zvezdin, A.K., Pyatakov, A.P.: Phase transitions and giant magnetoelectric effect in multiferroics. Physica **174**(4), 465–470 (2004)
157. Landau, L.D., Lifshits, E.M.: Statistical Physics. Part 1. Nauka, Moscow (1976) (in Russian)
158. Lines, M.E., Glass, A.M.: Principles and Application of Ferroelectrics and Related Materials. Clarendon Press, Oxford (1977)
159. Smolensky, G.A., et al.: Physics of Ferroelectric Phenomena, p. 396. Science, Leningrad (1985) (in Russian)
160. Strategy for the development of science and innovation in the Russian Federation for the period until 2015. Approved by the Interdepartmental Commission on Science and Innovation Policy (2006) (in Russian)
161. Quezel, S., et al.: Magnetic structure of the perovskite-like compound $TbMnO_3$. Physica B+C, Part 2, **86–87**, 916–918 (1977)
162. Cipriani, J., et al.: Raman scattering by two-magnon excitations in CoF_2. Phys. Lett. A **34**(3), 187–188 (1971)
163. Slyusar, V.: Metamaterials in antenna structures. Electron. Sci. Technol. Bus. (8), 66–70 (2009)
164. Noginov, M.A., et al.: Demonstration of a spacer-based nanolaser. Nature **460**, 1110–1113 (2009)
165. Ozbay, E.: Plasmonics: merging photonics and electronics at nanoscale dimensions. Science **311**, 189–193 (2006)
166. Pendry, J.B.: Extremely low frequency plasmons in metallic mesostructures. Phys. Rev. Lett. **76**(25), 4773–4776 (1996)
167. Mikheeva, O.P., Sidorov, A.I.: Absorption and scattering of infrared radiation by vanadium dioxide nanoparticles with a metal shell. J. Tech. Phys. **73**(5), 79–83 (2003) (in Russian)
168. Simakov, I.G., et al.: Acoustoelectric interaction in planetary systems of nanosized metal particles. Condensed States, 317–319 (2007) (in Russian)
169. Miyashita, T.: Sonic crystals and sonic waveguides. Measurem. Sci. Technol. (16), 47 (2005)
170. Joannopoulos, J.D., et al.: Photonic Crystals: Molding the Flow of Light. Princeton University Press, Princeton (2008)

171. Nikitov, S.A., et al.: Magneto-photonic and magnonic crystals based on ferrite films—new types of magnetic functional materials. Adv. Sci. Technol. **45**, 1355–1363 (2006)
172. Leonhardt, U.: To invisibility and beyond. Nature **471**, 292–293 (2011)
173. Lisenkov, I.V.: Features of the propagation of acoustic waves in liquid and solid composite media with cylindrical inhomogeneities. Ph.D. Thesis (Physical and Mathematical Sciences), Moscow (2010) (in Russian)
174. Russell, P.S.J.: Photonic crystal fibers. Science **299**, 358–362 (2003)
175. Reznichenko, L.A., et al.: Modeling of high temperature ferropiezoelectric ceramic materials. Inorg. Mater. **31**(6), 840–842 (1995)
176. Kuzminov, Yu.S.: Lithium Niobate and Tantalate. Science, Moscow (1975). (in Russian)
177. Verbenko, I.A.: Multicomponent multifunctional electroactive media with different thermodynamic background. PhD Thesis (Physical and Mathematical Sciences). Southern Federal University Press, Rostov-on-Don (2009) (in Russian)
178. Avvakumov, E.G., Gusev, A.A.: Mechanical Activation Methods in the Processing of Natural and Man-Made Raw Materials. Geo, Novosibirsk (2009). (in Russian)
179. Chao, W., et al.: Sol–gel synthesis and characterization of lead-free LNKN nanocrystalline powder. J. Crystal Growth **310**, 4635–4639 (2008)
180. Kumar, M., et al.: Large magnetization and weak polarization in sol-gel derived $BiFeO_3$ ceramics. Mater. Lett. **62**, 1159–1161 (2008)
181. Bogatin, A.S., Bulanova, A.L., Andreev, E.V., et al.: Identifying dielectric spectra: cole–cole distribution, constant phase element, and lognormal distribution. Bull. Russ. Acad. Sci. Phys. **84**, 1425–1426 (2020)
182. Bondarenko, E.I., et al.: The influence of the granulometric state of niobium pentoxide on the phase diagram of the (Na, Li)NbO_3 system and the thermodynamic characteristics of its solid solutions. In: Proceedings of XIII Conference on Physics of Ferroelectrics. Tver **2**, 88 (1992) (in Russian)
183. Fesenko, E. G., et al.: Preparation, properties and application of niobate piezoceramics. In: Proceedings Increasing the Efficiency of Using Niobium in the National Economy. Part 1: Raw Materials Sources, Raw Material Processing Technology, Use of Niobium). Giredmet, Moscow, 91–95 (1986) (in Russian)
184. Kovalenko, M.I., et al.: Dependence of the strength properties of niobate ferroelectric ceramics on the dispersion of niobium pentoxide. Lett. Tech. Phys. **26**(24), 1–6 (2000) (in Russian)
185. Kovalenko, M.I., et al.: Strength and electrical properties of fluorine-containing niobate ferroelectric ceramics. Lett. Tech. Phys. **26**(33), 30–37 (2000) (in Russian)
186. Kuznetsova, E.M., et al.: Polymorphism of niobium pentoxide and properties of alkali metal niobates—the basis of ferropiezoelectric materials. Tech. Phys. Lett. **27**(5), 36–41(2001) (in Russian)
187. Reznichenko, L.A., et al.: Development of technology for producing niobium pentoxide for niobate piezoceramics. In: Proceedings on Increasing the Efficiency of Using Niobium in the National Economy. Pt. 1: Raw Materials Sources, Raw Material Processing Technology, Use of Niobium), pp. 96–100. Giredmet, Moscow (1986) (in Russian)
188. Reznichenko, L.A., et al.: Dependence of the properties of niobate piezoceramics on the impurity composition of niobium pentoxide. Inorg. Mater. **24**(10), 1708–1715 (1988)
189. Reznichenko, L. A., et al.: Dependence of the properties of niobate piezoceramics on the granulometric composition of Nb_2O_5. VINITI, Moscow, No. 7660. B-88, 10.25.88 (1988) (In Russian).
190. Reznichenko, L.A., et al.: Influence of the dispersity of the initial niobium pentoxide on the properties of piezoceramics. Inorg. Mater. **26**(10), 2184–2189 (1990)
191. Reznichenko, L.A., et al.: Sintering of niobate piezoceramics from fine powders. Inorg. Mater. **26**(10), 2190–2193 (1990)

192. Reznichenko, L.A., et al.: Influence of dispersion characteristics of raw materials on the properties of niobate piezoceramics. Inorg. Mater. **29**(6), 862–867 (1993)
193. Reznichenko, L.A., et al.: Dependence of the properties of niobate piezoceramics on the dispersion of Nb_2O_5 of various grades. Inorg. Mater. **29**(7), 1004–1007 (1993)
194. Reznichenko, L.A., et al.: To determine the requirements for the granulometric composition of niobium pentoxide of various qualifications. VINITI, Moscow, No. 343-B93, 02.11.93 (1993) (in Russian)
195. Reznichenko, L.A., et al.: Liquid phase in alkali metal niobates. In: Proceedings on 7th International Conference on Physics of Ferroelectric Semiconductors (IMFS-7), pp. 149–151. Rostov-on-Don (1996) (in Russian)
196. Reznichenko, L.A., et al.: The phenomenon of secondary discontinuous recrystallization in niobate ceramics. In: Proceedings on 7th International Conference on Physics of Ferroelectric Semiconductors (IMFS-7), pp. 170–171. Rostov-on-Don (1996) (in Russian)
197. Reznichenko, L.A., et al.: Secondary discontinuous recrystallization as a manifestation of the self-organization effect in niobate ferroelectric ceramics. In: Procedings on 8th International Conference on Physics of Ferroelectric Semiconductors (IMFS-8), pp. 159–160. Rostov-on-Don (1998) (in Russian)
198. Reznichenko, L.A., et al.: Ordering and annihilation of defects by rearrangement of coordination polyhedra with the formation of block structures niobium oxides of various compositions. In: International Symposium "Ordering in Minerals and Alloys" ("OMA-2000"), pp. 111–126. Rostov State University Press, Rostov-on-Don (2000) (in Russian)
199. Reznichenko, L.A., et al.: Secondary discontinuous recrystallization as a manifestation of the effect of self-organization in niobate ferroelectric ceramics. Ferroelectrics **247**(1–3), 95–105 (2000)
200. Reznichenko, L.A., et al.: Features of recrystallization processes in complex oxide ceramics based on alkali metal niobates. In: Proceedings on International Symposium of Order, Disorder and Properties of Oxides (ODPO-2001), Sochi, 261–272 (2001) (in Russian)
201. Reznichenko, L.A., et al.: The mechanism of compaction of fluorine-containing oxides. In: Proceedings on International Symposium of Order, Disorder and Properties of Oxides (ODPO-2001), Sochi, pp. 273–278 (2001) (in Russian)
202. Reznichenko, L.A., et al.: Kinetics of sintering of complex oxide materials of various chemical compositions. In: Proceedings on International Symposium of Order, Disorder and Properties of Oxides (ODPO-2001), Sochi, pp. 279–288 (2001) (in Russian)
203. Reznichenko, L.A., et al.: Crystallographic shift in niobium oxides of various compositions. Crystallography **49**(5), 909–916 (2004) (in Russian)
204. Reznichenko, L.A., Razumovskaya, O.N., Ivanova, L.S.: Optimization of raw materials for niobate piezomaterials. News of North-Caucasus Scientific Center of High School. Nat. Sci. (2), 92–97 (1991) (in Russian)
205. Razumovskaya, O.N., et al.: Optimization of the processes of synthesis and sintering of bismuth ferrite and its solid solutions with ferrites of rare earth elements. In: Proceedings of VII International Science-Technology Conference Fundamental Problems of Radio-Electronic Instrument Making (INTERMATIC–2009), pp. 155–160. MIREA, Moscow (2009) (in Russian)
206. Farhadi, S., Zaidi, M.: Bismuth ferrite ($BiFeO_3$) nanopowder prepared by sucrose-assisted combustion method: a novel and reusable heterogeneous catalyst for acetylation of amines, alcohols and phenols under solvent-free conditions. J. Mol. Catal. A: Chem. **299**, 18–25 (2009)
207. He, X., Gao, L.: Synthesis of pure phase $BiFeO_3$ powders in molten alkali metal nitrates. Ceram. Int. **35**, 975–978 (2009)
208. Mazumder, R., et al.: Spark plasma sintering of $BiFeO_3$. Mater. Res. Bull. **44**, 555–559 (2009)
209. Xu, J.-H., et al.: Low-temperature synthesis of $BiFeO_3$ nanopowders via a sol-gel method. J. Alloy. Compd. **472**, 473–477 (2009)

210. Verbenko, I.A., et al.: Dielectric spectra and behavior when heated under a constant (polarization) electric field of lead-free ferroelectric ceramics of the multicomponent system $[(K_{0.5}Na_{0.5})_{1-x}Li_x](Nb_{1-y-z}Ta_ySb_z)O_3$, modified with oxides of calcium, strontium, cadmium, copper, titanium. In: Proceedings on the 11th International Conference on Physics of Dielectrics (Dielectrics-2008), vol. 1, pp. 194–197. Saint Petersburg (2008) (in Russian)
211. Verbenko, I.A., et al.: Optimization of the properties of lead-free ferroelectric niobate ceramics by means of their superstoichiometric modification with oxide compounds of d-metals. In: Proceedings on All-Russian Scientific and Practical Conference Science and Education in the Chechen Republic: State and Development Prospects, pp. 76–79. Grozny (2011) (in Russian)
212. Verbenko, I.A.: Nanotechnological process of producing lead-free ferroelectric ceramics. In: Proceedings on International Science Technology Conference Part 1. Taganrog, pp. 214–215 (2010) (in Russian)
213. Andryushin, K.P., et al.: Thermal stability and electrical conductivity of $BiFeO_3$/REE multiferroics. Izv. RAS. Ser. Physical **75**(8), 1137–1139 (2011) (in Russian)
214. Rybyanets, A.N., et al.: Porous piezoceramics: production technology, mathematical models, experiment. Universities News, North-Caucasus Region. Tech. Sci. Special issue, 82–90 (2004) (in Russian)
215. Rybyanets, A.N., et al.: Porous piezoceramics based on lead magnoniobate for ultrasonic applications. In: Proceedings of the International Scientific and Practical Conference Fundamental Problems of Piezoelectric Instrument Making, pp. 105–108. MIREA, Moscow (2003) (in Russian)
216. Rybyanets, A.N., et al.: Technology for producing porous piezoceramics of lead titanate and lead metaniobate. In: Proceedings of the International Scientific and Practical Conference Fundamental Problems of Piezoelectric Instrument Making, pp. 109–111. MIREA, Moscow (2003) (in Russian)
217. Banno, H.: Effect of shape and volume fraction of closed pores on dielectric, elastic and electromechanical properties of dielectric and piezoelectric ceramics—a theoretical approach. Amer. Ceram. Soc. Bull. **66**(9), 1332–1337 (1987)
218. Lopatin, S.S., Lupeiko, T.G.: Properties of porous piezoelectric ceramics such as lead zirconate titanate. Inorg. Mater. **27**(9), 1948–1951 (1991) (in Russian)
219. Munehiko, K., Hirol, F.: Porous ceramics for a piezoelectric element. Application **61**, 13673 (1986)
220. Reznichenko, L.A., et al.: Features of the crystal structure of porous ceramics $NaNbO_3$. In: Proceedings of the 7th International Symposium Phase Transformations in Solid Solutions and Alloys (OMA-2004), Sochi, pp. 247–250 (2004) (in Russian)
221. Miller, A.I., et al.: The influence of mechanochemical activation on the phase formation and reactivity of various groups of ferroactive materials. In: Proceedings of the XV International Science Conference of Young Scientists, Postgraduates and Students "PERSPECTIVE-2011", pp. 209–212. KBR, Elbrus, KBSU EEC (2011) (in Russian)
222. Miller, A.I., et al.: Dielectric spectroscopy of solid solutions of a multicomponent system including ferroelectric relaxors prepared from mechanically activated initial mixtures. In: Proceedings on XIV International Interdisciplinary Symposium Order, Disorder and Properties of Oxides (ODPO-14), vol. 2, pp. 30–34. Rostov-on-Don–Loo (2011) (in Russian)
223. Miller, A.I., et al.: Evolution of dielectric spectra of mechanically activated relaxor materials. In: Proceedings on XLV School of PNPI RAS on Physics of Condensed Matter (PKM-2011), p. 121. Saint Petersburg (2011) (in Russian)
224. Miller, A.I., et al.: Methods of chemical and mechanochemical influences aimed at eliminating the reasons that impede the practical use of $BiFeO_3$. In: Proceedings on III International, Interdisciplinary Symposium Media with Structural and Magnetic Ordering (Multiferroics-3), pp. 101–105. Rostov-on-Don-Loo (2011) (in Russian)
225. Reznichenko, L.A., et al.: Methods of Synthesis, Hot Pressing, Optimization of Backfill Compositions, Identification of the Microstructure of Niobate Ferroelectric Ceramics. USSR Patents No. 857075, 1981; No. 1087489 1983 (in Russian)

226. Reznichenko, L.A., et al.: Piezoelectric Ceramic Materials for High-frequency Converters for Various Purposes. USSR Patents No. 968011, 1982; No. 1296549, 1987 (in Russian)
227. Reznichenko, L.A., et al.: Highly Sensitive Materials. USSR Patents No. 1096251, 1984; No. 1294790, 1987 (in Russian)
228. Reznichenko, L.A., et al.: High-temperature Ferropiezoelectric Materials. Russian Patents No. 2712081, 2020; No. 2712083, 2020 (in Russian)
229. Reznichenko, L.A., et al.: Materials for High Voltage Sources. USSR Patent No. 642273 (1979) (in Russian)
230. Reznichenko, L.A., et al.: Glass-Ceramic Materials. USSR Patents No. 854914, 1981; 863560, 1981 (in Russian)
231. Reznichenko, L.A. et al.: Environmentally friendly posistor materials based on lithium metaniobate, including iron-modified ones. In: Ninth International Interdisciplinary Youth Symposium Physics of Lead-free Piezoactive and Related Materials. Modeling of Ecosystems (Analysis of the Current State and Development Prospects), vol. 1, pp. 158–159. SFedU Press, Rostov-on-Don (2020) (in Russian)
232. Reznichenko, L.A., et al.: Semiconductor Material. Russian Patents No. 1574094, 1994; No. 1618190, 1995 (in Russian)
233. Reznichenko, L.A., et al.: Capacitor materials. Russian Patents No. 667526, 1979; No. 1138397, 1985 (in Russian)
234. Reznichenko, L.A., et al.: Pyroelectric Materials. Russian Patent No. 2413186, 2011 (In Russian).
235. Reznichenko, L.A., et al.: Materials for Low- and Medium-Frequency Applications. USSR Patents No. 619470, 1978; No. 1294791, 1987 (in Russian)
236. Reznichenko, L.A., et al.: Material for Ferroelectric Substrates. USSR Patents No. 854915, 1981; No. 962261, 1982 (in Russian)
237. Reznichenko, L.A., et al.: Material for Frequency-Selective Devices. USSR Patent No. 1000439, 1983 (in Russian)
238. Panchenko E.M. et al.: On the possibility of creating efficient electrets based on complex niobium oxides that do not contain lead. In: Ninth International Interdisciplinary Youth Symp. Physics of Lead-free Piezoactive and Related Materials. Modeling of Ecosystems (Analysis of the Current State and Development Prospects), vol. 1, pp. 152–153. SFedU Press, Rostov-on-Don (2020) (in Russian)
239. Reznichenko, L.A., et al.: Electret Materials. USSR Patent No. 975673 (1982) (in Russian)
240. Turik, A.V., et al.: Heterogeneous multiferroics: magnetoelectricity and piezoelectric effect. Phys. Solid State **51**(8), 1580–1583 (2009) (in Russian)
241. Chernobabov, A.I., et al.: Chaotic dynamics in piezoactive statistical mixtures. Phys. Solid State **51**(7), 1419–1421 (2009) (in Russian)
242. Radchenko, G.S., Turik, A.V.: Giant piezoelectric effect in layered ferroelectric-polymer composites. Phys. Technol. **45**(9), 1676–1679 (2003)
243. Turik, A.V., et al.: Magnetoelectricity in two-dimensional statistical mixtures. Phys. Solid State **51**(7), 1395–1397 (2009) (in Russian)
244. Kovalenko, M.I.: Features of structure formation and destruction phenomena in polycrystalline materials based on sodium niobate. Ph.D. Thesis (Physical and Mathematical Sciences). Rostov State University Press, Rostov-on-Don (2001) (in Russian)
245. Klevtsov, A.N., et al.: Preparation of high-strength superconducting ceramics of the composition $YBa_2Cu_3O_{7-\delta}$. Prob. HTSC, Rostov-on-Don **2**, 32–51 (1990). (in Russian)
246. Parinov, I.A.: Microstructure and Properties of High-Temperature Superconductors, 2nd edn. Springer, Heidelberg, New York, Dordrecht, London (2013)
247. Razumovskaya, O.N., et al.: Solid solutions $Li_xLa_2CuO_{4\pm y}$, $Li_xNd_2CuO_{4\pm y}$, $Li_xLa_{2-z}Nd_zCuO_{4\pm y}$. Inorg. Mater. **32**(2), 185–187 (1996)

248. Titov, S.V., et al.: Ionic conductivity, structural features and multifractal properties of grain boundaries in $CuCr_{1-x}V_xS_2$. Izv. RAS. Ser. Physical **71**(5), 743–744 (2007)
249. Tretyakov, Yu.D.: Nanotechnologies and nanomaterials in the past, present and future. Altern. Energy Ecol. **86**(6), 10–19 (2010)
250. Hong, Ch.-H., et al.: Lead-free piezoceramics—Where to move on? J. Materiomics, 1–24 (2016)
251. Koruza, J., et al.: Requirements for the transfer of lead-free piezoceramics into application. J. Materiomics **4**(1), 13–26 (2018)
252. Leontsev, S.O., Eitel, R.E.: Progress in engineering high strain lead-free piezoelectric ceramics. Sci. Technol. Adv. Mat. **11**(4), 044302 (2010)
253. Priya, S., Nahm, S.: Lead-Free Piezoelectrics. Springer, New York (2012)
254. Roedel, J., et al.: Perspective on the development of lead-free piezoceramics. J. Am. Ceram. Soc. **92**(6), 1153–1177 (2009)
255. Roedel, J., et al.: Transferring lead-free piezoelectric ceramics into application. J. Eur. Ceram. Soc. **35**(6), 1659–1681 (2015)
256. Shrout, T.R., Zhang, S.J.: Lead-free piezoelectric ceramics: alternatives for PZT? J. Electroceram. **19**(1), 113–126 (2007)
257. Shvartsman, V.V., Lupascu, D.C.: Lead-free relaxor ferroelectrics. J. Am. Ceram. Soc. **95**(1), 1–26 (2011)
258. Reznichenko, L.A., et al.: Development of lead-free ferropiezoelectric materials science at the turn of the millennium. In: Mater. first Int. Interdisciplinary Symp. Lead-free Ferroelectric Piezoelectric Ceramics and Related Materials: Production, Properties, Applications (Retrospective—Modernity—Forecasts) (LFFC—2012). Rostov-on-Don, 9–19 (2012) (in Russian)
259. Verbenko, I.A.: Lead-free ceramics. Development of the direction in 2013–2014. In: Proceedings on Third International Youth Symposium on Physics of Lead-free Piezoactive and Related Materials (Analysis of the Current State and Development Prospects) (LFPM-2014), Rostov-on-Don, Tuapse, vol. I (3), pp. 1–12 (2014) (In Russian)
260. Verbenko, I.A., Reznichenko, L.A.: Lead-free ferroelectric ceramics based on alkali metal niobates: history, technology, prospects. In: Proceedings on Second International Youth Symposium on Physics of Lead-free Piezoactive and Related Materials (Analysis of the Current State and Development Prospects) (LFPM-2013), Rostov-on-Don, Tuapse vol. I(2), 52–64 (2013) (in Russian)
261. Reznichenko, L.A., et al.: Lead-free long ago. In: Parinov, I.A. (ed.) Piezoceramic Materials and Devices, pp. 1–69. Nova Science Publishers, New York (2010)
262. Reznichenko, L.A., et al.: Ferroelectrics, relaxors, multiferroics with different thermodynamic prehistory. In: Parinov, I.A. (ed.) Piezoceramic Materials and Devices, pp. 1–50. Nova Science Publishers, New York (2012)
263. Reznichenko, L.A., et al.: Lead-free ceramic of third millennium. In: Parinov, I.A. (ed.) Piezoceramic Materials and Devices, pp. 107–144. Nova Science Publishers, New York (2012)
264. Directive 2011/65/EU on the restriction of the use of certain hazardous substances in electrical and electronic equipment. European Parliament and the Council. Official Journal of the EU (2011)
265. Measures for the Administration on Pollution Control of Electronic. Information Products, Ministry of Information Industry China Order, 39 (2006)
266. Regulation (EC) concerning the Registration, Evaluation, Authorisation and Restriction of Chemicals (REACH). European Parliament and the European Council. Official Journal of the European Union, 1907/2006 (2006)
267. Solid Waste: Hazardous Electronic Waste. U.S. California Senate Bill (2004)
268. Malic, B., et al.: Lead-free piezoelectrics based on alkaline niobates: synthesis, sintering and microstructure. Acta Chim. Slov. **55**(4), 719–726 (2008)

269. Wu, J.G., et al.: Potassium-sodium niobate lead-free piezoelectric materials: past, present, and future of phase boundaries. Chem. Rev. **115**(7), 2559–2595 (2015)
270. Jaffe, B., et al.: Piezoelectric Ceramics. Academic Press, London, New York (1971)
271. Tennery, V.J., Hang, K.W.: Thermal and x-ray diffraction studies of $NaNbO_3$–$KNbO_3$ system. J. Appl. Phys. **39**(10), 4749–4753 (1968)
272. Cross, L.E., Nicholson, B.J.: The optical and electrical properties of single crystals of sodium niobate. Philos. Mag. **46**(376), 453–466 (1955)
273. Vousden, P.: The structure of ferroelectric sodium niobate at room temperature. Acta Crystallogr. **4**(6), 545–551 (1951)
274. Koruza, J., et al.: Phase transitions of sodium niobate powder and ceramics, prepared by solid state synthesis. J. Appl. Phys. **108**(11), 113509 (2010)
275. Megaw, H.D.: 7 phases of sodium niobate. Ferroelectrics **79**(1–4), 87–89 (1974)
276. Shirane, G., Danner, H., Pavlovic, A., Pepinsky, R.: Phase transitions in ferroelectric $KNbO_3$. Phys. Rev. **93**(4), 672–673 (1954)
277. Tellier, J., et al.: Crystal structure and phase transitions of sodium potassium niobate perovskites. Solid State Sci. **11**(2), 320–324 (2009)
278. Baker, D.W., et al.: A comprehensive study of the phase diagram of $K_xNa_{1-x}NbO_3$. Appl. Phys. Lett. **95**(9), 091903 (2009)
279. Egerton, L., Dillon, D.M.: Piezoelectric and dielectric properties of ceramics in the system potassium-sodium niobate. J. Am. Ceram. Soc. **42**(9), 438–442 (1959)
280. Fukada, M., et al.: Fabrication of lead-free piezoelectric $(Na_{0.5}K_{0.5})NbO_3$ ceramics by a modified solid-state reaction method. IEEE T. Ultrason. Ferr. **55**(5), 988–993 (2008)
281. Lee, Y.H., et al.: Piezoelectric properties and densification based on control of volatile mass of potassium and sodium in $(Na_{0.5}K_{0.5})NbO_3$ ceramics. Jpn. J. Appl. Phys. **47**(6), 4620–4622 (2008)
282. Zhang, B.P., et al.: Effect of sintering temperature on electrical properties of $(Na_{0.5}K_{0.5})NbO_3$ lead-free piezoelectric ceramics prepared by normal sintering. Ferroelectrics **358**, 188–195 (2007).
283. Damjanovic, D.: Contributions to the piezoelectric effect in ferroelectric single crystals and ceramics. J. Am. Ceram. Soc. **88**(10), 2663–2676 (2005)
284. Akdogan, E.K., et al.: Origin of high piezoelectric activity in ferroelectric $(K_{0.44}Na_{0.52}Li_{0.04})(Nb_{0.84}Ta_{0.1}Sb_{0.06})O_3$ ceramics. Appl. Phys. Lett. **92**(11) (2008)
285. Pavlenko, A.V., et al.: Ferropiezoelectric ceramic materials based on sodium and potassium niobates. Dielectric and piezoelectric characteristics at temperatures from 0 °C to 100 °C. Russian GOST R 8.937-2017 (2017) (in Russian)
286. Reznichenko, L.A., et al.: Lead-free piezoelectric ceramic material. Russian Patent No. 2580538, 03.16.2016 (in Russian)
287. Reznichenko, L.A., et al.: Ferropiezoelectric ceramic materials based on sodium and potassium niobates. Dielectric and piezoelectric characteristics at temperatures (0–100) °C. Russian Standard Certificate No. 322. Federal State Unitary Enterprise "VNIIMS", Moscow, GSSSD 322–2017 (2017) (in Russian)
288. Wang, K., et al.: Temperature-insensitive (K, Na)NbO_3-based lead-free piezoactuator ceramics. Adv. Funct. Mater. **23**(33), 4079–4086 (2013)
289. Li, J.F., et al.: (K, Na)NbO_3-based lead-free piezoceramics: fundamental aspects, processing technologies, and remaining challenges. J. Am. Ceram. Soc. **96**(12), 3677–3696 (2013)
290. Verbenko, I.A., et al.: Correlation connections composition (chemical composition)–structure (phase diagram)–properties (dielectric, piezoelectric, mechanical) in a multicomponent system based on sodium-potassium niobates. Ecol. Indus. Prod. **1**, 51–59 (2009). (in Russian)
291. Pavlenko, A.V., et al.: The influence of nickel oxide on the dielectric, piezoelectric parameters of composite materials based on alkali metal niobates. Constr. Compos. Mater. **1**, 60–64 (2015). (in Russian)

292. Sadykov, Kh.A., et al.: The influence of 3d transition metal ions on the formation of electrical properties of polycrystalline materials based on alkali metal niobates. Izv. RAS. Ser. Phys. **77**(9), 1253–1255 (2013)
293. Sadykov, Kh.A., et al.: Effects of simple and combined modification of niobate materials that do not contain toxic elements. Constr. Compos. Mater. **3**, 45–55 (2013). (in Russian)
294. Sadykov, Kh.A., et al.: Features of dielectric spectra of niobate materials modified with manganese and copper oxides. Izv. RAS. Ser. Phys. **78**(1), 108–109 (2014)
295. Zhang, S.J., et al.: Characterization of lead free $(Na_{0.5}K_{0.5})NbO_3$–$LiNbO_3$ piezoceramic. Solid State Commun. **141**(12), 675–679 (2007)
296. Wang, R.P., et al.: Tuning the orthorhombic-rhombohedral phase transition temperature in sodium potassium niobate by incorporating barium zirconate. Phys. Status Solidi **3**(5), 142–144 (2009)
297. Wang, R.P., et al.: Temperature stability of lead-free niobate piezoceramics with engineered morphotropic phase boundary. J. Am. Ceram. Soc. **98**(7), 2177–2182 (2015)
298. Wang, X.P., et al.: Giant piezoelectricity in potassium-sodium niobate lead-free ceramics. J. Am. Chem. Soc. **136**(7), 2905–2910 (2015)
299. Bernard, J., et al.: Low-temperature sintering of $K_{0.5}Na_{0.5}NbO_3$ ceramics. J. Am. Ceram. Soc. **91**(7), 2409–2411 (2008)
300. Matsubara, M., et al.: Sinterability and piezoelectric properties of (K, Na)NbO_3 ceramics with novel sintering aid. Jpn. J. Appl. Phys. **43**(10), 7159–7163 (2004)
301. Ryu, J., et al.: Sintering and piezoelectric properties of KNN ceramics doped with KZT. IEEE T. Ultrason. Ferr. **54**(12), 2510–2515 (2007)
302. Verbenko, I.A., et al.: Glass ceramics based on alkali metal niobates. Int. J. Appl. Eng. Res. **10**(2), 4007–4010 (2015)
303. Matsubara, M., et al.: Processing and piezoelectric properties of lead-free (K, Na)(Nb, Ta)O_3 ceramics. J. Am. Ceram. Soc. **88**(5), 1190–1196 (2005)
304. Matsubara, M., et al.: Synthesis and characterization of $(K_{0.5}Na_{0.5})(Nb_{0.7}Ta_{0.3})O_3$ piezoelectric ceramics sintered with sintering aid $K_{5.4}Cu_{1.3}Ta_{10}O_{29}$. Jpn. J. Appl. Phys. **44**(9a), 6618–6623 (2005)
305. Matsubara, M., et al.: Sintering and piezoelectric properties of potassium sodium niobate ceramics with newly developed sintering aid. Jpn. J. Appl. Phys. **44**(1a), 258–263 (2005)
306. Han, H.-S., et al.: Hardening behavior and highly enhanced mechanical quality factor in $Na_{0,5}K_{0,5}NbO_3$-based ceramics. J. Eur. Ceram. Soc. **37**(5), 2083–2089 (2017)
307. Park, B.C., et al.: Highly enhanced mechanical quality factor in lead-free $Na_{0,5}K_{0,5}NbO_3$ piezoelectric ceramics by codoping with $K_{5.4}Cu_{1.3}Ta_{10}O_{29}$ and CuO. Mater. Lett. **64**(14), 1577–1579 (2010)
308. Eichel, R.A., et al.: Interactions of defect complexes and domain walls in CuO-doped ferroelectric (K, Na)NbO_3. Appl. Phys. Lett. **102**(24), 242908 (2013)
309. Sadykov, Kh.A., et al.: Phases, microstructure, dielectric and piezoelectric properties of solid solutions of the $NaNbO_3$–$CuNb_2O_6$ system, prepared from Nb_2O_5 of various qualifications. In: Proceedings on Second International Youth Symposium Physics of Lead-free Piezoactive and Related Materials. (Analysis of the Current State and Development Prospects) (LFPM-2013), vol. II(2), pp. 164–183. Rostov-on-Don, Tuapse (2013) (in Russian)
310. Pardo, L., et al.: Ecological, lead-free ferroelectrics. In: Magnetic, Ferroelectric, and Multiferroic Metal Oxides, pp. 201–219 (2018)
311. Zhang, S., et al.: Piezoelectric materials for high power, high temperature applications. Mat. Lett. **59**, 3471–3475 (2005)

312. Gavrilyachenko, V.G., Turik, A.V.: Educational and methodological development for the lecture courses “Crystal Physics”, “Physics of Ferroelectrics” and for the laboratory work on the topic “Piezoelectric Effect in Ferroelectrics”. Ed. CMC “Kolibri”, Rostov-on-Don, p. 24 (2000) (in Russian)
313. Berlincourt, D., et al.: Piezoelectric and piezomagnetic materials and their application in converters. In: W. Meson (ed.) Physical Acoustics, vol. 1, part A. World, Moscow, 204–326 (1966) (In Russian)
314. Danciger, A.Ya.: Ferroelectric solid solutions of multicomponent systems of complex oxides and highly efficient piezoelectric materials based on them. DrSc Thesis (Physical and Mathematical Sciences). Rostov State University Press, Rostov-on-Don (1985) (in Russian)
315. Shuvalov, L.A., Likhacheva, Yu.S.: Damping of oscillations of resonators made of ferroelectric single crystals. Izv. Acad. Sci. USSR. Ser. Phys. **24**(10), 1216–1224 (1960) (in Russian)
316. Shirokov, A.M., Shuvalov, L.A.: On the temperature dependence of internal friction in crystals of Rochelle salt. Crystallography **8**(5), 733–737 (1963) (in Russian)
317. Russian GOST 12370–80. Piezoceramic Materials. Test Methods. Group E. Introduced 07.01.81, 30 (1981) (in Russian)
318. Russian OST 11 0444-87. Piezoceramic materials. Technical conditions. Group E10. Introduced 01.01.88, 140 (1988) (in Russian)
319. Jaffe, H., Berlincourt, D.A.: Piezoelectric transducer materials. Proc. IEEE **53**, 1372–1386 (1965)
320. Dauman, R.: Bone conduction: an explanation for this phenomenon comprising complex mechanisms. Eur. Ann. Otorhinolaryngol. Head Neck Dis **130**(4), 209–213 (2013)
321. Russian Patents. Piezoelectric Ceramic Material. No. 2358953, 20.06.2009; No. 2440954, 27.01.2012; No. 2440955, 27.01.2012; No. 2498958, 20.11.2013; No. 2498959 20.11.2013; No. 2498960, 20.11.2013; No. 2498961, 20.11.2013; No. 2542004, 20.02.2015; No. 2542008, 20.02.2015; No. 2542009, 20.02.2015; No. 2542012, 20.02.2015; No. 2547875, 10.04.2015; No. 2548278, 20.04.2015; No. 2551156, 20.05.2015; No. 2561439, 27.08.2015 (in Russian)
322. Hong, C.-H., et al.: Ring-type rotary ultrasonic motor using lead-free ceramics. J. Sens. Sci. Technol. **24**(4), 228–231 (2015)
323. Tou, T., et al.: Properties of $(Bi_{0.5}Na_{0.5})TiO_3$–$BaTiO_3$-$(Bi_{0.5}Na_{0.5})(Mn_{1/3}Nb_{2/3})O_3$ lead-free piezoelectric ceramics and its application to ultrasonic cleaner. Jpn. J. Appl. Phys. **48**, 07GM03.234 (2009)
324. Feahler, S., Reobler, U.K., Kastner, O., Eckert, J., Eggeler, G., Emmerich, H., et al.: Caloric effects in ferroic materials: new concepts for cooling. Adv. Eng. Mater. **14**(1–2), 10–19 (2012)
325. Yurasov, Yu.I. Environmentally friendly ultrasonic parking sensor and detonation sensor based on lead-free ferroelectric piezoceramics. In: Proceedings on All-Russian Seminar on Radiation and Industrial Ecology. Southern Federal University, Rostov-on-Don, pp. 394–397 (2016) (in Russian)

Chapter 2
Regularities of the Change of Physical Properties of Solid Solutions Based on Sodium Niobate During Their Modification

2.1 Introduction

One of the most effective methods for modifying the physical properties of solar cells is the introduction of small (no more than 10 at%) additives of monoxides of various elements (modification). The considerable potential of this method is attributable to the presence of a considerable quantity of metal monoxides and their combinations that can dissolve in complex oxides, particularly in ferroelectric PTS, across a wide range of concentrations, as well as oxides of elements that form "fluxes." (B_2O_3) is introduced into the original objects, namely SiO_2, V_2O_5, and finally, additives of combined action that influence the properties of the original systems due to the formation of the liquid phase and cation-exchange interaction with the base (modifying glasses). The method's unquestionable advantage lies in its capacity to, while maintaining a set of specific properties of the modified objects, selectively alter the values of certain parameters in a desired direction by slightly varying their composition. In the event of a substantial alteration (improvement) in the ceramic (ρ, $\overline{D}$) and technological ($T_{\text{sint.}}$, $\Delta T_{\text{sint.}}$) characteristics of the modified materials, as well as such modifications in dielectric and piezoelectric parameters that can be considered as the "finishing" of these target values, modification appears to be a parameter optimization method. The introduction of significant alterations to the characteristics, resulting in the emergence of a novel set of properties, provides a rationale for the conceptualization of modification as a means of generating materials with qualitatively novel attributes and adjustable parameters.

From a theoretical standpoint, the identification of general patterns of change in physical properties when modifying PTS provides insight into the composition–structure–property relationships. This allows us to consider it as a method for studying the ferroelectric state. Consequently, the outcomes of modification have a broad spectrum of applications, ranging from fundamental research to technological processes at the atomic-molecular level. In this context, the significance of the problem of modifying ferroelectric PTS becomes evident.

I. A. Verbenko et al., *Environmentally Friendly Smart Materials with Special Electrical and Magnetic Properties*, Engineering Materials,
https://doi.org/10.1007/978-3-031-90387-8_2

As indicated by bibliographic data, the modification of SS based on the PZT system has been the subject of the most comprehensive study, while the potential for modifying the system itself has been exhausted. Multicomponent lead-containing solids have also been examined in depth in this regard. The analysis and generalization of a vast quantity of experimental data (more than 300 combinations of modifying oxides are known for the PZT system alone) enabled the classification of all known modifiers of these systems and the establishment of general patterns of change in their physical properties. This approach has been successfully employed in the prediction of the properties of new FPC materials and the identification of optimal methods for their fabrication [1].

The potential for modifying SS based on NN has been comparatively understudied, with the focus exclusively on the $NaNbO_3$–$KNbO_3$ system [2–8]. Table 2.1 provides a list of classifying modifiers of the system, indicating the method of introducing additives and the presumed location of the modifiers within the crystal lattice. The data presented herein pertains to the influence of modifiers on the ceramic ($T_{sint.}$, ρ), dielectric ($\varepsilon_{33}^{T}/\varepsilon_0$, T_c), piezoelectric (d_{31}, K_p), and elastic (V_R, E_{yu}) characteristics of the SS system.

The concentration ranges of modifiers (in mol%) in which the described changes in parameters are observed are indicated.

No modifications have been made to other NN-based systems.

In our study of piezoceramics with the composition (Na, Li)NbO_3, we set ourselves the following tasks:

(i) to establish the influence of introduced modifiers on the structure and properties of solid solutions. Additionally, the study aims to identify patterns of changes in physical properties during heterovalent substitutions of ions and to clarify the mechanism for modifying the $NaNbO_3$ system.
(ii) to optimize the main dielectric and piezoelectric characteristics of the SS (in particular $\varepsilon_{33}^{T}/\varepsilon_0$, to reduce or increase d_{31}, T_c, K_p, Q_M, V_R) in order to bring their values closer to the specified ones, which are determined by the requirements of developers of equipment in which ceramics of this composition are used.
(iii) to enhance the manufacturability of the SS system, which entails reducing $T_{sint.}$, $\overline{D}$, expanding $\Delta T_{sint.}$, increasing ρ and σ.
(iv) to expand the range of possible applications of FPC materials based on alkaline niobates by developing a set of properties that allow them to be used not only in their conventional fields of electronic technology.

The resolution of these issues will facilitate the development of materials with a diverse range of applications based on modified SS of alkali metal niobates, which can be produced in an industrial setting.

Table 2.1 Classification of modifiers of the system $NaNbO_3$–$KNbO_3$

Type of substitutions	Modifier	Method of introduction	Modifier content (in mol%)	Modification effects								References
				$\Delta\rho$	$\varepsilon_{33}^{T}/\varepsilon_0$	Δd_{31}	ΔK_p	ΔE_{yu}	ΔV_R	ΔT_c	$\Delta T_{sint.}$	
Heterovalent $n_{mA} > n_A$	Pb^{2+}	In the form of $Pb_{0.5}NbO_3$ in a stoichiometric ratio according to the formula $0 < x \leq 0.1$	Up to 10	+	+	+	–	+	+	+	–	[9, 10]
			Up to 5[a]	+	+	+	+ (0–4) – (4–5)	+ (0–3) – (3–5)	+ (0–3) – (3–5)	–		[11]
			Up to 3[b]	+	+	+	+ (0–0.5) – (0.5–1) + (1–3)	– (0–1) + (1–3)	– (0–1) + (1–3)	–		
Heterovalent $n_{mA} > n_A$	Cd^{2+}	In the form of $Cd_{0,5}NbO_3$ in a stoichiometric ratio according to the formula $(K_{1-y}Na_y)_{1-x}A_{0.5x}NbO_3$, with $0 < x \leq 0.1$	Up to 10	+	+	+ (0–3.5) – (3.5–10)	+ (0–3) – (3–10)	0	+	–		[9, 10]
Heterovalent $n_{mA} > n_A$	Mg^{2+}	Compensated by $(Na_{1-y-0.5x}K_{y-0.5x}A_x)NbO_3$ $0 < x \leq 0.05$	Up to 5	–	+	+	+				–	[12–15]
	Ca^{2+}			+	+	+	+				–	
	Sr^{2+}			+	–	+	+				–	
	Ba^{2+}			+	+	+	+				–	
Heterovalent $n_{mA} > n_A$	Bi^{3+}	Not specified	Up to 1.5[a]	+ (0–1) – (1–1.5)	+	+	+ (0–1) – (1–1.5)	+	+	–		[11]
			Up to 2[b]	+	+	+ (0–1) – (1–2)	+ (0–0.5) + (0.5–2)	– (0–1) + (1–2)	–	–		
	La^{3+}		Up to 1.5[a]	+	+	+	–	+	+	–		

(continued)

Table 2.1 (continued)

Type of substitutions	Modifier	Method of introduction	Modifier content (in mol%)	Modification effects								References
				$\Delta\rho$	$\varepsilon_{33}^{T}/\varepsilon_0$	Δd_{31}	ΔK_p	ΔE_{yu}	ΔV_R	ΔT_c	$\Delta T_{sint.}$	
Heterovalent $n_{mA} > n_A$	Ge^{4+}	In the form of GeO_2 oxide, above the stoichiometric ratio	Up to 1	+	+	+	– (0–0.5) + (0.5–1)	+	+	–		
			Up to 5	+	+	+	+	+	+	+		[9, 10]

Notes $+\Delta A$ ($-\Delta A$) is the increase (decrease) of parameters; ($A = \rho$, $\varepsilon_{33}^{T}/\varepsilon_0$, d_{31}, K_p, E_{yu}, V_R, T_c, $T_{sint.}$); the parameter A do not change, when $\Delta A = 0$; n_A (n_B) is the valence of (Na, K) and Nb; n_{mA} (n_{mB}) is the valence of a modifier replacing (Na, K) or Nb

[a], [b] Results of modification of SS, obtained from two batches of Nb_2O_5, differing in the content of impurities

2.2 Heterovalent Ion Substitutions in the Sodium-Lithium Niobate System

The object of study was a solid solution of composition $Na_{0.875}Li_{0.125}NbO_3$, which lies on the Rh boundary of the MO system $NaNbO_3$–$LiNbO_3$ and exhibits extreme values of piezoelectric parameters.

The modification was executed in accordance with the established formulae:

$$(Na_{0.875}Li_{0.125})_{1-nx}M_x\square_{x(n-1)}NbO_3 \quad (2.1)$$

$$(Na_{0.875}Li_{0.125})_{1-x}M_x[Nb_{1-(n-1)/5x}\square_{(n-1)/5x}]O_3 \quad (2.2)$$

The symbol M represents the cationic species, which are the ions Mg^{2+}, Sr^{2+}, La^{3+} and Sb^{3+}, with valence n. The symbol $\square$ indicates vacancies.

The concentration of the introduced modifiers, designated as x, was found to be in the range from 0.3 to 5 at%.

The SS(1) and (2) types will be designated as SS(1) and SS(2), respectively.

It is well established that heterovalent modification maintains the balance of valencies in solid solution through the creation of vacancies. In this instance, the physical properties of a solid solution are contingent upon both the number and the position of vacancies within the crystal structure. The analysis of formulae (2.1) and (2.2) indicates that in this case, the valence balance in the solid solution is preserved due to the formation of vacancies in the A- or B-sublattices. The light color of the samples and the high values of $\rho_V \sim 10^{10}$ Ω m indicate the absence of significant oxygen deficiency. Furthermore, the maintenance of equilibrium in SS(1) necessitates the creation of a greater number of vacancies than in SS(2) when the concentration of modifiers is identical in both cases. As the valence of the modifiers increases, the concentration of vacancies also increases. It is also important to note that the number of vacancies in the A- and B-sublattices differs due to the inequality of A-O and B-O bonds. The B-O bonds are more covalent than the A-O bonds. Given the disparity in the number and nature of vacancies created in SS(1) and SS(2), it can be reasonably assumed that differences in their physical properties will also be observed. A study of the PZT system revealed that it is not possible to obtain a solid solution (without the presence of foreign phases) solely through the introduction of B-vacancies. Instead, the implementation of these vacancies must occur in a manner that redistributes them among both B- and A-sites [16]. Given the close structural relationship between the PZT and $NaNbO_3$–$LiNbO_3$ systems, it can be reasonably assumed that a similar redistribution of vacancies will occur. Consequently, the actual number of B-vacancies in SS(2) will be less than that calculated by formula (2.2).

Let us consider the properties of SS(1) and SS(2). Figure 2.1a–e illustrates the dependence of perovskite cell parameters on the concentration of Mg^{2+}, Sr^{2+}, La^{3+}, and Sb^{3+} in SS(1) (Fig. 2.1a, b) and SS(2) (Fig. 2.1c, e).

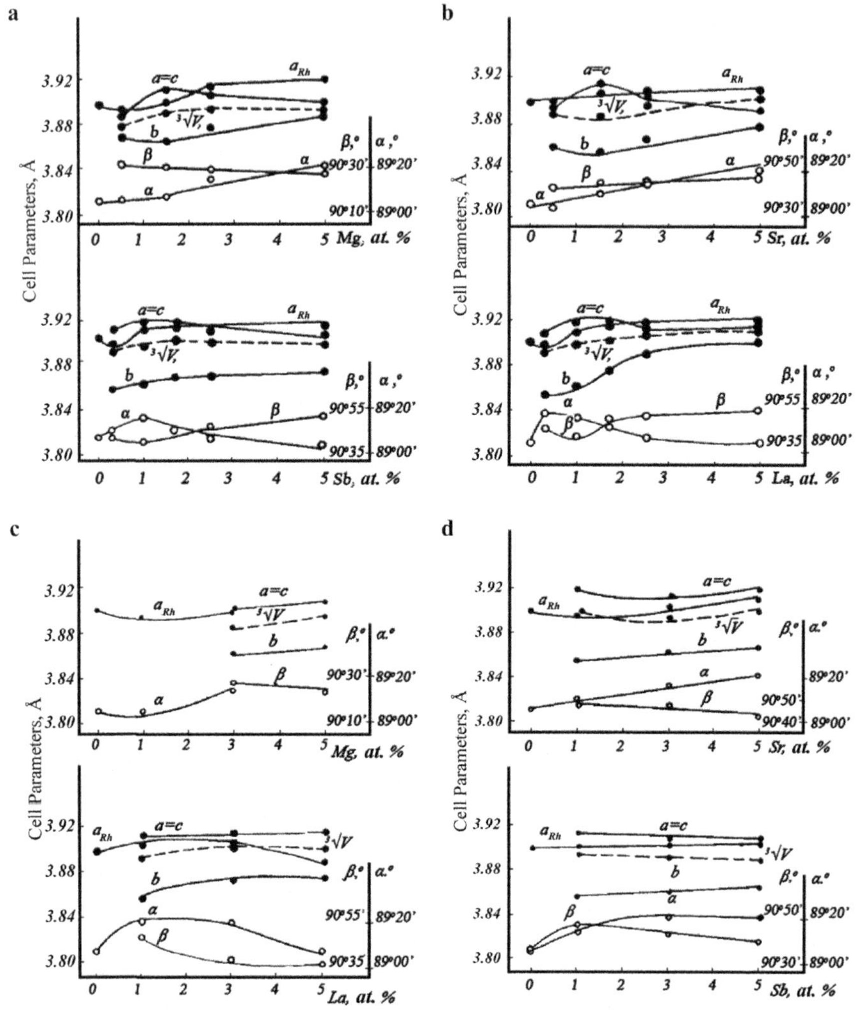

Fig. 2.1 **a** Dependencies of the parameters of perovskite cells on the content of Mg^{2+}, Sb^{3+} in SS(1). **b** Dependencies of the parameters of perovskite cells on the content of Sr^{2+}, La^{3+} in SS(1). **c** Dependencies of the parameters of perovskite cells on the content of Mg^{2+}, La^{3+} in SS(2). **d** Dependencies of the parameters of perovskite cells on the content of Sr^{2+}, Sb^{3+} in SS(2)

X-ray diffraction revealed the formation of pure SS with a mixture of R and Rh phases, which indicates a shift of MO in the $NaNbO_3$–$LiNbO_3$ system to the Rh region during modification.

The SS(1) composition exhibits a finer grain size (average grain size $\overline{D} = 2$–$6\ \mu m$) than the unmodified composition ($\overline{D} = 18\ \mu m$). The introduction of donor modifiers (Mg^{2+}, Sr^{2+}, La^{3+}, Sb^{3+}) results in a decrease in the specific volume resistivity ρ_V (Fig. 2.2a), which corroborates the previously stated assumption about the *n*-type

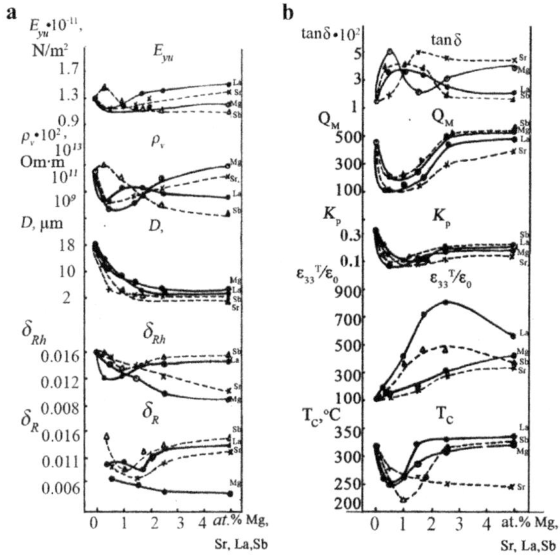

Fig. 2.2 **a** Dependences of structural, microstructural and electrical parameters on the content of Mg^{2+}, Sr^{2+}, La^{3+}, Sb^{3+} in SS(1). **b** Dependencies of dielectric and piezoelectric parameters on the content of Mg^{2+}, Sr^{2+}, La^{3+}, Sb^{3+} in SS(1)

conductivity in the $NaNbO_3$–$LiNbO_3$ system. SS(1) are distinguished by a lower value of spontaneous deformation (SD) or homogeneous deformation parameter δ, increased values of dielectric constant ($\varepsilon_{33}^T/\varepsilon_0$), mechanical ($1/Q_M$) and dielectric losses (tan δ) (Fig. 2.2a, b) in comparison to the original composition.

The effect of Mg^{2+}, Sr^{2+}, La^{3+}, and Sb^{3+} on the indicated parameters can be classified as "soft ferroferric" (SF) modifiers, as described in Ref. [17]. The Mo displacement results in a reduction in the electromechanical coupling coefficient (K_p) (Fig. 2.2b), as its maximum in the $NaNbO_3$–$LiNbO_3$ system aligns with the Rh boundary. The behavior of Q_M and T_c upon the introduction of Sr^{2+}, La^{3+}, and Sb^{3+} correlates with changes in SD. Specifically, a decrease in Q_M and T_C is associated with a decrease in $\delta_{Rh}(p)$. The minimum values of T_C at 0.3 at% La^{3+} and 1 at% Sb^{3+} correspond to the minimum values of δ. When Mg^{2+} is modified within the range of 0.5–5 at%, an increase in Q_M and T_C is observed, despite the simultaneous decrease in the SD of Rh- and R-cells. One potential explanation for this phenomenon is a notable reduction in grain size, which has been shown to result in a decline in the number of domain walls [16] and a weakening of their interaction with vacancies. Additionally, an increase in the space charge at grain boundaries [18] has been observed. The incorporation of trivalent modifiers (La^{3+}, Sb^{3+}) in the range of up to 2.5 at% has been observed to result in a greater increase $\varepsilon_{33}^T/\varepsilon_0$ than that observed with the incorporation of Mg^{2+} and Sr^{2+}. This phenomenon is likely attributable to the formation of a greater number of vacancies. In this instance, the most pronounced alterations are observed in the presence of La^{3+}.

The analysis of the properties of SS(2) revealed that, similar to SS(1), they do not contain impurity phases, are fine-grained, and exhibit increased conductivity (Fig. 2.3a). The nature of the change in SD, Q_M, tan δ, $\varepsilon_{33}^T/\varepsilon_0$, V_R, and E_{yu} in

SS(2) indicates that the Mg^{2+}, La^{3+}, and Sb^{3+} can still be classified as SF-modifiers. At fixed modifiers concentrations, the Q_M decreases in the sequence: Sb^{3+}, Mg^{2+}, La^{3+}. The value of Q_M reaches an extreme at 1 at% of Mg^{2+}, Sb^{3+}, and 3 at% of La^{3+} and corresponds to the modified compositions at the Rh-Mo boundary (see Fig. 2.3b). SS(2) with Sr^{2+}, in contrast to SS(2) with Mg^{2+}, Sb^{3+}, and La^{3+}, is distinguished by diminished values of, tan δ, $1/Q_M$, augmented values of V_R and E_{yu}. This phenomenon is likely attributable to a pronounced reduction in grain size at low concentrations of Sr^{2+}. Consideration of the behavior of the relative change in mechanical goodness $\frac{\Delta Q_M}{Q_{Mo}}$ (where Q_{Mo} is the mechanical quality factor of an unmodified material) of fit depending on the individual characteristics of equally acting SF modifiers in SS(1) and SS(2) (see Fig. 2.4) showed that the value of Q_{Mo} increases with decreasing unstressed cation (modifier), namely oxygen distance (l_o) [19] and increasing electronegativity (EN) of modifiers in the corresponding oxidation degree [20]. Furthermore, in SS(2), the dependencies of $\frac{\Delta Q_M}{Q_{Mo}}$ on EN are more pronounced than in SS(1). The data obtained are consistent with the previously documented results of modifying the SS of the PZT system [21]. The observed correlation between ferrohardness (HF) (estimated value $\frac{\Delta Q_M}{Q_{M\rho}}$ [21, 22]) and the nature of the chemical bond may be attributed to the following factors. The introduction of Mg^{2+}, La^{3+}, and Sb^{3+} ions results in an increase in the degree of covalence of *A*–O bonds, which favors an increase in their directionality and inequality [19]. This, in turn, contributes to an increase in atomic displacements, which consequently leads to an increase in SD [19] and the associated HF [22]. Considering the previously established correlation between the stability of HF and the degree of covalence between *A* and O, it is possible to draw parallels between the properties of SS(1) and SS(2). Given the analogous nature of the alterations in parameters when introducing La^{3+}, Sb^{3+}, and Mg^{2+} (Figs. 2.1a–d and 2.2a, b), a comparison will be made using the example of modification with La^{3+} (Fig. 2.5a, b).

At low concentrations of modifiers (up to 1.5 at%), SS(1) in terms of parameters δ_{Rh}, T_C, Q_M is more SF than SS(2). However, at concentrations exceeding 1.5 at%, SS(2) becomes more SF. The unequal degree of HF (SS(1) and SS(2)) is due to the inequality of the created vacancies. At low concentrations of modifiers, the quantity of vacancies appears to be a decisive factor (SS(1) with many vacancies is more "ferrosoft" than SS(2)), while at increased concentrations, the position in the structure (SS(2) with vacancies in the more covalent *B*-sublattice becomes more "ferrosoft"). The greater ferroelectric softness of SS(2) is likely to result in more pronounced changes in the value of $\frac{\Delta Q_M}{Q_{Mo}}$ from EN in them. This is associated with the increased compliance of soft systems to an increase in Q_M during modification [21].

The analysis of the data obtained allows us to draw a conclusion about the implementation of *B*-vacancies in the system. This is evidenced by the formation of pure SS(2), which exhibits distinct properties from SS(1), as well as a more pronounced dependence on the individual characteristics of the modifiers. The latest data suggests that the concentration changes in Q_M, in SS(2) may be more sharp than in SS(1), yet this hypothesis was not supported by the experimental results. One potential explanation for this phenomenon is the redistribution of vacancies in SS(2) across *B*- and *A*-sites, a process similar to that previously observed in the PZT system [16]. It is

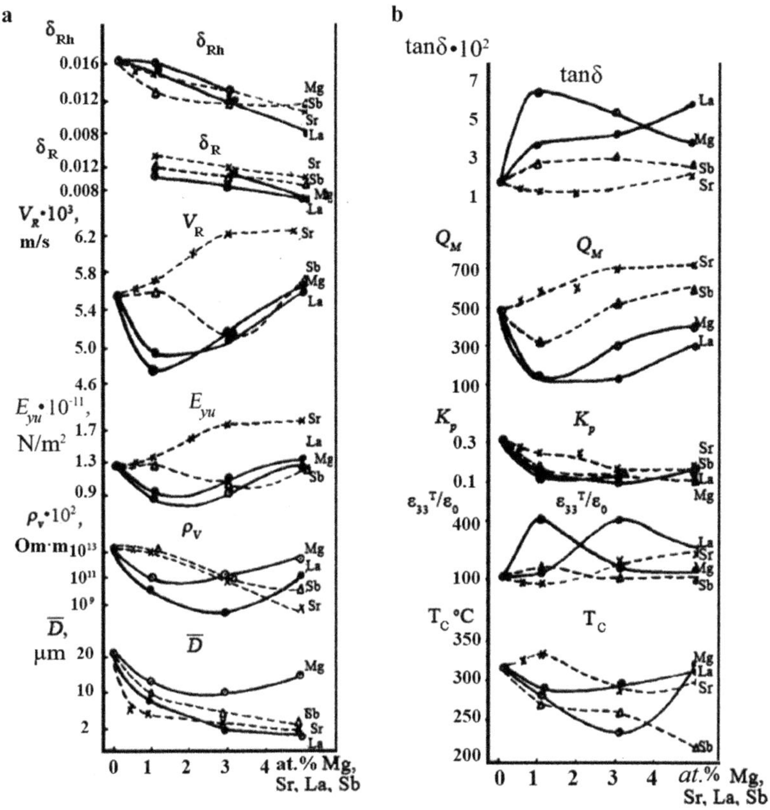

Fig. 2.3 **a** Dependencies of structural, microstructural and electrical parameters on the content of Mg^{2+}, Sr^{2+}, La^{3+}, Sb^{3+} in SS(2). **b** Dependencies of dielectric and piezoelectric parameters on the content of Mg^{2+}, Sr^{2+}, La^{3+}, Sb^{3+} in SS(2)

important to note that at high concentrations of modifiers, the nature of the change $\varepsilon_{33}^{T}/\varepsilon_0$ does not agree with the nature of the change in SD, T_C and Q_M depending on the type of SS. Consequently, at 1.5–5 at% La^{3+} SS(1) are more HF in terms of the values of SD, T_c, Q_M, and SF in the value of $\varepsilon_{33}^{T}/\varepsilon_0$ than SS(2). It is possible that the lower value of $\varepsilon_{33}^{T}/\varepsilon_0$ for SS(2) is due to the predominant 90° reorientation of domains during polarization, as previously observed in studies [12, 23]. This assumption is corroborated by the fulfillment of the relation: $\varepsilon_{33}^{T}/\varepsilon_0 < \varepsilon/\varepsilon_0$ in SS(2). The modification by formulae (2.1) and (2.2) yielded SS with low values of $\varepsilon_{33}^{T}/\varepsilon_0$ and high values of V_R, Q_M, and K_P. These materials have the potential to serve as the basis for the development of FPC materials for microwave engineering. A series of author certifications [13, 14, 24–28] has validated the practical value of these materials.

Concurrently, a Sr-containing material with a $\varepsilon_{33}^{T}/\varepsilon_0 = 90$, a vaporization rate of 5.9×10^3 m/s, a $Q_M = 1400$, and a $K_P = 0.29$ was introduced at the All-Union

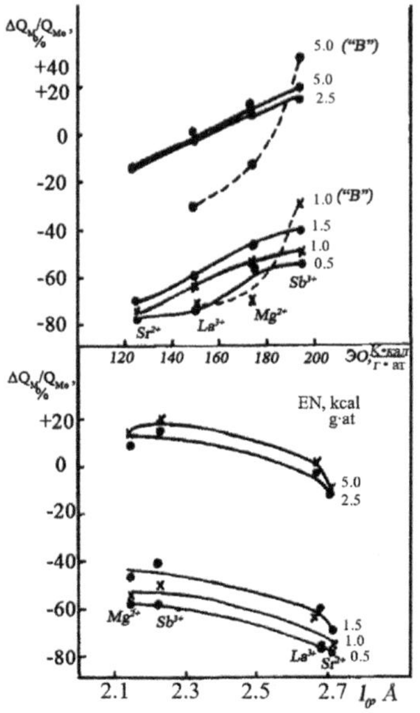

Fig. 2.4 Dependencies of the value $\frac{\Delta Q_M}{Q_{Mo}}$ on the individual characteristics of modifiers (l_0, EN); the dotted lines represent SS(2), and the numbers on the curves indicate the concentration of the introduced modifiers

Scientific Research Institute for use in surfactant devices and recommended for serial production between 1980 and 1982.

The results of the modification are also reflected in the publications [15, 29–31].

2.3 Effect of Modifying Glasses on the Properties of Solid Solutions of $NaNbO_3$–$LiNbO_3$ System

The utilization of the glass modification method [32] permitted a notable reduction in temperature and an expansion of the sintering temperature range, accompanied by an enhancement in mechanical strength and the generation of a fine-grained, homogeneous, pore-free structure of niobate materials, which was not achieved when they were produced by the HP method [33, 34]. The advantage of glass–ceramic technology is the intensification of ceramic formation processes under conditions of liquid-phase sintering [11, 35, 36], which provides, first and foremost, an improvement in the ceramic properties of the material. The outcomes of numerous investigations into FC have also demonstrated that the incorporated glass additives are not inert with respect to electrical parameters, thereby influencing the comprehensive array of physical properties intrinsic to the original systems. Several studies have demonstrated an improvement in dielectric and piezoelectric characteristics in a variety

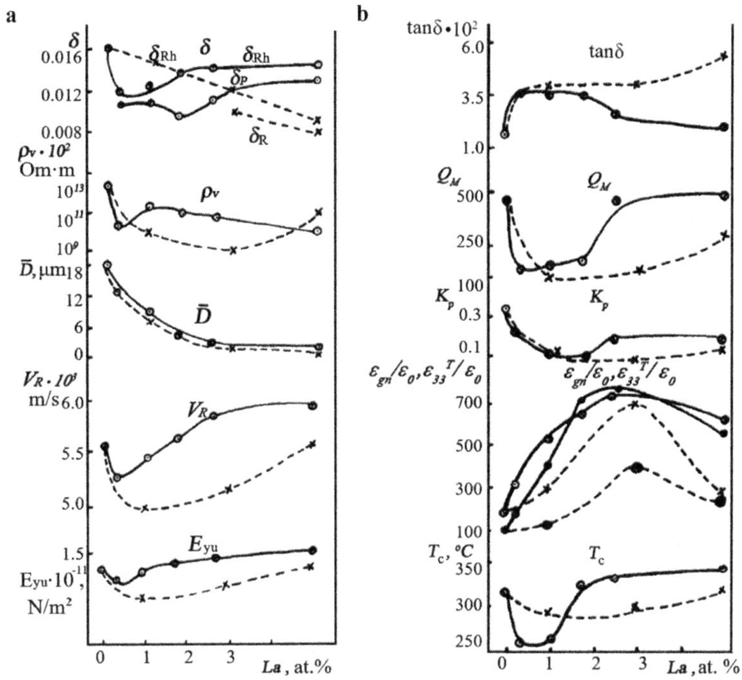

Fig. 2.5 **a** Dependencies of structural, microstructural and electrical parameters on the lanthanum content in SS(1) and SS(2) designated by dotted lines. **b** Dependencies of structural, microstructural and electrical parameters on the lanthanum content in SS(1) and SS(2) designated by dotted lines

of materials, including glass–ceramics, which has led to their practical use in engineering applications [9]. Despite the enumerated advantages, the inability to broadly regulate the parameters of FC through technological means (in many cases, FC were produced through conventional ceramic technology) constrained their scope of application. In this context, studies were conducted in [37, 38] to investigate the impact of small glass additives on the sintering and properties of FPC materials obtained by HP. The focus of these studies was SS based on PZT materials. The experimental results indicated that the production of denser hot-pressed FC (in comparison to conventional FC) was achieved with increased values of piezoelectric parameters, as well as improved stability of the latter when exposed to strong electric fields [37]. It appeared reasonable to attempt to obtain a FC based on sodium niobate SS by the hot-pressing method to improve its ceramic characteristics ($T_{sint.}$, ΔT_{sint}, ρ, $\overline{D}$, σ) and electrical properties. In this case, the $NaNbO_3$–$LiNbO_3$ SS system was chosen as the basis. It is important to note that the combined effect of introducing glasses and GP in the case of PZT compositions can to some extent be foreseen. This is because each operation separately leads to the same effects, namely a reduction of $T_{sint.}$ and $\overline{D}$, and an expansion of $\Delta T_{sint.}$. It is not evident that the use of HP in the production of niobate FC will yield a discernible outcome, given that HP does not result in a reduction in $T_{sint.}$ and $\overline{D}$ in niobate materials. Moreover, the synthesis of SC in the $NaNbO_3$–$LiNbO_3$ system has not been previously achieved. The inability to predict the effects of modifying glasses in HP ceramics based on sodium-lithium niobates is also due to the specificity of phase transformations in the binary oxide systems that make up the initial SS. These transformations can complicate the interaction processes between the glass and the base, and as a result, hinder the development of recrystallization during sintering. The production of niobate glass ceramics also necessitates a more rigorous selection of the composition, quantity, and method of introducing glass additives. In addition to the undesirable effects that may arise during normal firing, the formation of uncontrolled metastable compounds (products of the interaction of the sintered material with the substance of the intercrystalline liquid phase) further complicates the process of forming piezoceramics, as this may result in severe deformation of workpieces containing glass under the influence of pressure during HP.

The solids of the "pure" and modified $NaNbO_3$–$LiNbO_3$ system were selected as objects of study: $Na_{0.875}Li_{0.125}NbO_3$ (A), $Na_{0.85}Li_{0.122}$ $W_{0.028}NbO_3$ (B), $Na_{0.86625}Li_{0.12375}Sr_{0.01}Nb_{0.998}O_3$ (C), in which glasses of composition (in weight percent) were introduced: Na_2O—7.97; Nb_2O_5—34.1; WO_3—44.7; SiO_2—3.9; B_2O_3—9.0; Al_2O_3—0.33(α); Na_2O—9.9; Nb_2O_5—53; SrO—20.6; SiO_2—5.0; B_2O_3—11.1; Al_2O_3—0.4(β); Na_2O—9.1; Nb_2O_5—48.6; SrO—18.9; SiO_2—4.6; H_3BO_3—18.4; Al_2O_3—0.4 (γ) in quantity 0.75–5 wt% in excess of the stoichiometry of the basic composition. The A-base was modified with all glasses (α, β, γ), the B-base with β-glass, and the C-base with α-glass. Table 2.2 presents a summary of the composition of glasses and their content in the initial SS.

The glasses were prepared in the following manner. Samples of a mixture of glass oxides were combined in a porcelain mortar, placed into porcelain crucibles, and heated in the furnace. The fusion process was conducted for a period of one hour at a temperature of 1250 °C. The glass formed in the result of fusion was poured onto a steel plate, then subjected to a thorough crushing process. This powder in the indicated quantities was subsequently introduced into a previously synthesized and crushed powder of composition A, B, or C. As an essential element in the process of liquid sintering is the granulometric composition of the crystalline phase [10], the introduction of glass additives was preceded by a meticulous grinding of the synthesized product before passing through a 10,000 holes/cm^2 sieve. Additionally, considerable focus was placed on the homogenization of the mixture following the introduction of glasses. The resulting powder was then mixed in jasper mortars for a period of two hours in ethyl alcohol, after which it was dried and subjected to HP.

The formation of single-phase solid solutions with a perovskite-type structure in FC systems has been confirmed by X-ray diffraction. The introduction of initial SS glasses did not result in the detection of additional impurity phases at any of the specified concentrations of modifying glasses. Furthermore, throughout the entire concentration range of glass additives, all FCs retain the structural characteristics of the original SS (see Figs. 2.6, 2.7 and 2.8).

The exception to this is the C-α system, in which the SS shifts into the region with a mixture of R and Rh phases upon the introduction of glasses (Fig. 2.9a).

Table 2.2 Composition of modifying glasses and their content in initial SS

Composition of modifying glasses System: initial SS-glass (see text)	Na_2O	Nb_2O_5	WO_3	SiO_2	B_2O_3	Al_2O_3	SrO	H_3BO_3	Glass content in initial SS			
										mass%		
									0.75	1.5	3.0	5.0
A-α	7.97[a] 20.3[b]	34.1 20.3	44.7 29.7	3.9 9.4	9.0 20.3	0.33 0.50			1.20	mol% 2.41	4.76	7.85
A-β	9.9 19.9	53.0 24.9		5.0 9.45	11.1 19.9	0.4 24.9			0.89	1.86	3.73	6.19
A-γ	9.1 16.49	48.6 20.53		4.6 8.6		0.4 4.49	18.9 20.48	18.4 33.44	1.039	2.08	4.123	6.89
B-α	9.9 19.9	53.0 24.9		5.0 9.95	11.1 19.9	0.4 4.98	20.6 24.9		0.95	1.86	3.75	6.37
C-α	7.97 20.3	34.1 20.3	44.7 29.7	3.9 9.4	9.0 20.3	0.33 0.50			1.10	2.21	4.65	7.25

[a] mass%
[b] mol%

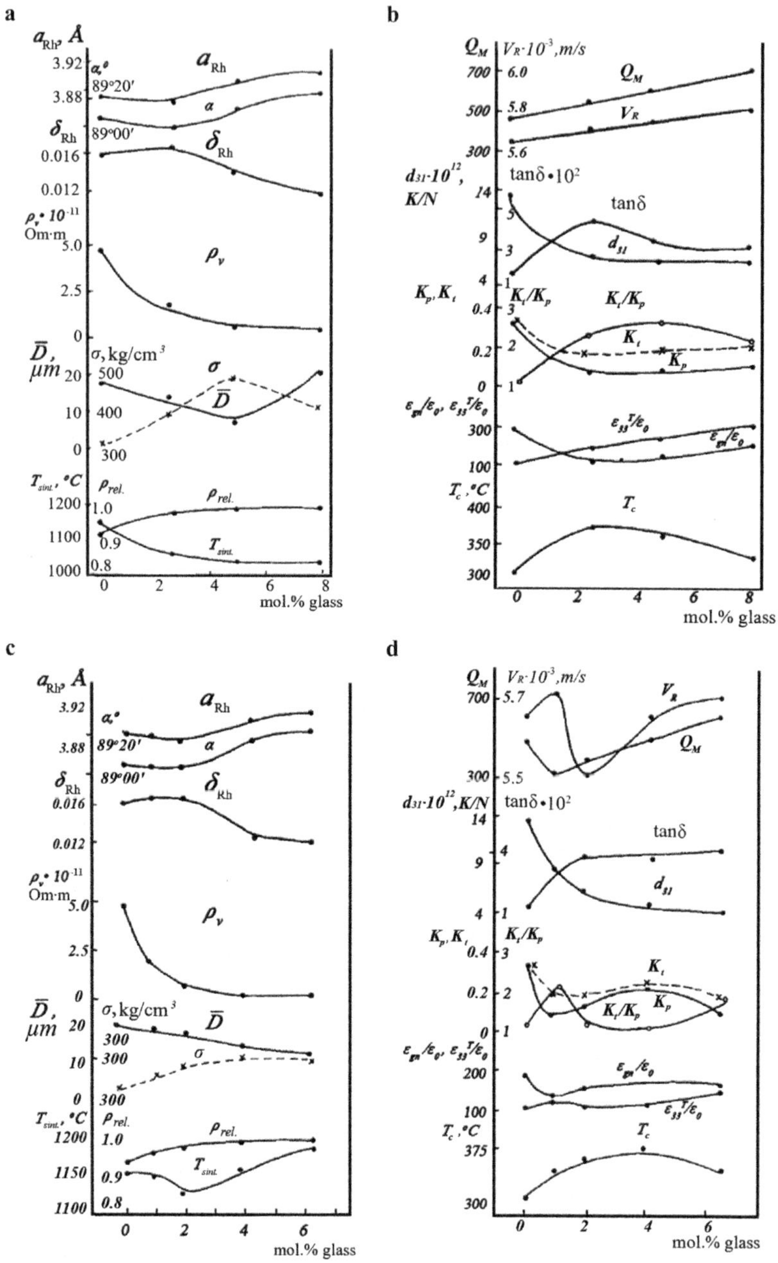

Fig. 2.6 **a** Dependencies of structural, microstructural, electrical and mechanical characteristics on the glass content in the A-α system. **b** Dependencies of dielectric, piezoelectric and mechanical characteristics on the glass content in the A-α system. **c** Dependencies of structural, microstructural and electrical and mechanical characteristics on the glass content in the A-β system. **d** Dependencies of dielectric, piezoelectric and mechanical characteristics on the glass content in the A-β system

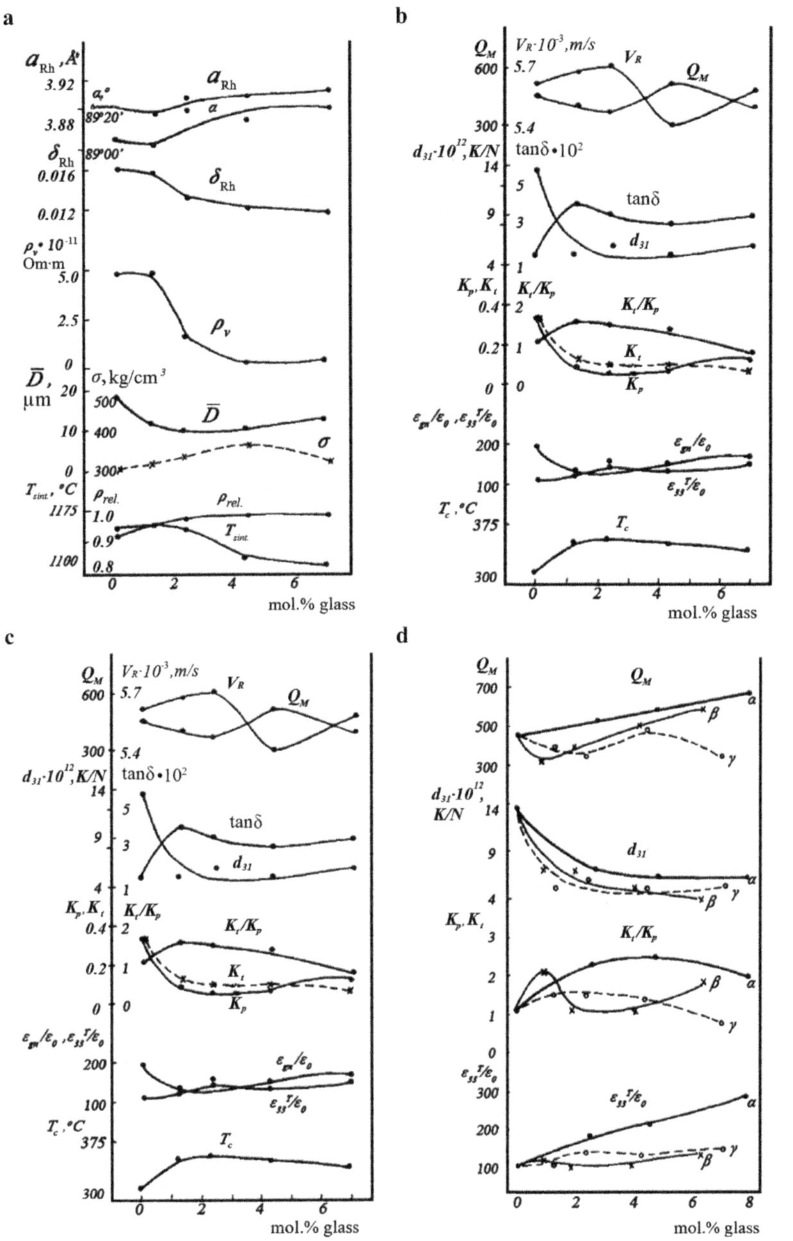

Fig. 2.7 **a** Dependencies of structural, microstructural, electrical and mechanical characteristics on the glass content in the A-γ system. **b** Dependencies of dielectric, piezoelectric and mechanical characteristics on the glass content in the A-γ system. **c** Dependencies of structural, microstructural, electrical and mechanical characteristics on the glass content in A-α, A-β, A-γ systems. **d** Dependencies of dielectric, piezoelectric and mechanical characteristics on glass content in A-α, A-β, A-γ systems

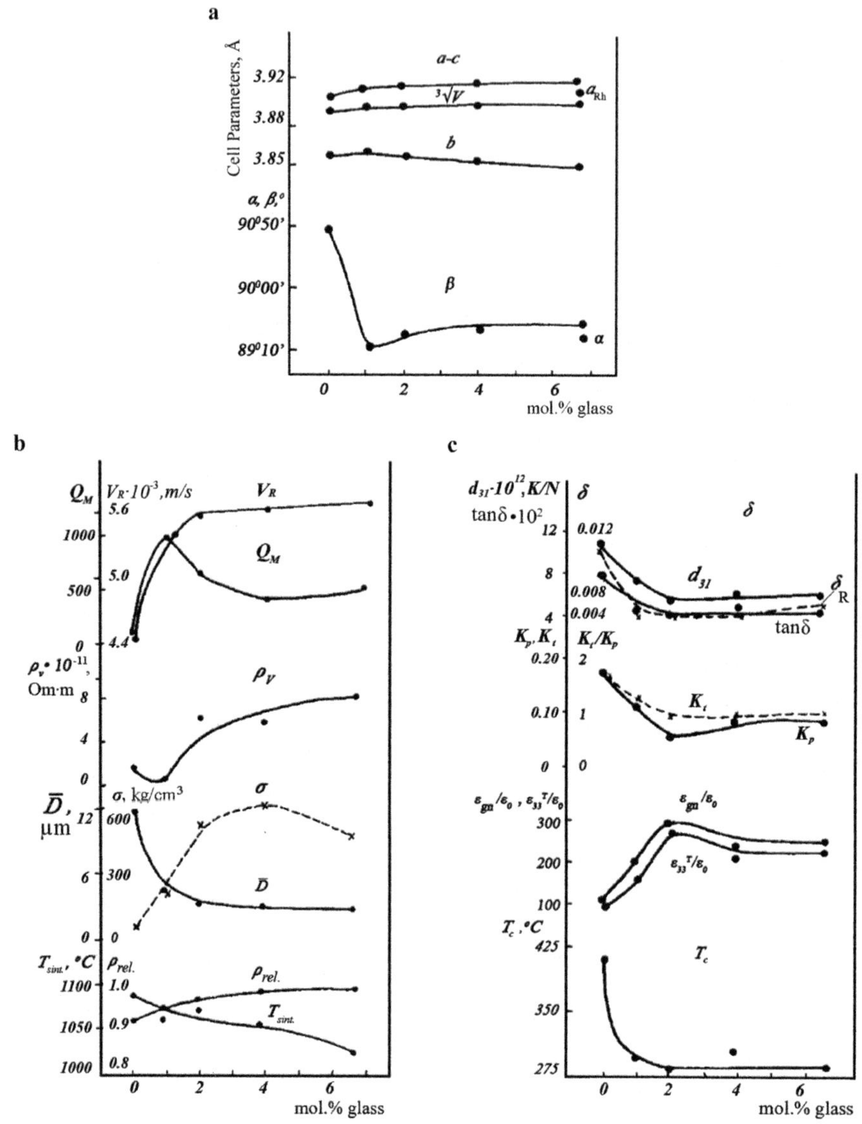

Fig. 2.8 **a** Dependencies of structural characteristics on glass content in the B-β system. **b** Dependencies of microstructural, electrical and mechanical characteristics on the glass content in the B-β system. **c** Dependencies of dielectric, piezoelectric and mechanical characteristics on the glass content in the B-β system

Furthermore, electron microscopic studies of the samples did not reveal the presence of any discernible amount of X-ray amorphous phase, whether in the form of thickened intercrystalline layers or as inclusions (Fig. 2.10b–d). This contrasts with observations in FC based on PZT systems, where such phases have been reported [10, 37, 39, 40]. However, an increase in the glass content in some systems (B–β) beyond the specified limits results in the precipitation of side crystalline phases (large quadrangular grains in Fig. 2.10d).

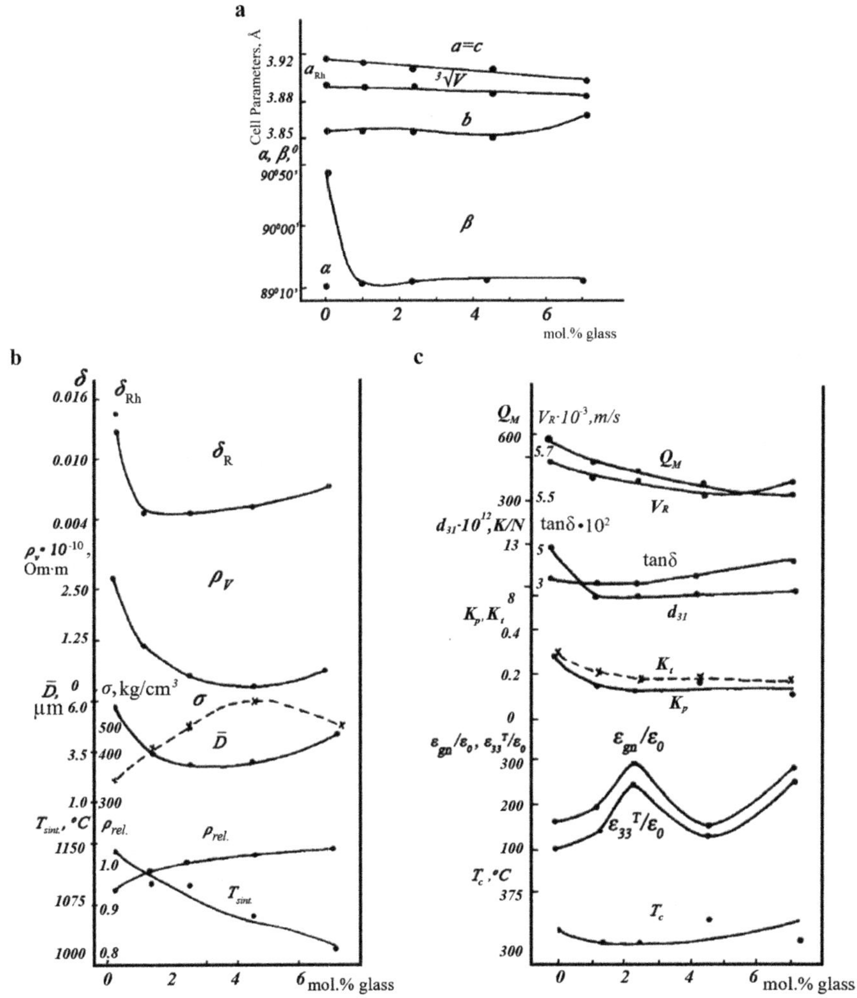

Fig. 2.9 **a** Dependencies of structural characteristics on glass content in the C-α system. **b** Dependencies of microstructural, electrical and mechanical characteristics on the glass content in the C-α system. **c** Dependencies of dielectric, piezoelectric and mechanical characteristics on the glass content in the C-α system

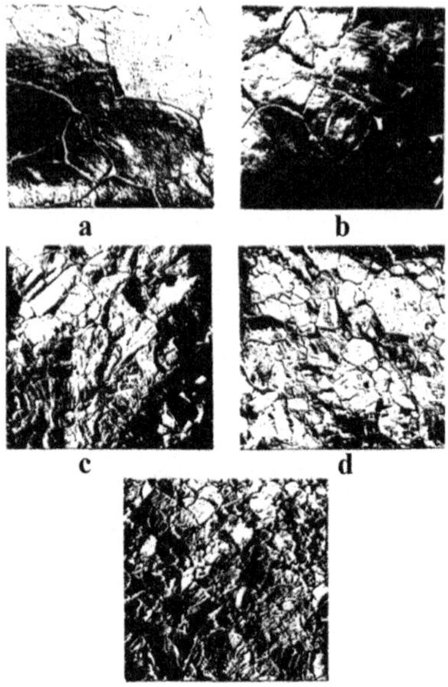

Fig. 2.10 Microstructure of FC in system B-β at different glass contents: **a** 0; **b** 1.10; **c** 2.65; **d** 7.25; **e** > 7.25 (mol%) (magnification 1500×)

A comparison of the influence of α, β, and γ glasses on $T_{sint.}$, $\overline{D}$, and σ of SS composition $Na_{0.875}Li_{0.125}NbO_3$ is presented here. In the A-α system (Fig. 2.6a), a decrease in $T_{sint.}$ is observed in the entire concentration range of the introduced glass, with the most pronounced temperature changes (by 100 °C) occurring in the region with a low content (up to 1.5 mass%) of the modifier. In the range of 1.5–3 mass% glass, $T_{sint.}$, decreases by 25–30 °C. Furthermore, an increase in the amount of α-glass in the system by more than 3 mass% leads to only a slight change in $T_{sint.}$.

The average grain size ($\overline{D}$) decreases almost threefold with the introduction of glass to 3 mass% and further increases in the glass content in ceramics lead to enlargement of the grains, resulting in an unevenly grained structure (the maximum grain size reaches $\overline{D} = \sim 30–40$ μm). The grains lose their regular geometric cut, becoming rounded, and the grain boundaries bend and thicken, resulting in a chaotic packing of grains.

A correlation exists between changes in $\overline{D}$ and changes in σ. A decrease in $\overline{D}$ is associated with an increase in the mechanical strength of the material, while the increasing of grains is indicative of a decrease in strength of the ceramic. Consequently, as the concentration of the introduced additive increases, σ initially rises, reaches a maximum, and then declines.

The maximum value of σ observed with the introduction of 3 mass% α-glass (497 kg/cm^2) is significantly greater than the σ value of the original composition (307 kg/cm^2) and corresponds to ceramics with the smallest grain size ($D \sim 7.6$ μm).

The introduction of a small amount (up to 1.5 mass%) of β-glass (Fig. 2.6c) into the same initial SS leads, as in the previous case, to a decrease in T_{sint}. Nevertheless, the decline in T_{sint} is considerably less pronounced. The average grain size in the entire concentration range of modifying glasses tends to decrease by approximately twofold. The strength of ceramics increases, reaching a maximum value of approximately 400 kg/cm^3 at a glass concentration of 3 mass%. Thereafter, σ remains essentially constant.

γ-glass in SS (Fig. 2.7a) exhibits a distinct behavior when compared to α- and β-glasses. Unlike the glasses, γ-glass does not shift T_{sint} to the region of lower temperatures, even at glass concentrations lower 1.5 mass%. A decrease in T_{sint} of approximately 50 °C is observed when the glass concentration is greater than 1.5 mass% (in the range of 1.5–3 mass%). However, this decrease is insignificant when the glass concentration is further increased. $\overline{D}$ reduces in FC with a low content (up to 0.75 mass%) of γ-glass is approximately half. $\overline{D}$ changes insignificantly in the range from 0.751 to 1.5 mass%. In the range from 1.5 to 3 mass% with the average grain size remaining relatively constant. At $\gamma > 3$ mass%, a slight increase in $\overline{D}$ is observed. The changes in $\overline{D}$ are correlated with the changes in σ: fine-grained ceramics are characterized by increased values of σ, while coarse-grained ceramics are characterized by decreased values of σ. Furthermore, the strength of ceramics when modified with γ-glass increases (in the region with a low glass content) but less affected than when α- and β-glasses are introduced into the same base.

The density of FC in all the cases described increases with increasing glass concentration (most significantly for small amounts of glass) (Fig. 2.7c), while water absorption decreases. In this case, glasses, despite the difference in composition, lead to an almost identical increase in density (Fig. 2.7c).

A study of the dependencies of the electrical characteristics of FC on the concentration of introduced additives demonstrated that the nature of their changes is consistent, regardless of the composition of the modifying glasses (Fig. 2.7d). In this case, small additions of glass (up to 1.5 mass%) lead to sharp changes in $\varepsilon/\varepsilon_0$, $\varepsilon_{33}^{\mathrm{T}}/\varepsilon_0$; an increase in tan δ and a decrease in ρ_V, a noticeable deterioration in piezoelectric properties (a decrease in d_{31}, K_P, K_t), an increase in their concentration (1.5–5 mass%) contributes to the stabilization of some characteristics (practically constant d_{31}, K_P, K_t) and insignificant changes in $\varepsilon/\varepsilon_0$, $\varepsilon_{33}^{\mathrm{T}}/\varepsilon_0$, ρ_V, tan δ. The behavior of Q_M, V_R, E_{yu} FC containing various (α, β, γ) glasses differs somewhat from that observed in other systems. An increase in the amount of α-glass leads to a linear increase in these parameters (Fig. 2.7d). In contrast, the changes in A-β and A-γ systems are non-monotonic.

Thus, the addition of glass in small quantities (less than 0.75 mass% for β-glass and less than 1.5 mass% for γ-glass) results in an increase in V_R and E_{yu}, while a decrease in Q_M is observed. In the range from 0.75 to 1.5 mass% for β-glass and from 1.5 to 3 mass% for γ-glass, the opposite trend is observed: for γ-glass, V_R and E_{yu} decrease, while Q_M increases. At $\beta > 1.5$ mass% and $\gamma > 3$ mass%, V_R and E_{yu} increase, while Q_M increases for β-glass and decreases with the introduction of γ-glass. A noteworthy aspect of all SCs is the disparate rate of decline in the planar (K_P) and thickness (K_t) electromechanical coupling coefficients. This phenomenon is responsible for the observed increase in the K_t/K_P ratio at a specific glass concentration, which is significantly higher than the corresponding value in the original SS ($K_t/K_P \sim 1$). Consequently, at glass concentrations from 1.5 to 3 wt% K_t/K_P, the values reach from

2.5 to 2.69, respectively. The highest value (2.69) is observed in the A-α system with 3 mass% of the glass.

Comparative analysis of the obtained data (Fig. 2.7c, d) allowed one to identify the composition and concentration of the introduced glass additives, which provide the most effective change in the parameters of the initial solid solution. Such a better modifier turned out to be α-glass, 3 mass% of which in the A-base leads to a significant decrease in T_{sint}, an increase in σ and the K_t/K_P ratio.

To optimize the properties of the FC material in the A-α system, a series of experiments were conducted to ascertain the most effective method of introducing glasses into the SS. In this instance, α-glass (in an amount of 3 mass%) was introduced either into the charge (i), or into the pre-fired one (at $T_1 = 800$ °C, $\tau_1 = 5$ h) (ii), or into the synthesized one ($T_1 = 800$ °C, $\tau_1 = 5$ h; $T_2 = 850$ °C, $\tau_2 = 5$ h) (iii) product. The data obtained (Table 2.3) indicate a significant decrease in T_{sint}, $\overline{D}$ and an increase in σ, V_R, Q_M, K_t/K_P FC compared to the original material. The effects are observed to be at their greatest when 3 mass% (4.8 mol%) α-glass is introduced using methods (ii) and (iii).

The introduction of glass into B-bases (Figs. 2.8 and 2.9) and C-bases (Figs. 2.8 and 2.9) produced similar results.

The following is an explanation of the patterns of change observed in the physical properties of solids when modified with glasses of different compositions.

The sintering of FC occurs in the presence of a liquid phase, resulting from the introduction of glasses into the initial composition, which then forms a low-melting eutectic [41]. If the base ceramic material is adequately wet, the liquid phase facilitates the formation of a compressive capillary force at the junction of solid particles, thereby exerting a pulling force on the particles [36] (this force is greatest when the wetting is full [42, 43]). Consequently, the sintering process in the presence of a wetting liquid phase is markedly enhanced [11, 35, 36]. The intensification of the sintering process of ceramics containing glass results in a decrease in their $T_{\text{sint.}}$ and an increase in ρ. The increase of ρ observed in systems with increasing glass concentration is also the result of the "binding" of crystallite grains of ceramics by the residual glassy phase after recrystallization [44].

To ascertain the impact of modifying glasses on the sintering process of piezoceramics, we recorded curves demonstrating the correlation between the shrinkage of samples of the original SS and the resulting FC in relation to $T_{\text{sint.}}$. The experimental results, as illustrated in Figs. 2.11 and 2.12, indicate a shift in the temperature at which plastic flow is initiated in FC systems towards lower temperatures in comparison to unmodified SS. Consequently, the introduction of α-glass into the initial SS (A), which undergoes plastic flow at 900 °C, results in a temperature shift of 100 °C. Similarly, the introduction of β- and γ-glasses into the A-base results in a temperature shift of 200 °C (Figs. 2.11 and 2.12).

Furthermore, the incorporation of the glass into the system results in an acceleration of the compaction process, with the A-α system exhibiting the highest rate of compaction when 5 mass% glass is introduced. This leads to a notable reduction in the T_{sint} of the corresponding FC.

Table 2.3 Parameters of SS composition $Na_{0.875}Li_{0.125}NbO_3$, glass-modified composition $0.203NaO–0.203Nb_2O_5–0.282WO_3–0.005Al_2O_3–0.203B_2O_3–0.094SiO_2$ at different stages of preparation of press powder

Composition (mol%)			Method of glass introduction (see text)	Parameters								
$NaNbO_3$	$LiNbO_3$	Glass		T_{sint}, °C	$\overline{D}$, μm	σ, kg/cm^2	K_t	K_P	K_t/K_P	$\varepsilon_{33}^T/\varepsilon_0$	V_R, km/s	Q_M
87.5	12.5	0		1150	18.5	307	0.350	0.334	1.041	107	5.60	465
85.4	12.2	2.4	(iii)	1060	15.3	386	0.174	0.078	2.24	190	5.68	540
			(i)	1070	8.2	400	0.180	0.074	2.50	215	5.74	447
83.3	11.9	4.8	(ii)	1060	7.6	492	0.250	0.094	2.69	165	5.80	509
			(iii)	1040	7.6	497	0.192	0.077	2.53	232	5.71	*591*
80.6	11.5	7.9	(iii)	1040	20.6	404	0.199	0.094	2.12	274	5.80	600

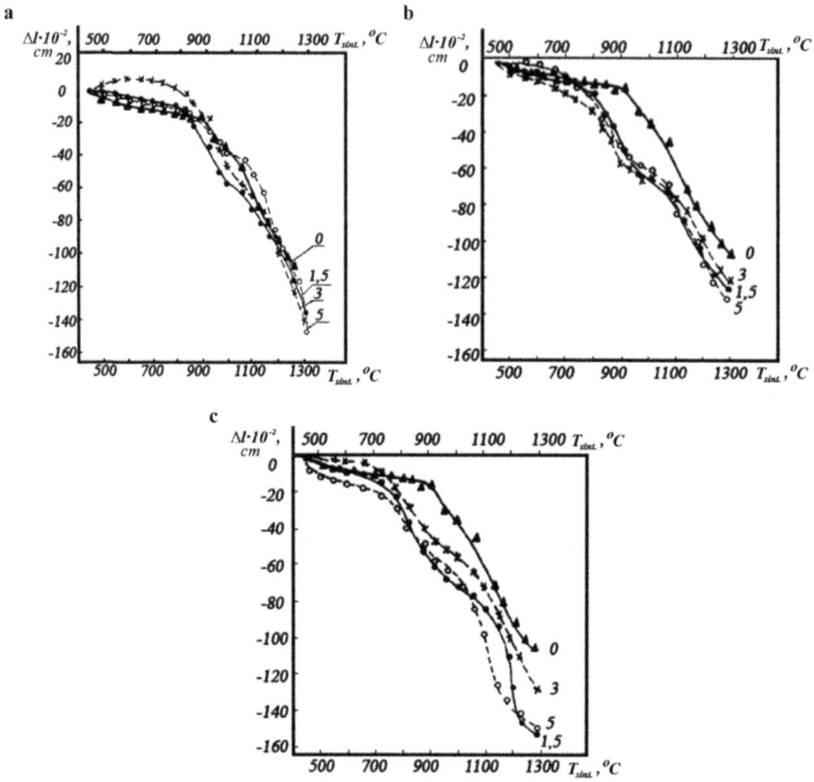

Fig. 2.11 **a** Dependencies of FC shrinkage in A-α systems on sintering temperature at different glass contents; the numbers in the curves denote the concentration of introduced glasses. **b** Dependencies of FC shrinkage in A-β systems on sintering temperature at different glass contents; the numbers in the curves denote the concentration of introduced glasses. **c** Dependencies of FC shrinkage in A-γ systems on sintering temperature at different glass contents; the numbers in the curves denote the concentration of introduced glasses

The observed reduction in the average grain size with the introduction of modifying glasses is associated with a deceleration in their growth during the sintering process. This is due to the melt of glass, which fills the space between crystallite grains, promoting the recrystallization process through the liquid phase [39].

An increase in $\overline{D}$ at specific glass concentrations (Figs. 2.6a, 2.7a and 2.9b) may be attributed to both the inhomogeneity of grain sizes within the microstructure of samples containing a substantial amount of glass, and secondary recrystallization, during which there is an accelerated growth of individual grains situated in energetically more favorable conditions [45].

The strengthening of FC, associated with their fine grain size can be attributed to a number of factors: a reduction in stresses at the boundaries between grains and phases (crystalline and glass phase) with a reduction in $\overline{D}$ [45]; an increase in the magnitude of the fracture stress is proportional to $(\sqrt{D})^{-1}$ [46]; the barrier effect

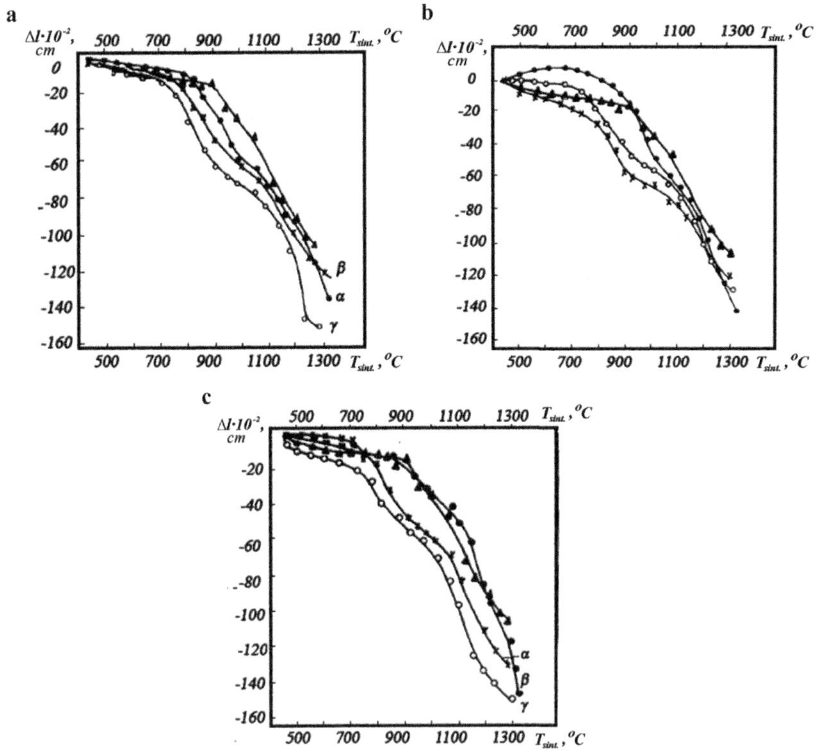

Fig. 2.12 **a** Dependencies of FC shrinkage in systems A-α, A-β, A-γ on sintering temperature at glass content of 1.5 mass%. **b** Dependencies of FC shrinkage in the A-α, A-β, A-γ systems on the sintering temperature at a glass content of 3 mass%. **c** Dependencies of FC shrinkage in systems A-α, A-β, A-γ on sintering temperature at a glass content of 5 mass%

identified as strengthening role of grain and interphase boundaries in the result of their capacity to act as powerful obstacles to moving dislocations; a strengthening of the processes of segregation of impurity atoms at grain boundaries, pinning dislocations and increasing grain-boundary relaxation; increasing the rate of proliferation of dislocations and, in connection with this, strengthening the interaction between them [47]. Another factor contributing to the strengthening of ceramics may be the cementation of crystalline phase grains with a melt that fills the pores and intergranular voids [39, 40]. In addition, the glass phase itself is to some extent a stress reliever, reducing the sensitivity of FC to surface damage that leads to stress concentration [48].

The research yielded a piezoceramic material containing Na_2O, Li_2O, Nb_2O_5, characterized in that to increase mechanical strength and the ratio of the thickness coefficient of electromechanical coupling to the planar, as well as to reduce the sintering temperature by the inclusion of the oxides WO_3, B_2O_3, and SiO_2, along with Al_2O_3, in the following ratio of components (in mol%): The Na_2O content ranges from 41.088 to 43.184, while the Li_2O content varies from 5.69 to 6.10, the Nb_2O_5 content is between 47.096 and 49.284, while the WO_3 content ranges from 0.704 to 2.292, the B_2O_3 content is between 0.189 and 1.594, while the SiO_2 content is between 0.0994 and 0.740 for Al_2O_3 from 0.0121 to 0.317. The material containing, in addition to the listed oxides SrO ranges from 0.237 to 1.586 mol% results in a significant reduction in the average size of crystallite grains.

In addition, a method has been developed for producing the piezoceramic materials mentioned above, including the synthesis of part of the oxides: Na_2O, Li_2O; Nb_2O_5,WO_3 taken in the following ratio (in mol%): 39.82–42.695 (Na_2O); 5.69–6.10 (Li_2O); 45.51–48.795 (Nb_2O_5); 0–2.79 (WO_3) with subsequent hot pressing, characterized in that a pre-fused and then crushed mixture of oxides: Na_2O, Nb_2O_5, WO_3, B_2O_3, Al_2O_3 was added to the synthesized oxide system in an amount of 0.95–7.85 mol% with the following mixture ratio: 0.189–1.594 (Na_2O); 0.237–1.594 (Nb_2O_5); 0.094–0.740 (SiO_2); 0–2.292 (WO_3); 0.189–1.594 (B_2O_3); 0.0121–0.317 (Al_2O_3); 0–1.586 (SrO).

Figures 2.11 and 2.12 demonstrate dependencies of FC shrinkage in the systems considered on sintering temperature at different glass contents. The reduction of the average grain size and the enhancement of the strength of ceramics permit the fabrication of extremely thin piezoelectric sensors, with an operating frequency range that is considerably higher than that of transducers, based on conventional HP ceramics.

The parameters of the compositions investigated are presented in Tables 2.4 and 2.5. For purposes of comparison, we present the characteristics of the opposing materials (A and B):

(A) $T_{sint.} = 1150\,°C$; $\overline{D} = 18\,\mu m$; $\sigma = 307\,kg/cm^3$; $K_t/K_P = 1.045$;
(B) $T_{sint.} = 1085\,°C$; $\overline{D} = 12\,\mu m$; $\sigma = 53\,kg/cm^3$; $K_t/K_P = 0.96$.

The reduction of $T_{sint.}$ of the proposed materials contributes to the preservation of stoichiometry of the given composition. This is due to the lower sublimation of components, the impossibility of formation of side crystalline phases, and the absence of thermal dislocation of SS in the region of reduced FC.

Table 2.4 Parameters of the investigated compositions

Composition (mol%)				Parameters							
Na_2O	Li_2O	Nb_2O_5	WO_3	T_{sint}, °C	$\overline{D}$, μm	K_t	K_P	$V_R \times 10^{-3}$, m/s	$\varepsilon_{33}^T/\varepsilon_0$	σ, kg/cm^3	K_t/K_P
43.184	6.10	49.284	0.704	1060	15.3	0.174	0.078	5.68	190	386	2.22
42.616	5.95	48.566	1.39	1040	7.6	0.192	0.077	5.71	232	497	2.53
41.909	5.76	47.669	2.292	1040	20.6	0.199	0.094	5.80	274	404	2.13
42.299	6.015	48.368	2.79	1065	4.6	0.21	0.109	5.68	151	235	1.92
42.090	5.960	48.143	2.78	1070	3.6	0.19	0.057	5.63	272	535	3.33
41.661	5.845	47.694	2.71	1050	3.6	0.20	0.080	5.67	215	604	2.50
41.088	5.69	47.096	2.64	1020	3.6	0.175	0.070	5.62	230	486	2.51

Table 2.5 Parameters of the investigated compositions

Composition (mol%)				Parameters							
B_2O_3	SiO_2	Al_2O_3	SrO	T_{sint}, °C	$\overline{D}$, μm	K_t	K_P	$V_R \times 10^{-3}$, m/s	$\varepsilon_{33}^{T}/\varepsilon_0$	σ, kg/cm³	K_t/K_P
0.489	0.227	0.0121	0	1060	15.3	0.174	0.078	5.68	190	386	2.22
0.966	0.447	0.024	0	1040	7.6	0.192	0.077	5.71	232	497	2.53
1.594	0.740	0.0392	0	1040	20.6	0.199	0.094	5.80	274	404	2.13
0.189	0.0994	0.0473	0.237	1065	4.6	0.21	0.109	5.68	151	235	1.92
0.370	0.185	0.093	0.463	1070	3.6	0.19	0.057	5.63	272	535	3.33
0.746	0.373	0.187	0.934	1050	3.6	0.20	0.080	5.67	215	604	2.50
1.268	0.6338	0.317	1.586	1020	3.6	0.175	0.070	5.62	230	486	2.51

Furthermore, $\overline{D}$ plays a role in the inhibition of the destructive effects characteristic of the studied compositions. This results in an extension of the sintering temperature range, which, in turn, improves the reproducibility of the characteristics. The high K_t/K_P ratio ($\geq$ 2.5) enables the materials to be employed in single frequency resonators, operating on thickness vibrations, as well as in high-frequency accelerometers with selective sensitivity. In such devices, a significant excess of longitudinal sensitivity over transverse sensitivity is necessary to determine the exact direction of mechanical action.

The results of the study of FC in niobate systems have been published in [49].

Based on the analysis of the data in this section, the following observations can be made:

1. The heterovalent modification mechanism is identical in both PZT-type and systems based on sodium niobate. The key factor influencing the formation of a set of physical properties in modified solid solutions is the mobility of domain walls, which is determined by SD and their interaction with vacancies introduced by the appropriate modifiers. The degree of change in parameters is dependent on the concentration of vacancies and the nature of their distribution in the crystal structure.
2. The relationship between HF and the nature of the chemical bond in SS of the modified $NaNbO_3$–$LiNbO_3$ system is analogous to that observed in SS of the based on PZT-system. This is caused by a change in the degree of covalency of the bonds, their direction and unequal value.
3. A modified SS with parameters of practical importance has been developed, combining high piezoelectric (d_{31}, K_P) and elastic (V_R, E_{yu}) characteristics with relatively low values of $\varepsilon_{33}^{T}/\varepsilon_0$. This material can be used as the basis for microwave devices. This confirms the effectiveness of heterovalent modification as a method of targeted change in the physical properties of SS of the FE systems.

4. The physical properties of SS based on $NaNbO_3$ can be significantly altered by the introduction of glasses, with optimal results being achieved when obtaining SS using a method that combines elements of glass–ceramic technology and HP.
5. The selection of a specific qualitative and quantitative composition of modifying glasses allows one to produce materials with promising applications in a variety of electronic devices.

2.4 Conclusion

This section has been presented a comprehensive analysis and generalization of the extensive experimental data. The classification of the utilized modifiers, the order and method of their introduction, and the influence of different modifiers on the ceramic, dielectric, piezoelectric and elastic characteristics of solid solutions of the systems $NaNbO_3$–$KNbO_3$ and $NaNbO_3$–$LiNbO_3$ were evaluated. This study demonstrated the advantages of glass–ceramic technology in obtaining solid solutions of the system $NaNbO_3$–$LiNbO_3$. The impact of α-, β-, and γ-glasses on different characteristics of solid solutions of the system $NaNbO_3$–$LiNbO_3$ was assessed. It was demonstrated that the selection of the optimal qualitative and quantitative composition of modifying additives and glasses allows the production of materials with promising applications in various types of radio-electronic devices.

References

1. Grineva, L.D.: Study of the patterns of changes in physical properties when modifying ferroelectric solid solutions of complex oxides with a perovskite-type structure. Ph.D. Thesis (Physical and Mathematical Sciences), Rostov State University Press, Rostov-on-Don (1975) (in Russian)
2. Ananyeva, A.A., et al.: Piezo- and dielectric properties of modified potassium-sodium niobate ceramics. In: Methods for Obtaining and Studying Piezo- and Ferroelectric Materials and Feedstock for Them, pp. 89–95. MDNTP, Moscow (1966) (In Russian)
3. Dambekalne, M.Ya., et al.: The influence of modification on the piezoelectric properties and texture of ceramics of solid solutions of potassium and sodium niobate. Electron. Technol. Ser. 14. Mater. (Inorganic Dielectrics) **8**, 33–35 (1969) (in Russian)
4. Freidenfeld, E.Zh.: Physico-chemical principles of synthesis and electrical properties of ceramic and glass-crystalline piezoelectrics. DrSc Thesis (Technical Sciences), Moscow (1972) (In Russian)
5. Ugryumova, M.A.: Research and development of new piezoceramics based on solid solutions of barium and lead niobate, potassium and sodium niobate. Ph.D. Thesis (Technical Sciences), Moscow (1968) (in Russian)
6. Ugryumova, M.A.: Piezoceramics based on (K, Na)NbO_3 solid solutions. Electron. Technol. Ser. 14. Mater. (Inorganic Dielectrics) **14**(20), 130–138 (1969) (In Russian)
7. Yanson, G.D., et al.: Ceramic solid solutions of potassium and sodium metaniobates modified with alkaline earth metal ions. In: Piezo- and Ferroelectric Materials and Their Application, pp. 36–38. MDNTP, Moscow (1972) (In Russian)
8. Yanson, G.D., et al.: Kinetics of sintering of ferroelectric ceramics based on alkali niobates modified with oxides of alkaline earth elements. In: Proceedings of IV Intersectoral Meeting, Moscow, pp. 40–45 (1972) (In Russian)

9. Macmillan, P.W.: Glassceramics. Ph.D. Thesis (Technical Sciences). Mir, Moscow (1967) (in Russian)
10. Yanson, G.D., Evdokimova, O.G.: Glass-ceramics based on alkaline metaniobates. In: Proceedings of the Piezo- and Ferroelectric Materials and Their Application, pp. 14–18. MDSTP, Moscow (1972) (in Russian)
11. Skorokhod, V.V.: Rheological Bases of the Theory of Sintering. Naukova Dumka, Kiev (1972) (in Russian)
12. Balash, V.A., Kramarov, O.P., Shpoliansky, Y.A.: Investigation of the direct and inverse piezoelectric effect of a segmented ceramic. Izv. USSR Acad. Sci. Ser. Phys. **29**(11), 2086–2090 (1965) (in Russian)
13. Fesenko, E.G., et al.: Piezoelectric ceramic material. USSR Patent 591435, Invention Bulletin No. 5 (1978) (in Russian)
14. Fesenko, E.G., et al.: Piezoelectric ceramic material. USSR Patent 694478, Invention Bulletin No. 40 (1979) (in Russian)
15. Reznichenko, L.A., Grineva, L.D., Razumovskaya, O.N.: Effect of modification on the properties of solid solutions based on alkali metal niobates. In: Proceedings of V International Conference on State and Prospects of Development of Methods for Obtaining and Analyzing Ferrite Ferro-, Piezoelectric and Capacitor Materials and Raw Materials for Them, pp. 87–88. Donetsk (1975) (in Russian)
16. Harate, K.H., Hennings, D.: Distribution of A-site and B-site vacancies in (Pb, La)(Ti, Zr)O_3 ceramics. J. Am. Am. Ceram. Soc. **55**(5), 230–234 (1972)
17. Jhomann, H.: Piezoelectric mechanism in bleizirconat-titanat. J. Angew. Phys. **20**(6), 554–557 (1966)
18. Okazaki, K, Hirama, K: Effect of grain size and Porosity on electrical and optical properties of PLZT ceramics. J. Am. Ceram. Soc. **56**(2), 82–86 (1973)
19. Fesenko, E.G.: The Perovskite Family and Ferroelectricity. Atomizdat, Moscow (1972) (In Russian)
20. Povarennykh, A.S.: Crystal Chemical Classification of Minerals. Springer, New York, NY (1972)
21. Grineva, L.D., Danziger, A.Ya., Fesenko, E.G.: Some features of modification of solid solutions of the type $PbTiO_3$–$PbZrO_3$–$PbB_{(1-\beta)}{}' B_\beta{}'' O_3$. In: Proceedings of VIMI RIPORT, vol. 6 (1975) (in Russian)
22. Fesenko, E.G., Danziger, A.Ya., Razumovskaya O.N. Dielectric and piezoelectric properties of solid solutions of multicomponent systems of complex oxides. Izv. AN USSR. Ser. Inorganic Mater. **14**(5), 928–931 (1978) (in Russian)
23. Jaffe, B., et al.: Piezoelectric Ceramics. Academic Press, London, New York (1971)
24. Fesenko, E.G., et al.: Piezoelectric ceramic material. USSR Patent 587129, Invention Bulletin No. 1 (1978) (in Russian)
25. Fesenko, E.G. et al.: Piezoelectric ceramic material. USSR Patent 619468, Invention Bulletin No. 30 (1978) (in Russian)
26. Fesenko, E.G., et al.: Piezoelectric ceramic material. USSR Patent 642272, Invention Bulletin No. 2 (1979) (in Russian)
27. Fesenko, E.G., et al.: Piezoelectric ceramic material. USSR Patent 642273, Invention Bulletin No. 2 (1979) (in Russian)
28. Fesenko, E.G., et al.: Piezoelectric ceramic material. USSR Patent 657463, Invention Bulletin No. 42 (1979) (in Russian)
29. Grineva, L.D., et al.: Study of the patterns of changes in physical properties when modifying solid solutions based on alkali metal niobates. In: Crystallization and Properties of Crystals, Novocherkassk, vol. 6, pp. 57–65 (1979) (in Russian)
30. Reznichenko, L.A., Grineva, L.D.: Influence of production conditions on physical and chemical properties of materials based on alkali metal niobates. In: Proceedings of V Seminar Ferro- and Piezomaterials and Their Application, Moscow, pp. 144–147 (1978) (in Russian)

31. Reznichenko, L.A., Grineva, L.D.: Investigation of the regularities of physical properties changes at modification of solid solutions based on the alkali metal niobates. In: Proceedings of VI All-Union Conference on State and Prospects of the Development of Methods for Obtaining and Analyzing Ferrite, Ferro-Piezoelectric Capacitor and Resistive Materials and Raw Materials for Them, Donetsk, vol. 68 (1978) (in Russian)
32. Bondarenko, V.S., et al.: Development, Research and Application of Ferroelectric Materials in Ultraacoustics. Barium Titanate, pp. 195–201. Science, Moscow (1973) (in Russian)
33. Hardiman, B., et al.: Hot pressing of sodium-lithium niobates ceramics with perovskite-type structures. Ferroelectrics **12**(1–4), 157–159 (1976)
34. Jaeger, R.E., Egerton, L.: Hot pressing of potassium-sodium niobates. J. Am. Ceram. Soc. **45**(5), 209–213 (1962)
35. Eremenko, V.N., Naidich, Y.V., Lavrynenko, I.A.: Sintering with Participation of Liquid Phase, p. 252. Naukova Dumka, Kiev (1967) (in Russian)
36. Kainarsky, I.S.: Processes of Refractories Technology. Metallurgy, Moscow (1969) (in Russian)
37. Flidlider, G.M., Balkevich, V.D., Kholodkova, L.A.: Effect of glass additives on the properties of piezoceramics during hot pressing. Glass Ceram. **10**, 30–32 (1977) (in Russian)
38. Kinoshita, M., Wakabayashi, H.: Hot pressing of glass powder and PZT-glass powder compacts. Bull. Govt. Ind. Res. Inst. Osaka **24**, 208–213 (1973)
39. Kleine, R.Z., Freidenfeld, E.J.: Some regularities of modification of ferroelectric ceramic materials by glass. In: Proceedings of the Piezo- and Ferroelectric Materials and Their Application, pp. 95–97. MDSTP, Moscow (1975) (in Russian)
40. Ugryumova, M.A. Influence of glass-forming oxides on the mechanical strength of piezoceramics PZT-13. In: Proceedings of the Piezo- and Ferroelectric Materials and Their Application, pp. 19–23. MDSTP, Moscow (1971) (in Russian)
41. Kleine, R.Z., Freidenfeld, E.J., Kutuzova, T.K.: Sintering and properties of lead zirconate-titanate ceramics with glass addition. In: Proceedings of the Piezo- and Ferroelectric Materials and Their Application, pp. 11–13. MDSTP, Moscow (1972) (in Russian)
42. Naidich, Yu.V., Kolesnichenko, G.A.: Study of the wetting of diamond and graphite by liquid metals. Powder Metall. **1**, 35–38 (1964) (in Russian)
43. Naidich, Yu.V., Lavrinenko, I.A.: Capillary adhesion forces between solid particles with an intermediate liquid layer at the contact. Powder Metall. **9**, 709–710 (1967) (in Russian)
44. Freidenfeld, E.Zh., Selection of glass for modification of PZT piezoceramics. In: Proceedings of V Interbranch Conference on Methods of Obtaining and Analyzing Ferrite, Ferroelectric, Piezoelectric and Capacitor Materials and Raw Materials for Them, Moscow, pp. 46–50 (1973) (in Russian)
45. Kingery, W.D.: Introduction to Ceramics. Stroyizdat, Moscow (1967) (in Russian)
46. Orowan, E.: Fracture and strength of solids. Rep. Prog. Phys. **12**, 185–232 (1949)
47. Krasovsky, A.Y.: Physical Foundations of Strength. Naukova Dumka, Kiev (1977) (in Russian)
48. Zatsarinniy, V.P.: Strength of Piezoceramics, p. 208. Rostov State University Press, Rostov-on-Don (1978) (in Russian)
49. Raevskiy, I.P., Reznichenko, L.A., Panich, A.E.: Peculiarities of obtaining segmented ceramics based on alkali metal niobates by methods of hot pressing and glass-ceramic technology. In: Proceedings of IX All-Union Council on Ferroelectricity, Rostov-on-Don, vol. 2, p. 133 (1979) (in Russian)

Chapter 3
Regularities of Changes in Physical Properties of Oxides with Perovskite-Type Structure During Modification

3.1 Modification Effects in (Na, Li) NbO_3 and $LiNbO_3$ System

Table 3.1 provides a summary of the scheme, formulae, and quasi-chemical equations, associated with cationic and vacancy modification of one of the SS of the (Na, Li)NbO_3 system. The table illustrates the position of modifiers within the crystal lattice of the matrix, as well as the nature and concentration of vacancies that have been formed. The SS of composition $Na_{0.875}Li_{0.125}NbO_3$, situated at the Rh boundary of the MO of the (Na, Li)NbO_3 system and exhibiting exceptional values of piezoelectric parameters [1, 2], was selected as the subject of investigation.

3.1.1 Schemes, Formulae and Quasi-chemical Equations of Modification of Solid Solutions of (Na, Li)NbO_3 System

In the context of cationic modification, the term refers to the introduction of elements that are not part of the original structure. In contrast, vacancy modification pertains to alter the defect state of the base without a significant change in its chemical composition. This is achieved by modifying the concentration of the *AO* component or the oxidation degree of the *B*-component. Despite the conditionality of this division (in the first case, vacancies are formed, and in the second case, chemical composition is altered), it seems legitimate to posit that in each of the options, the leading mechanism of formation of a new set of properties is either primarily due to the influence of crystallochemical characteristics of modifiers (Fig. 3.1a, b—I–V) or due to the number and "quality" of vacancies (Fig. 3.1c—VI–VIII).

I. A. Verbenko et al., *Environmentally Friendly Smart Materials with Special Electrical and Magnetic Properties*, Engineering Materials,
https://doi.org/10.1007/978-3-031-90387-8_3

Table 3.1 Formulae and quasi-chemical equations of cationic stoichiometric modification

Replacement options			Formulae and equations
Substi-tution in *A*-sublat-tice	Isovalent, $n_{MA} = n_A$		$n = 1$ $(Na, Li)_{1-x}NbO_{3-x/2} + xMO_{0.5} \rightarrow (Na, Li)_{1-x}M_xNbO_3$ $M_2O \leftrightarrow 2M^x{}_{Na} + O^x{}_O$
	Heterova-lent, $n_{MA} > n_A$	Vacancies in *A*-substrate	$(Na, Li)_{1-nx}NbO_{(3-n)/2x} + xMO_{n/2} \rightarrow (Na, Li)_{1-nx}M_x^n\square_{x(n-1)}NbO_3$ $n = 2$ $(Na, Li)_{1-2x}NbO_{3-x} + xMO \rightarrow (Na, Li)_{1-2x}M_x\square_xNbO_3$ $MO \leftrightarrow M^{\cdot\cdot}{}_{Na,Li} + V'{}_{Na,Li} + O^x{}_O$ $n = 3$ $(Na, Li)_{1-3x}NbO_{-1.5x} + xMO_{1.5} \rightarrow (Na, Li)_{1-3x}M_x\square_{2x}NbO_3$ $M_2O_3 \leftrightarrow 2M^{\cdot\cdot}{}_{Na,Li} + 4V'{}_{Na,Li} + 3O^x{}_O$
	Heterova-lent, $n_{MA} > n_A$	Vacancies in *B*-substrate	$(Na, Li)_{1-x}M_xNb_{1-(n-1)x/5}O_{3-n/2x} + xMO_{n/2} \rightarrow (Na, Li)_{1--x}M_xNb_{1-(n-1)x/5}\square_{1-(n-1)x/5}O_3$ $N = 2$ $(Na, Li)_{1-x}M_xNb_{1-1x/5}O_{3-x} + xMO \rightarrow (Na, Li)_{1-x}M_xNb_{1-1x/5}\square_{x/5}O_3$ $5MO \leftrightarrow 5M^{\cdot}{}_{Na,Li} + V^{5'}{}_{Na} + 5O^x{}_O$ $n = 3$ $(Na, Li)_{1-x}M_xNb_{1-2x/5}O_{3-1.5x} + xMO_{1,5} \rightarrow (Na, Li)_{1-x}M_xNb_{1-2x/5}\square_{2/5}O_3$ $2.5M_2O_3 \leftrightarrow 5M^{\cdot\cdot}{}_{Na,Li} + 2V^{5'}{}_{Na} + 7.5O^x{}_O$
Substi-tution in *B*-sublat-tice	Isovalent, $n_{MB} = n_B$		$(Na, Li)Nb_{1-x}O_{3-2.5x} + xMO_{2.5} \rightarrow (Na, Li)Nb_{1-x}M_xO_3$ $M_2O_5 \leftrightarrow 2M^x{}_{Nb} + 5O^x{}_O$
	Heterova-lent, $n_{MB} > n_B$	Vacancies in *A*-substrate	$(Na, Li)_{1-x(n--5)}Nb_{1-x}O_{3-n/2x} + xMO_{n/2} \rightarrow (Na, Li)_{1-x(n-5)}\square_{x(n-5)}Nb_{1-x}M_xO_3$ $n = 6$ $(Na, Li)_{1-x}Nb_{1-x}O_{3-3x} + xMO_3 \rightarrow (Na, Li)_{1-x}\square_xNb_{1-x}M_xO_3$ $MO_3 \leftrightarrow M^{\cdot}{}_{Nb} + V'{}_{Na} + 3O^x{}_O$

(continued)

Table 3.1 (continued)

Replacement options			Formulae and equations
	Heterova-lent, $n_{MB} > n_B$	Vacancies in B-substrate	$(Na, Li)Nb_{1-nx/5}O_{3-n/2x} + xMO_{n/2} \rightarrow (Na, Li)Nb_{1-nx/5}M_x\square_{x(n-5)/5}O_3$ $n = 6$ $(Na, Li)Nb_{1-6x/5}O_{3-3x} + xMO_3 \rightarrow (Na, Li)Nb_{1-6x/5}M_x\square_{x/5}O_3$ $5MO_3 \leftrightarrow 5M^{\cdot}_{Nb} + V'_{Nb} + 15O^x_O$
	Heterovalent, $n_{MB} < n_B$	Vacancies in O-substrate	$(Na, Li)Nb_{1-x}O_{3-5x/2} + xMO_{n/2} \rightarrow (Na, Li)Nb_{1-x}M_xO_{3-x(5-n)/2}$ $n = 3$ $(Na, Li)Nb_{1-x}O_{3-2.5x} + xMO_{1.5} \rightarrow (Na, Li)Nb_{1-x}M_xO_{3-x}\square_x$ $M_2O_3 \leftrightarrow 2M''_{Nb} + 2V^{\cdot\cdot}_O + 3O^x_O$ $n = 4$ $(Na, Li)Nb_{1-x}O_{3-2.5x} + xMO_2 \rightarrow (Na, Li)Nb_{1-x}M_xO_{3-x/2}\square_{x/2}$ $MO_2 \leftrightarrow M'_{Nb} + 0.5V^{\cdot\cdot}_O + 2O^x_O$
Substitution in AB-sublattices	Isovalent		$[(Na, Li)_{1-nx}M^n{}_x\,\square_{x(n-1)}]_{1-y(n'-5)}\square_{y(n'-5)}Nb_{1-y}\,Me^{n'}{}_yO_3$
Heterovalent	Substitution in A (vac. A) + substitution in B (vac. B)		$(Na, Li)Nb_{1-nx}\,M^n{}_x\,\square_{x(n-1)}Nb_{1-n'y/5}\,Me^{n'}{}_y\,\square_{y(n'-5)/5}O_3$
	Substitution in A (vac. B) + substitution in B (vac. A)		$[(Na, Li)Nb_{1-x}$ $M_x]_{1-y(n'-5)}\square_{y(n'-5)}(Nb_{1-(n-1)x/5}\square_{(n-1)x/5})_{1-y}Me^{n'}{}_yO_3$
	Substitution in A (vac. B) + substitution in B (vac. B)		$(Na, Li)_{1-x}\,M_x\square_{y(n'-5)}(Nb_{1-(n-1)x/5}\square_{(n-1)x/5})_{1-n'y/5}Me^{n'}{}_y\square$ $_{y(n'-5)/5}O_3$

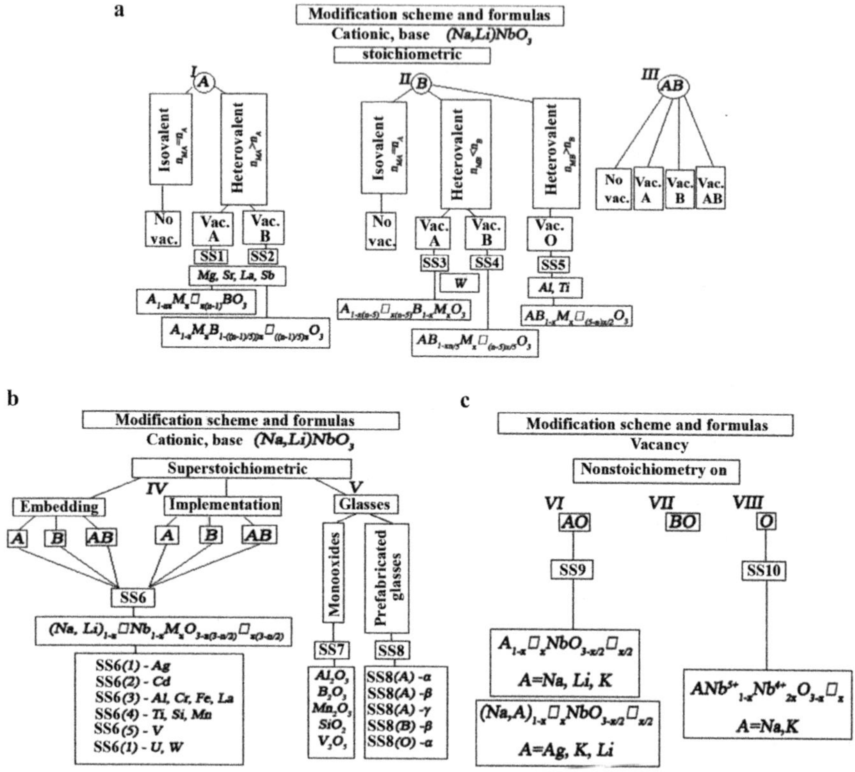

Fig. 3.1 **a** Schemes and formulae of cationic stoichiometric modifications of some research objects (I, II, III are the schemes, indicating types of modifications; □ is a conventional designation of vacancy). **b** Schemes and formulae of cationic super stoichiometric modification of some research objects (IV, V are the schemes indicating types of modifications; □ is a conventional designation of vacancy). **c** Schemes and formulae of vacancy modification of some research objects (VI, VII, VIII are the schemes indicating types of modification, □ is a conventional designation of vacancy)

The schemes present variants of stoichiometric (isovalent vacancy-free and heterovalent with the formation of vacancies, the position of which in the crystal structure can be different) and super stoichiometric isomorphic substitution of ions in the *A*-, *B*-, and *AB*-sublattices.

In the case of *A*-substitutions (I), only the scenario, in which $n_{MA} > n_A$* (i.e., $n_M = 2, 3$) is considered as relevant, given that the valence of alkali metal (*AM*) is 1. For *B*-substitutions (II) are described possibilities with $n_M <> n_B$ (i.e. $n_{MB} = 6, 4, 3$). The isovalent modification is achieved at $n_{MA} = 1$, $n_{MB} = 5$. At combined modification (III), when ions in both cationic sublattices are substituted, a compensatory variant is possible that excludes the appearance of vacancies ($n_{MA} = 2$ and $n_{MB} = 4$, $n_{MA} = 3$ and $n_{MB} = 3$) or variant with the appearance of vacancies in *A*-, *B*-, and *AB*-positions.

When introduction of modifiers beyond stoichiometry (IV) the options for modifications to the crystal lattice of the base may be greater variations than when they are introduced in the formula.

The cation-modifier can become incorporated into regular nodes of the matrix crystal, which leads to the formation of vacancies in the cationic and oxygen sublattices. In this process, cation *A* causes vacancies in the *B*- and O-sublattices, while cation *B* causes vacancies in the *A*- and O-sublattices [3]. The modifier cation may be incorporated into the lattice, displacing the major component; it may also be incorporated into the lattice itself or present as an impurity distributed in the intergranular space.

In the context of vacancy modification, it was deemed appropriate to consider a controlled deviation from stoichiometry in the *A*O(VI) and O(VIII) components. In the latter case, significant oxygen defectivity was achieved by sintering SS in a reduced atmosphere [4].

3.1.2 Effects of Stoichiometric Modification of the (Na, Li) NbO_3 System

$\mathbf{n_{MB}} < n_{\mathbf{B}}$

The modifiers Al^{3+} and Ti^{4+} with valences lower than their substituted Nb^{5+} function are acceptors. Figure 3.2a and b demonstrate the relationships between perovskite cell parameters and the concentrations of Al^{3+} and Ti^{4+} in SS-5. Starting from 0.2 at%, in both cases, SS with a mixture of R and Rh phases crystallize. In the initial case, heterophase is maintained across the entire range of concentrations of the introduced elements. In the second case, the area of phase coexistence is constrained to a range from 0.2 at% to 0.5 at%, and concentrations exceeding 0.5 at%, SS exhibit monophasic behavior, manifesting as R-distortion of the cell. This indicates a rightward shift of the MO in the (Na, Li) NbO_3 system during the modification process, with a stronger shift occurring when Ti^{4+} is introduced. Therefore, the addition of the modifier has a profound impact on the structure of the initial SS, resulting in the formation of SS with varying degrees of cell distortion at different concentrations. At concentrations below 0.2 at%, the SS exhibit Rh distortion, while at concentrations between 0.2 and 0.5 at%, they display R + Rh distortion, and at concentrations exceeding 0.5 at%, they manifest R distortion. In this regard, it can be observed that only the composition SS (Na, Li) $Nb_{1-(0.2-0.5)} \cdot M_{(0.2-0.5)} O_{(0.2-0.5)}$ is situated within the MO, while the remaining single-phase Ti-bearing SS are positioned externally, on disparate sides of the boundaries. In Ti^{4+}-containing SS, the alterations in unit cell parameters are more pronounced than in the case of Al^{3+}, which is distinguished by practical constancy $\overline{a}_R = \sqrt[3]{V}$ [3]V across the entire range of modifier concentrations and a limited growth in $a = c$, a_{Rh}, α, β at $x > 0.5$ at% Al^{3+} (Fig. 3.2). Furthermore, the linear dimensions of SS cells with Ti^{4+} exhibit the most pronounced variation at low concentrations of Ti^{4+} ($x > 1$ at%), reaching a maximum at $x = 0.5$ at%, which is associated with the transition of SS to a single-phase state at this concentration of Ti^{4+}. For $x > 1$ at% Ti^{4+}, the linear dimensions of cells exhibit a weak change.

The angular characteristics exhibit notable non-monotonic fluctuations throughout the entire phase existence range. At $x = 1.5$ at% Ti^{4+}, the parameter β is at its minimal value, and it increases on either side of this value (Fig. 3.2b). The

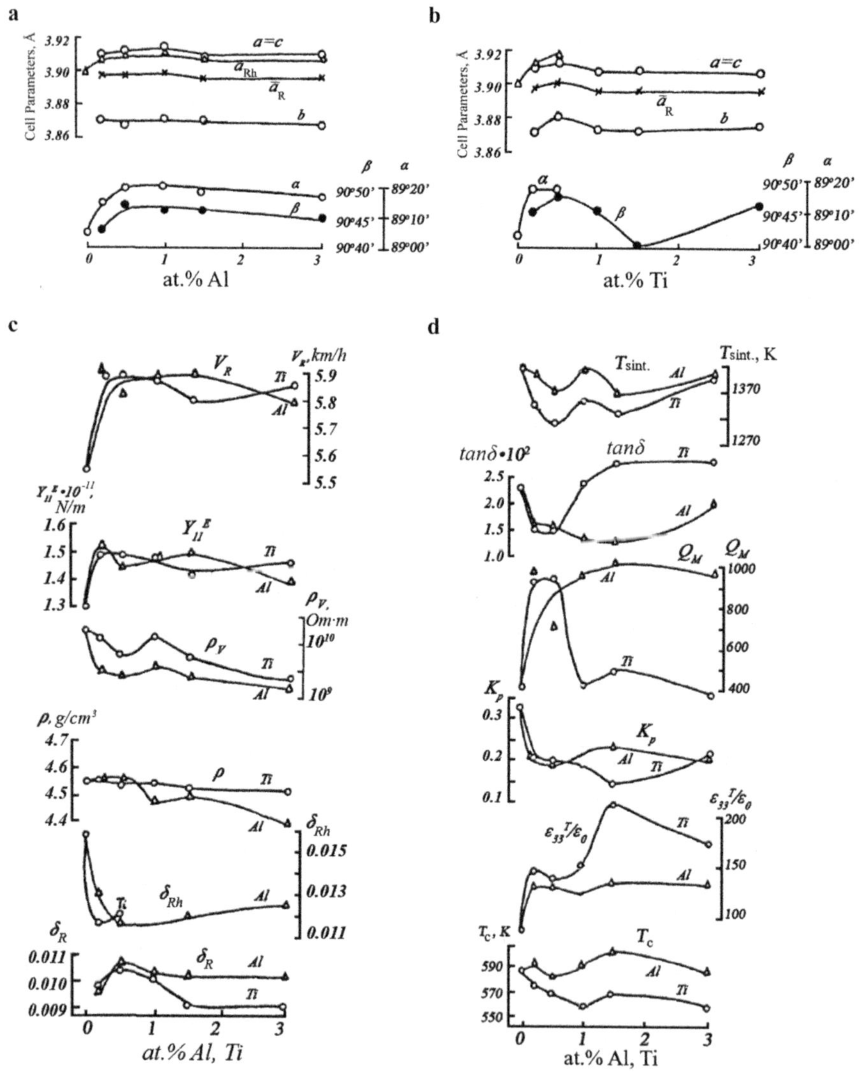

Fig. 3.2 **a** Dependencies of structural characteristics of SS on Al^{3+} content. **b** Dependencies of structural characteristics of SS on Ti^{4+} content. **c** Dependencies of structural and electrophysical characteristics of SS on the content of Al^{3+} and Ti^{4+}. **d** Dependencies of dielectric and piezoelectric characteristics of SS on the content of Al^{3+} and Ti^{4+}

course of the parameters δ_{Rh} and δ_R (Fig. 3.2c and d) differs significantly from one of another, yet exhibits a notable degree of similarity for SS with Al^{3+} and Ti^{4+}. A pronounced decline in the value of c is observed at $x \leq 0.2$ at% of modifiers, followed by a slight increase at $x > 0.3$ at.%. Concurrently, the values of δ_{Rh} remain markedly inferior to those observed in the original SS. δ_{Rh} exhibits anomalous changes in the concentration range of $0.2 < x < 1$ at% of both modifiers with a maximum at $x = 0.5$ at%. In the range of $1.0 < x < 1.5$ at%, the decrease observed in δ_R is greatest. Further increase of x leads only to very small changes in δ_R. The changes of δ_R in SS with Ti^{4+} are more significant than in SS with Al^{3+} (Fig. 3.2c). It is important to note that for δ_{Rh}, Al^{3+} remains relatively constant within the range of 1–3 at%, following significant fluctuations up to 1 at%. Considering this case, it can be reasonably presumed that the Al^{3+} is not incorporated into the structural composition of the SS, but rather present within the composition of internodes and vacant positions, or, more likely, in intergranular interlayers in the form of a liquid phase.

The introduction of acceptor modifiers Al^{3+} and Ti^{4+} resulted in an increase of ρ_V, in accordance with the previously established n-type conductivity of the SS composition (Na, Li) NbO_3 [5]. However, this phenomenon does not occur in any of the analyzed cases of modification. There is a notable decrease in ρ_V by approximately one order of magnitude in comparison with the initial SS. Additionally, the changes in ρ_V exhibit a stepwise character with a kink ("step") in the curves ρ_V (x) at $x = 1$ at% of Al^{3+} or Ti^{4+}.

It can be observed that SS with Ti^{4+} exhibit greater variations in $\rho_V(x) < 1.5$ at%, compared to SS with Al^{3+} (Fig. 3.2c). The explanation can be provided by considering the intrinsic disorder present in the system under study. We propose to consider the possibility of lithium being placed partially at Nb^{5+} sites in both the original SS and $LiNbO_3$ [6]. Subsequently, a certain quantity of B-modifiers will effectively substitute for Li, regarding which the values of Al^{3+} and Ti^{4+} valences are larger, and the elements will act as donors, thereby reducing ρ_V. The potential for Nb to occupy the Li sites results in the creation of vacancies within the B-sublattice. These vacancies have a negative charge relative to the crystal; they capture electrons. Embedding Al^{3+} and Ti^{4+} in the B-positions reduces the number of vacancies in this position, which also contributes to the shift of the donor-acceptor equilibrium of the system towards an increase in the concentration of n-type carriers. Some of the modifiers will undoubtedly replace niobium; however, it is probable that the associated ρ_V changes will be significantly less pronounced than those determined by the donor properties of Al^{3+} and Ti^{4+}.

The p-type conductivity of $LiNbO_3$ may also influence the behavior of ρ_V, making both conduction mechanisms equally probable and competing in the process of matrix SS modification. It should be noted that upon the introduction of Al^{3+} and Ti^{4+} into $LiNbO_3$, the character of the change in ρ_V in different concentration ranges of modifiers is distinct (see Table 3.2).

The donor properties of Al^{3+} and Ti^{4+}, as evidenced by the reduction of ρ_V and δ_{Rh} also influence the behavior of $\varepsilon_{33}^T/\varepsilon_0$ and T_C. Specifically, $\varepsilon_{33}^T/\varepsilon_0$ increases and T_C decreases in comparison to the original SS. Concurrently, the parameter alterations are most pronounced in SS with Ti^{4+}. At specific concentrations of Ti^{4+}, the $\varepsilon_{33}^T/\varepsilon_0$

Table 3.2 Modification effects in SS of PZT and (Na, Li) NbO_3 types

Valence of the modifier (M) (n_M) in relation to the valence of the A- or B-substituted ion ($n_{A,B}$)	Carriers of conduc-tivity	Effect of modifier (M) on ρ_v, structural and electrophysical characteristics			
		PZT	Type M by effect on ρ_v and electro-physical character-ristics	Alkaline metal niobates	Type M by effect on ρ_v and electro-physical character-ristics
		p-type of conductivity		n-type of conductivity	
$n_M > n_{A,B}$	Electrons	$-\rho_v$	Donor	$-\rho_v$	Donor
		$+\delta$ (SD), E_c, T_c, Q	SF	$-\varepsilon$, $\tan\delta$, $1/Q$	SF
		$-\varepsilon$, $\tan\delta$, $1/Q$		$+\delta$ (SD), T_c, Q	
$n_M < n_{A,B}$	holes	$-\rho_v$	acceptor	$-\rho_v$	acceptor
		$+\delta$ (SD), E_c, T_c, Q	HF	$-\varepsilon$, $\tan\delta$, $1/Q$	HF
		$-\varepsilon$, $\tan\delta$, $1/Q$		$+\delta$ (SD), E_c, T_c, Q	

is observed to increase by more than twice that observed in SS with Al^{3+} and the decrease in T_C is approximately equal to 20 %. However, the characteristics of SS with Al^{3+} exhibit a tendency to fluctuate around an average value, repeating the behavior of $\varepsilon_{33}^T/\varepsilon_0$ (x) and T_C (x) observed in Ti^{4+}-containing SS. In both the Al^{3+} and Ti^{4+} cases, the changes in $\varepsilon_{33}^T/\varepsilon_0$ and T_C are non-monotonic with extrema at $x = 0.2$ and 1.5 at% (Fig. 3.1d). Given the nature of the effect on the parameters, both modifiers can be classified within the SF group. This is analogous to the donor modification of PZT and (Na, Li) NbO_3 systems, as evidenced by SS-1 and SS-2. In this instance, as indicated by [7], mechanical $1/Q_M$ and dielectric $\tan\delta$ losses are expected to have increased, which occurs at $x > 0.5$ at% Ti^{4+}. In contrast, at lower modifier contents, Q_M increases and $\tan\delta$ decreases, conditioning the HF type of input elements. In this case, the Q_M changes follow the course of δ_P: the increase of δ_P, which causes a decrease in the mobility of domain walls, is accompanied by an increase in Q_M, and the decrease in the distortion of the unit cell, which promotes the active motion of domain walls, leads to a decrease in Q_M. Given the greater influence of oxygen vacancies on domain wall displacement (as previously demonstrated in [8]), it can be proposed that these vacancies should have a pivotal role in Q_M behavior. Since the introduction of modifiers of different valence changes the number of vacancies formed (Table 3.2), which compensate the charge of the introduced ion, the strengthening or weakening of their connection with the domain walls should change the degree of stiffening or softening of the system. As the quantity of introduced modifiers rises, the number of oxygen vacancies, as indicated in Table 3.3, also increases. These vacancies should interact with domain walls, impeding their motion and thereby increasing Q_M, that occurs at low concentrations of modifiers. The further decline of the Q_M can be elucidated by the ordering of single vacancies, whose elimination by crystallographic shift planes [9] is a phenomenon, corroborated by the data, obtained from the X-ray diffraction studies. This process occurs most rapidly (at lower concentrations) in SS with Ti^{4+}, as TiO_2 monoxide itself is prone to defect ordering [9]. It can be observed that vacancies play a significant role in the formation of Q_M, as demonstrated by the notable change in this parameter, observed for Al^{3+}, which forms a greater number of vacancies than Ti^{4+} (Table 3.3). The indirect confirmation of the occurrence of ordering defects is evidenced by the observed increase in sintering temperature, T_{sint}, following an initial decrease (at low concentrations of modifiers) and its extreme behavior in region x, where the process of annihilation of oxygen vacancies begins. Since sintering is a diffusion process proceeding by the vacancy mechanism [10], the role of vacancies in the formation of ceramics is substantial. An increase in their number at the initial stage will result in a reduction, and subsequent annihilation will lead to an increase in $T_{sint.}$ (Fig. 3.2d).

Table 3.2 presents the effects of modification at $n_M > n_{A,B}$ and $n_M < n_{A,B}$ in SS of PZT and (Na, Li)NbO_3 types.

The *MO* displacement upon the introduction of modifiers will result in a reduction in K_p, as its maximum value in the (Na, Li) NbO_3 system is situated at the Rh boundary of the MO. To establish the relationships between the composition (crystallochemical characteristics of ions), the structure (δ), defectivity (oxygen vacancies), and the properties (electrophysical characteristics) of SS-5, we introduce a new value called $\delta_{Paverage}$. This is the average (for the whole range of modifier concentrations) value of the homogeneous deformation parameter of the crystal cell is dominating in SS with rhombic phase. The characteristics of composition, structure, and defectivity are summarized in Table 3.3. The table presents the experimentally observed facts that confirm the known correlations between structure parameters, defectivity of SS, crystallochemical characteristics of modifiers, and electrophysical properties responsible for the HF of SS.

(i) Composition → structure:

$$\ni O_{Al} < \ni O_{Ti} \rightarrow \delta_{Al} > \delta_{Ti};\ Lo_{Al} < Lo_{Ti} \rightarrow \delta_{Al} > \delta_{Ti};\ \alpha_{Al} < \alpha_{Ti} \rightarrow \delta_{Al} > \delta_{Ti}$$

Table 3.3 Characteristics of modifiers (*M*): Al^{3+} and Ti^{4+}

Ion modifier	$\delta_{R(aver.)}$, SS-5 with 3 at% *M*	Crystallochemical characteristicsmodifiers			Number of oxygen vacancies in the formed SS, V''_O
		Electronega-tivity, EN, kJ/g-at	Unstressed distance (BO) *Lo*, Å	Polarisability $\alpha \times 10^{24}$ cm^3	
Al^{3+}	> 0.010	921.8	1.90	0.065	1
Ti^{4+}	< 0.0090	1089.4	1.97	0.272	0.5
Conclusion	$\delta_{Al} > \delta_{Ti}$	$EN_{Al} < EN_{Ti}$	$Lo_{Al} < Lo_{Ti}$ $1/Lo_{Al} > 1/Lo_{Ti}$	$\alpha_{Al} < \alpha_{Ti}$	$V''o_{Al} > V''o_{Ti}$

(ii) Structure → properties:

$$\delta_{\mathrm{Al}} > \delta_{\mathrm{Ti}}\ Q_{M\,(\mathrm{Al})} > Q_{M\,(\mathrm{Ti})};\ \tan\delta_{\mathrm{Al}} > \tan\delta_{\mathrm{Ti}};\quad \varepsilon_{\mathrm{Al}} < \varepsilon_{\mathrm{Ti}}.$$

(iii) Defectivity → properties

$$V''_{O(\mathrm{Al})} > V''_{O(\mathrm{Ti})} \to Q_{M\,(\mathrm{Al})} > Q_{M\,(\mathrm{Ti})};\ \tan\delta_{\mathrm{Al}} > \tan\delta_{\mathrm{Ti}}$$

It is noteworthy that, in contrast to SS-1, which are characterized by a direct dependence of δ_R on EN, in this case the transition to more EN ion Ti^{4+} leads to a decrease in δ_R (Fig. 3.3a). This is a consequence of the influence of other crystallochemical characteristics of elements on this value.

$n_{MB} > n_B$

The modifier W^{6+}, with a valence one unit higher than that of Nb^{5+}, is a donor. The introduction of the modifier was implemented in two distinct ways: firstly, by the organization of vacancies in the *A*-sublattice (SS-3), and secondly, by the organization of vacancies in the *B*-sublattice (SS-4).

Modification formulae are as follows: (Na, Li)$_{1-x}\square_x$Nb$_{1-x}$W$_x$O$_3$ (SS-3), (Na, Li) Nb$_{1-6/5x}$W$_x\square_{x/5}$O$_3$ (SS-4) (x = 0.001–0.05). Figures 3.3a, b demonstrate the relationships between the W^{6+} concentration and the parameters of perovskite SS cells in SS-3 and SS-4. It is evident that even a minimal introduction of the W^{6+} (x = 0.001) results to shift in the SS into the region, where both the R- and Rh phases coexist. Moreover, the amount of the Rh phase declines significantly (from 100 % in the initial SS to 5–15 % in SS-3, SS-4). The homogeneous deformation parameter of the R-cell demonstrates an increase with increasing x, whereas that of the Rh-cell exhibits a decrease. The predominant R-phase (and Rh-phase in SS-4) exhibits a decrease in cell volume at $x > 0.03$; this decrease is attributed to steric effects, whereby replacement of the larger Nb^{5+} (0.64 Å) by the smaller W^{6+} (0.60 Å) occurs.

The behavior of the electrophysical parameters (Fig. 3.4) is consistent with changes in the structure characteristics. In this instance, the primary influence on the dielectric and piezoelectric properties is exerted by the Rh-sub cell. With a decrease in δ_{Rh}, as follows from the known relationships, $\varepsilon_{33}^{\mathrm{T}}/\varepsilon_0$ increases, because $\varepsilon_{33}^{\mathrm{T}}/\varepsilon_0 \sim 1/\delta$); when $\tan\delta$ increases ($\tan\delta \sim 1/\delta$), T_C decreases ($T_C \sim \delta$). A sharp decline in K_P is associated with the departure of SS from the exclusively Rh region, which is typified by the maximum K_P, and its transition to the region of coexistence of the phase, where all characteristics exhibit diminished values.

The value of E_{yu} is essentially constant, except for the initial non-monotonic behavior, observed in the region of $0.00 < x < 0.01$. Interestingly, the introduction of donor W^{6+} does not lead to major changes in ρ_V, excluding a stretch of initial small drop and rise to the initial value. It appears that non-monotonic changes in characteristics in the region of small x are a consequence of the redistribution of W^{6+} on regular and irregular (vacancy) positions of SS.

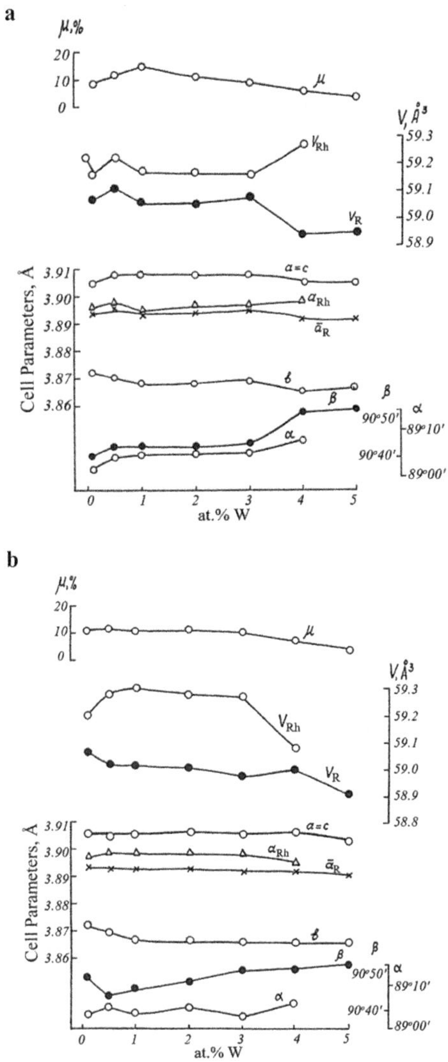

Fig. 3.3 **a** Dependencies of the perovskite cell parameters of SS-3 on the W^{6+} content. **b** Dependencies of the perovskite cell parameters of SS-4 on W^{6+} content

3.1.3 Regularities of Changes in Properties of Solid Solutions Based on Natrium Niobate at Super Stoichiometric Modification by One (Ag)-, Two (Cd)-, Three (Al, Cr, Fe, La)-, Four (Ti, Si, Mn)-, Five (V), Six (U, W) Valence Elements

The modification was carried out in accordance with the general formula (Na, Li)$NbO_3 + xMO_{n/2}$, where M represents the modifiers: Ag^{1+}, Cd^{2+}, Cr^{3+}, Fe^{3+}, La^{3+},

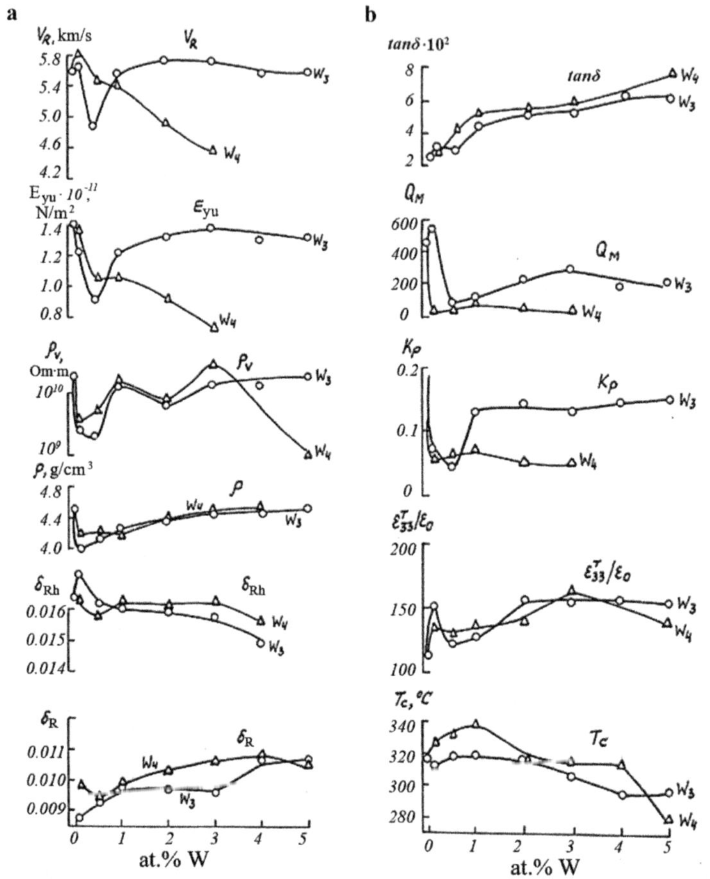

Fig. 3.4 **a** Dependencies of structural and electrophysical characteristics of SS-3 (W_3) on W^{6+} content. **b** Dependencies of dielectric and piezoelectric characteristics of SS4 (W_4) on W^{6+} content

Ti^{4+}, Si^{4+}, Mn^{4+}, V^{5+}, U^{6+}, W^{6+}; n denotes the valence of M, which is equal to 1–6; and x represents the concentration of M, which varies in the range of 0.001–0.03. By analyzing the properties of modified SS, the ionic radii of cations (r), lengths of unstressed interatomic bonds (L_{AO}, L_{BO}), and EN were used as arguments, as described in Table 3.4.

Furthermore, the crystal structure characteristics were considered, including the average parameters and homogeneous strain parameters of the unit cells of the coexisting R and Rh phases (respectively, $\overline{a}_R$, $\overline{a}_{Rh}$, δ_R, δ_{Rh}). The electronic configurations of the outermost layers of the M cations were also considered. X-ray diffraction studies demonstrated that, in almost all cases, the solubility of the introduced modifiers (i.e., the absence of impurity phases or a change in cell parameters) could be confirmed. In this instance, all components are integrated into the lattice, in accordance with size factor and the specific location within the structure results in a

Table 3.4 Crystallochemical characteristics of *A*-, *B*-, *M*-cations

Valences of *A*-, *B*-, *M*-cations	*A*-, *B*-, *M*-cations	Crystallochemical characteristics of *A*-, *B*-, *M*-cations [11]				
		r, Å Coordination number (c.n.) = 6	L_{AO}, Å	L_{BO}, Å	EN, kJ/g at	Electronic configuration of the outer layers
1	$(Na_{0.875}Li_{0.125})$	1.6 (c.n. = 12)	2.30 (sum of radius *A-O*)	–	499	S^1 (*Na*)
	Ag	1.52 (c.n. = 12)	2.51	–	733	$(d^{10})S^1$
2	Cd	1.11 (c.n. = 12)	–	–	195	$(d^{10})S^2$
3	Al	0.57	–	1.90	922	S^2P^l
	Cr	0.64	–	1.99	1048	d^5S^1
	Fe	0.67	–	2.03	1027	d^6S^2
	La	1.16 (c.n. = 12)	–	–	150	$(f^o)d^lS^2$
4	Ti	0.64	–	1.97	1089	d^2S^2
	Si	0.39	–	1.75 (sum of radius *B-O*)	1131	S^2P^2
	Mn	0.52	–	1.91	1299	d^5S^2
5	Nb	0.66	–	2.01	1006	d^4S^1
	V	0.40	–	1.76 (sum of radius *B-O*)	1320	d^3S^2
6	U	0.73	–	2.12	901	$(f^o)d^4S^2$
	W	0.60	–	1.97	985	d^4S^2

monotonous increase or decrease in the $\overline{a}_R$ of the R phase (Figs. 3.5, 3.6, 3.7, 3.8 and 3.9). Consequently, the replacement of (Na, Li) with Ag, Cd, La, which has a larger radius (*r*), results in an increase $\overline{a}_R$ (Figs. 3.5a, b and 3.6a). Similarly, the replacement of Nb with Fe and U, which also has a larger radius (*r*), results in the same behavior of $\overline{a}_R$ (Figs. 3.6b and 3.8b). Conversely, the replacement of *B*-cation with Cr and Ti, which has a smaller radius (*r*) than Nb, contributes to a decrease of $\overline{a}_R$ (Figs. 3.6c and 3.7c).

The exceptions are Mn, Si, and V, for which $\overline{a}_R$ has a negligible change. The very small sizes of these elements, which are close to the maximum permissible (r_B~ 0.51 Å), suggest that they could potentially be incorporated into the lattice, resulting in a sharp decrease in $\overline{a}_R$. This is in accordance with [8], which indicates that solubility should be stopped for *V* at ~1 at% and for Si, solubility should not occur at all. It is plausible that these cations are not incorporated into the lattice but rather into internodes or space of vacancies. One can also assume the formation with their

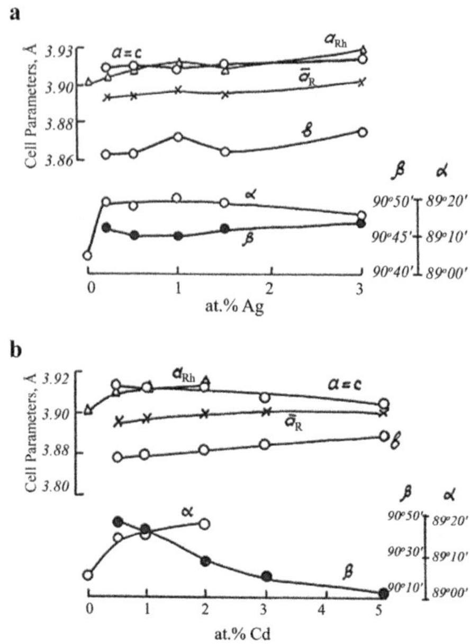

Fig. 3.5 **a** Dependencies of structural characteristics on Ag^{1+} content in SS-6 (1). **b** Dependencies of structural characteristics on Cd^{2+} content in SS-6 (2)

participation of liquid phases, localized in intercrystalline layers. In certain instances (Si), the distinction in the symmetry of valence orbitals may be a contributing factor in the non-integration in crystal structure phenomenon (Table 3.4). The markedly non-monotonic behavior of aluminum (Al) is worthy of note. The average parameter exhibits a notable decline to $x = 0.01$. The introduction of Al results to a slight increase in $\bar{a}_R$, which is then followed by a very slight change in it. It is most likely that at low concentrations of M, the Al is embedded in the lattice. At $x > 0.01$, however, it begins to fill irregular positions, embedding into the structure or forming a liquid phase. Figures 3.10, 3.11, 3.12, 3.13, 3.14, 3.15 and 3.16 demonstrate the dependent variations of the structural (δ_R, δ_{Rh}), ceramic (ρ, T_{sint}) and electrophysical characteristics on concentration of the investigated SS. It is interesting to consider the action of single modifiers (Ag, Cd, V), which have no other representatives in their valence group. The behavior of their electrophysical characteristics is generally consistent with the known logics of dependence on the structure parameters (δ_R, δ_{Rh}) (Fig. 3.11).

Small anomalies in the character of parameter changes at small x are associated with similar anomalies in the curves $S(x)$ and $p(x)$. In addition to structural characteristics, the density of ceramics has been found to affect the electrophysical properties. In turn, the fluctuations of the density at low M concentrations may be a consequence of the complex distribution of M in the structure. A comparison of the properties of SS, modified with trivalent elements, showed the following. The dependencies of the SS characteristics on the concentration with La "fall out" of all the regularities considered, as can be seen in Fig. 3.6. This is certainly a consequence of the different

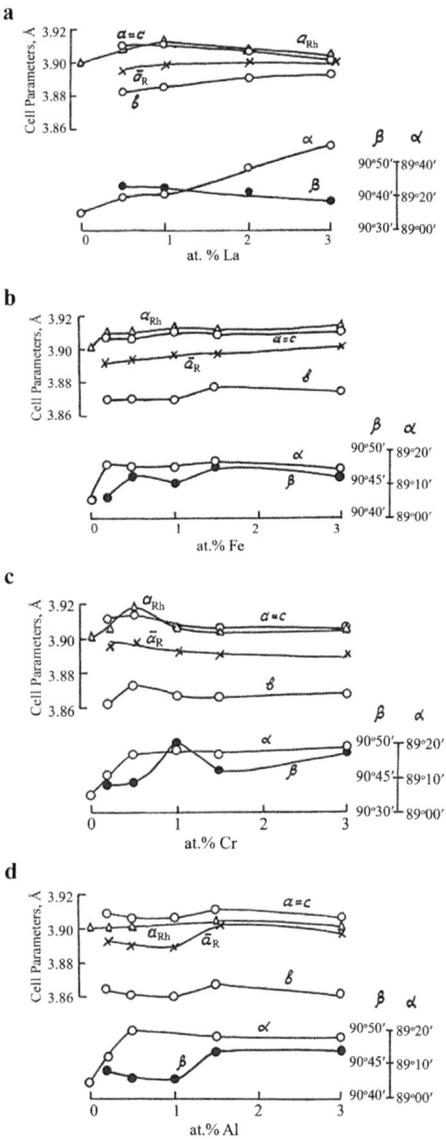

Fig. 3.6 **a** Dependencies of structural characteristics on La^{3+} content in SS-6 (3). **b** Dependencies of structural characteristics on Fe^{3+} content in SS-6 (3). **c** Dependencies of structural characteristics on Cr^{3+} content in SS-6 (3). **d** Dependencies of structural characteristics on Al^{3+} content in SS-6 (3)

arrangement of La in the lattice of the initial SS, compared to all other modifiers: it replaces the *A*-cation (Na, Li).

The introduction of other modifiers leads to similar changes in properties, with the greatest changes observed in the case of iron modification. This phenomenon is likely attributable to the fact that, of all the modifiers listed, iron, as previously noted, has a larger radius than Nb^{5+}, which results in distinct structural alterations compared

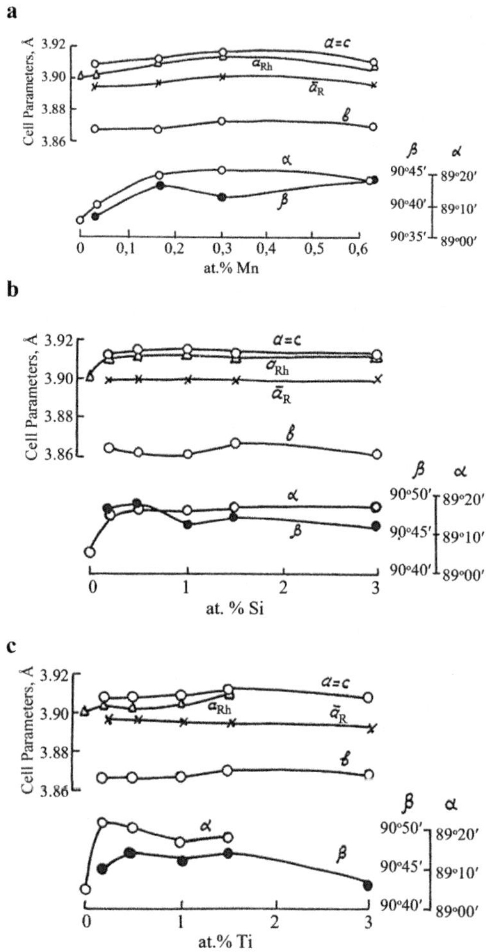

Fig. 3.7 **a** Dependencies of structural characteristics on Mn^{4+} content in SS-6 (4). **b** Dependencies of structural characteristics on Si^{4+} content in SS-6 (4). **c** Dependencies of structural characteristics on Ti^{4+} content in SS-6 (4)

to Cr^{3+} and Al^{3+}. As in previous cases, anomalies during parameters at low concentrations of modifiers are associated with similar anomalies S_p. A comparison of the action of Cr^{3+} and Al^{3+} reveals that the parameters in SS with Al^{3+} undergo a more pronounced change in the SS with Cr^{3+}, despite the similarity in size between Cr^{3+} and Nb^{5+}. However, the action of Al^{3+} is to some extent balanced by the formation of liquid phases (aluminates of different compositions) and by its complex distribution in the structure of the initial SS. To facilitate a more detailed comparison of the properties of SS with Cr^{3+} and Fe^{3+} embedded in the structure (in place *B*), let us to exclude La^{3+} (as located in place *A*) and Al^{3+} (with a combined character of distribution in the lattice) from consideration. The introduction of such acceptor modifiers at the *n*-type conductivity of the original system should be resulted in an increase in the ρ_V. Nevertheless, the ρ_V declines in both cases.

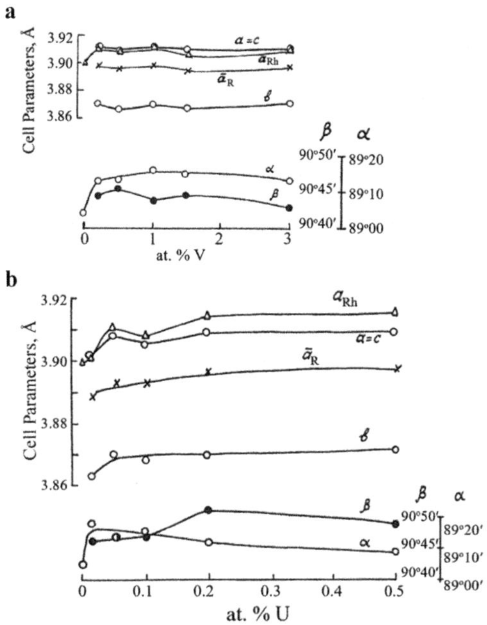

Fig. 3.8 **a** Dependencies of structural characteristics on V^{5+} content in SS-6 (5) and SS-6 (6). **b** Dependencies of structural characteristics on U^{6+} content in SS-6 (5) and SS-6 (6)

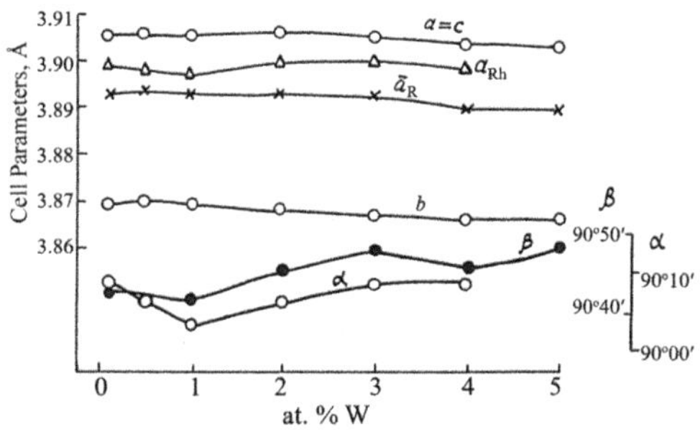

Fig. 3.9 Dependencies of structural characteristics on the W^{6+} content in SS-6 (6)

A similar explanation can be provided for this phenomenon, as was done previously for the stoichiometric modification of Al^{3+} and Ti^{4+}. The SS-6 phase, which contains Cr^{3+} and Fe^{3+}, has been found to exhibit greater values of the homogeneous strain parameter of the unit cell of the predominant rhombic phase. In the entire concentration range δ_R (Cr) > δ_R (Fe). The changes in δ_R are non-monotonic: up to 1 at% of modifier, δ_R increases; further (> 1 at%), δ_R decreases. The observed changes in Q_M are consistent with this behavior of δ_R. In both cases, the general tendency is to increase Q_M with up to 1 at% of modifier Q_M increasing $\left(Q_{M\,(\mathrm{Cr})} > Q_{M\,(\mathrm{Fe})}\right)$, and further (> 1 at% of modifier) Q_M decreasing $\left(Q_{M\,(\mathrm{Cr})} < Q_{M\,(\mathrm{Fe})}\right)$. In the case of Fe^{3+}, tanδ decreases, while in the case of Cr^{3+}, tanδ increases. Considering their impact on specific parameters (δ_R, Q_M, tanδ), these modifiers can be classified into the HF-group. However, an increase in $\varepsilon_{33}^{T}/\varepsilon_0$ (with a maximum at 0.2 at% of modifier) and

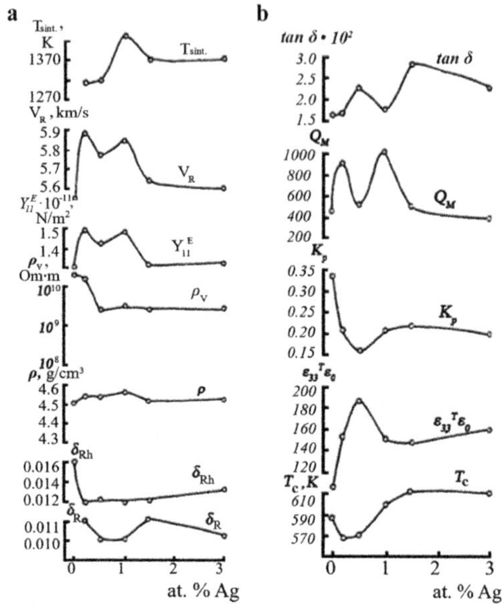

Fig. 3.10 **a** Dependencies of structural and ceramic characteristics on Ag^{1+} content in SS-6 (1). **b** Dependencies of electrophysical characteristics on Ag^{1+} content in SS-6 (1)

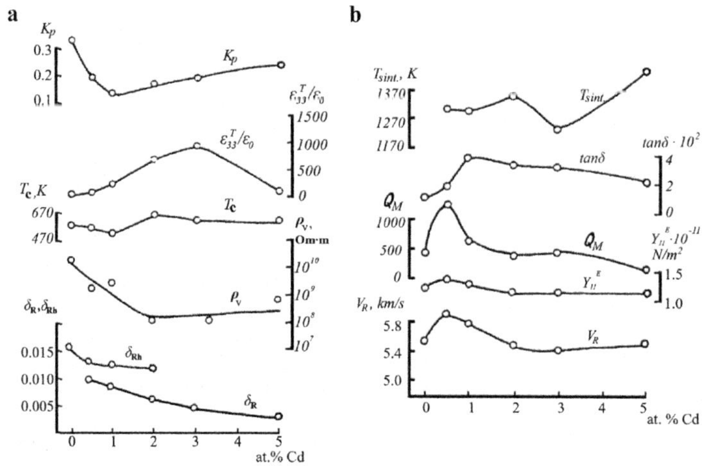

Fig. 3.11 **a** Dependencies of structural, ceramic dielectric characteristics on Cd^{2+} content in SS-6 (2). **b** Dependencies of electrophysical characteristics on the Cd^{2+} content in SS-6 (2)

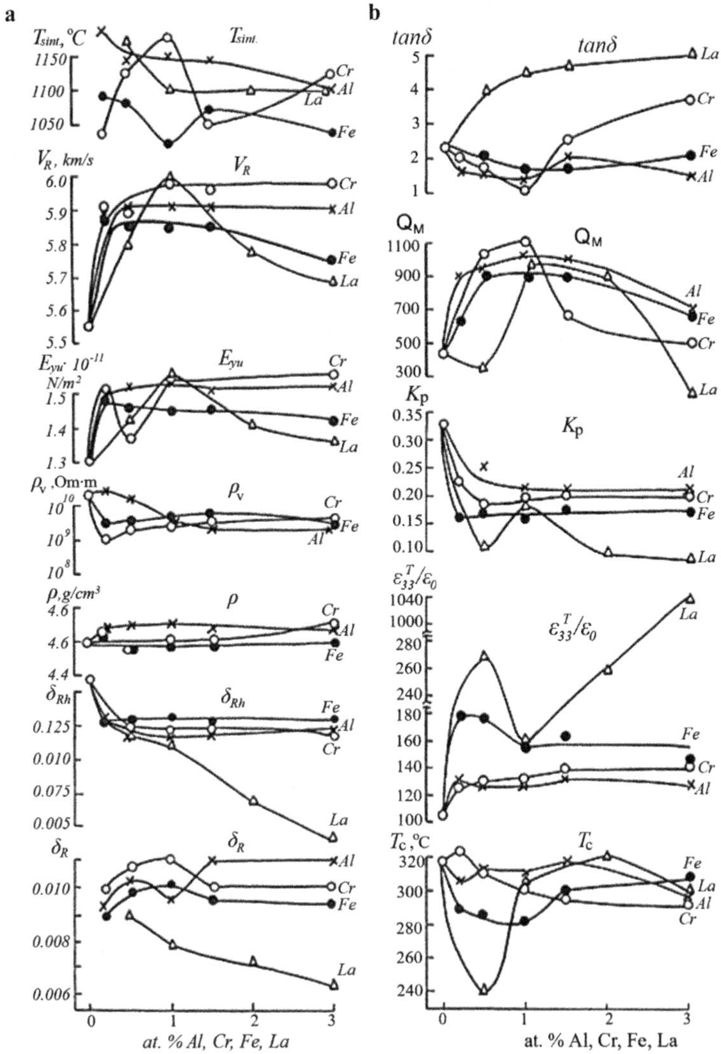

Fig. 3.12 **a** Dependencies of structural, ceramic and electrophysical characteristics on the content of Al^{3+}, Cr^{3+}, Fe^{3+}, La^{3+} in SS-6 (3). **b** Dependencies of dielectric and piezoelectric characteristics on the content of Al^{3+}, Cr^{3+}, Fe^{3+}, La^{3+} in SS-6 (3)

a decline in T_c indicate a "SF" of these elements. Consequently, the behavior of the analyzed modifiers is ambiguous. A certain "dualism" can be observed in the influence of the modifiers on two groups of electrophysical parameters. As previously observed, the reduction in K_p in both instances of modification can be attributed to the shift of SS in the phase diagram of the (Na, Li) NbO_3 system.

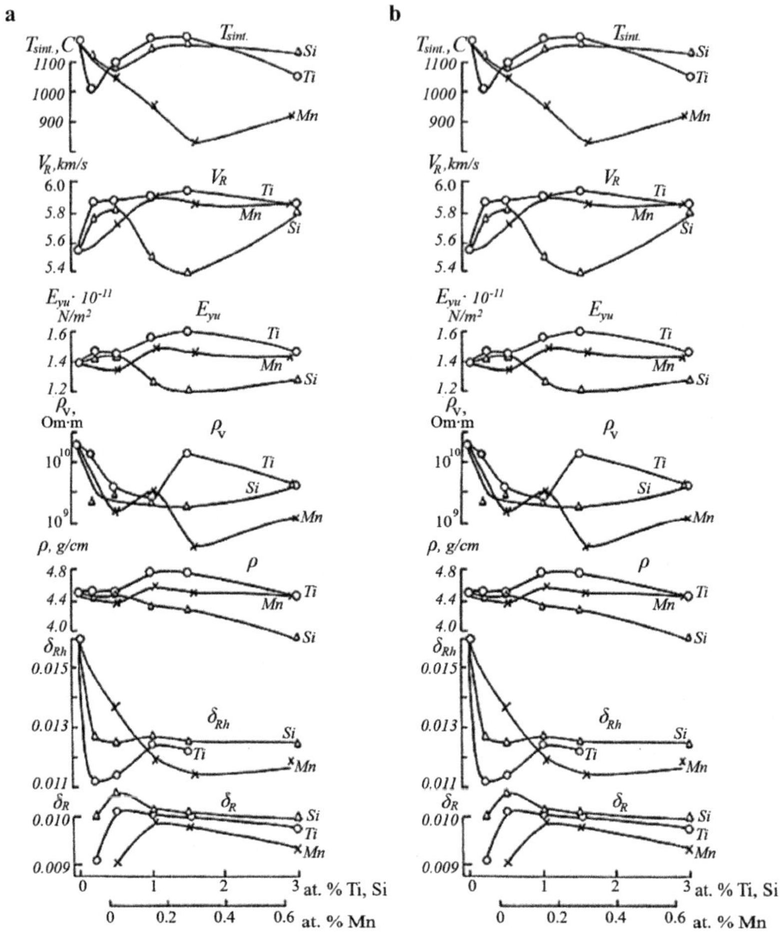

Fig. 3.13 **a** Dependencies of structural, ceramic and electrophysical characteristics on Ti^{4+}, Si^{4+}, Mn^{4+} content in SS-6 (4). **b** Dependencies of dielectric and piezoelectric characteristics on the content of Ti^{4+}, Si^{4+}, and Mn^{4+} in SS-6 (4)

A comparison of the properties of modified SS within the group of 4-valent modifiers is challenging due to the unique incorporation of introduced elements into the lattice, as previously discussed.

It is interesting to observe the alteration in the value of *c*, when transitioning from 3- to 6-valent modifiers. Figure 3.17 illustrates the variation of the Q_M in SS with 3 at% modifiers as a function of the number of oxygen vacancies in the SS. The obtained dependence is linear, thereby confirming the previously established fact of inhibition of domain wall motion by oxygen vacancies. A reduction in the mobility of domain walls leads to a decrease in mechanical losses (internal friction losses). This, in turn, contributes to an increase in Q_M.

Fig. 3.14 Dependencies of structural, ceramic and electrophysical characteristics on V^{5+} content in SS-6 (5)

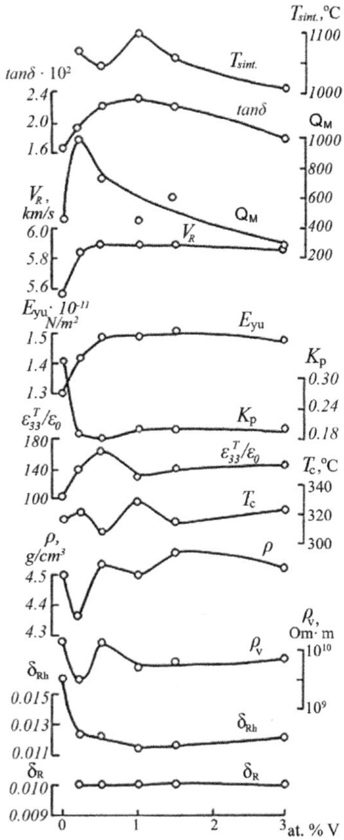

3.1.4 *Microstructure and Strength Characteristics of Modified Ceramics*

Given that one of the most significant attributes of ceramics is its strength, which is largely contingent upon the nature of the microstructure, it is imperative to monitor the alterations in this latter parameter when modifying SS. Table 3.5 presents the principal characteristics, and Figs. 3.18, 3.19, 3.20, 3.21 and 3.22 depict fragments of microstructure of some modified SS (based on $Na_{0.875}Li_{0.125}NbO_3$).

The samples were modified with Ti^{4+} (1 at%). Figure 3.18 presents etched surface and Fig. 3.18b shows no etched surface, ×350. Figure 3.18c presents the sample modified with Fe^{3+} (0.5 at%), Fig. 3.18d shows the sample with Fe^{3+} (1.5 at%), and Fig. 3.18e corresponds to modification with Al^{3+} (1 at%). These samples exhibited high strength characteristics. The initial three modifiers facilitate the formation of a fine-grained structure (see Table 3.5). The strengthening of ceramics is associated with their fine grain size and can be caused by a reduction in stresses at grain boundaries when $\overline{D}$ decreases [12].

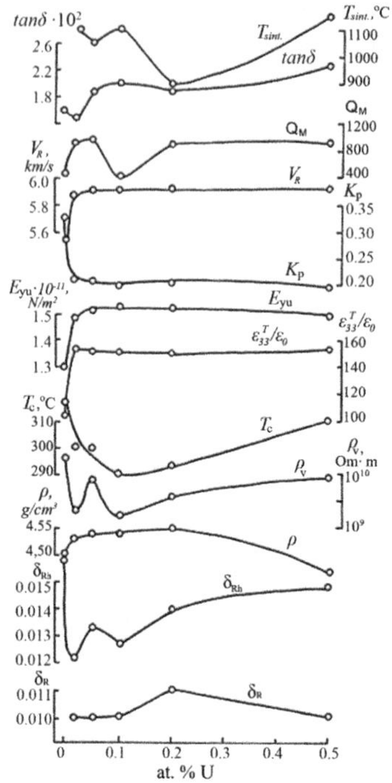

Fig. 3.15 Dependencies of structural, ceramic and electrophysical characteristics on the U^{6+} content of SS-6 (6)

The strengthening of ceramics is associated with their fine grain size and can be attributed to a number of factors, including: the reduction of stresses at grain boundaries when $\overline{D}$ decreases [12], the increase in the value of the failure stress proportional to $\sqrt{(\overline{D})}$ [13], the barrier effect (i.e. the strengthening role of inter-grain and interfacial boundaries as powerful obstacles for moving dislocations), the strengthening of the processes of segregation of impurity atoms at grain boundaries (which fix dislocations and increase grain boundary relaxation), the increase in the rate of dislocation multiplication (and, in this connection, the strengthening of the interaction between them) [14]. The role of aluminum in the formation of LF (aluminates of various compositions [15]) binding (or "cementing") the grains and thus strengthening the ceramics.

The introduction of Mn facilitates the formation of a heterogeneous microstructure, comprising small grains at the center of the samples and giant densely packed grains at the periphery (Fig. 3.19). This results in the formation of a "crust" at the edges. The formation of the "crust" is a characteristic of NN-based ceramics. This phenomenon is linked to the extrusion of F-containing LF onto the surface, which drives secondary discontinuous recrystallization (growth of ideomorphic grains). However, in contrast to unmodified ceramics, in which the surface layer is characterized by extreme friability (as evidenced by its desquamation, or flake-like spalling), the "crust" of SS with Mn is highly dense due to the bonding action of fusions, based on Mn oxides [16], formed at relatively low temperatures (~ 1270 K).

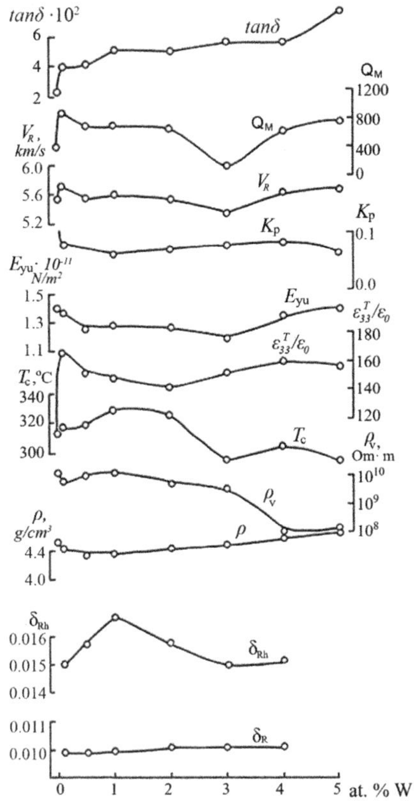

Fig. 3.16 Dependencies of structural, ceramic and electrophysical characteristics on the W^{6+} content of SS-6 (6)

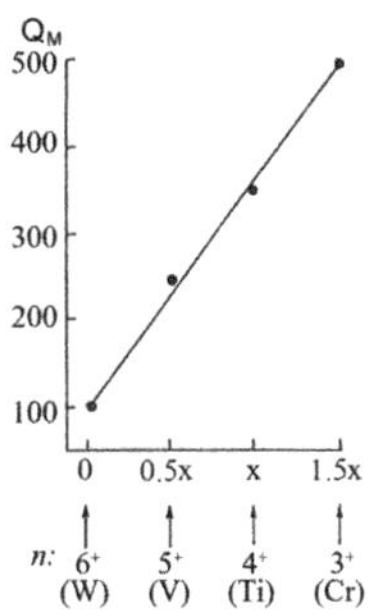

Fig. 3.17 Dependencies of mechanical goodness of fit Q_M on the number of oxygen vacancies in SS-6 (3-6)

The addition of Mn leads to the structure becoming more homogeneous; however, the "crust" effect remains. The strongest ceramics are formed when the minimum amount of Mn is present. It is typified by "constricting" internal stresses and minimum values of δ_R. As the amount of Mn increases, the internal stresses assume a "tensile" character, thereby causing the ceramics to become unstrengthened. At the Cr modification, a "crust" of giant grains is also observed. However, as the Cr content

Table 3.5 Characteristics of microstructure of modified SS

Modifier	Concentration at%	Average grain size, $\overline{D}$, μm	$\overline{D}_{max}$, μm	$\overline{D}_{min}$, μm	Features
Ag	0.20	12.2	35.0	1.0	
	1.50	230.0	500.0	1.0	Giant irregularly shaped grains in the middle of the samples
	3.00	I-4.0; II-120	II-400	1.0	Two subsystems of grain (I; II)
Mn	0.01	4.7	10.0	0.5	A “crust” of giant grains
	0.10	3.9	7.0	0.5	A “crust” of giant grains
	0.20	4.4	10.0	1.0	A “crust” of giant grains
Cr	0.20	4.7	70.0	1.0	A “crust” of giant grains, rarely—large in the center
	1.00	15.1	50.0	2.0	No “crust”
	3.00	11.4	30.0	1.0	No “crust”
Fe	0.20	6.6	40.0	0.5	Rare giant grains (up to 100–250 μm)
	1.50	4.8	20.0	0.5	Clusters of large and small grains
	3.00	6.2	10.0	0.5	Clusters of large and small grains
Al	0.20	18.1	50.0	2.0	Clusters of large and small grains
	1.00	12.6	50.0	2.0	
	3.00	13.9	40.0	2.0	
Ti	0.20	11.1	30.0	1.0	A lot of cubic grains
	1.00	12.5	30.0	3.0	The number of cubic grains is limited
	3.00	2.3	7.0	0.5	The number of cubic grains is limited
Si	0.20	12.5	40.0	2.0	
	3.00	12.8	30.0	2.0	Clusters of large and small grains

(continued)

Table 3.5 (continued)

Modifier	Concentration at%	Average grain size, $\overline{D}$, μm	$\overline{D}_{max}$, μm	$\overline{D}_{min}$, μm	Features
V	0.50	7.6	20.0	1.0	Clusters of large and small grains
	1.00	5.50	10.0	0.5	Clusters of large and small grains
	1.50	5.60	10.0	0,5	Clusters of large and small grains
?	0.01	9.00	40.0	1.0	Clusters of large and small grains
	0.10	7.00	70.0	1.0	The size variation is higher
	0.50	7.80	25.0	1.0	More homogeneous

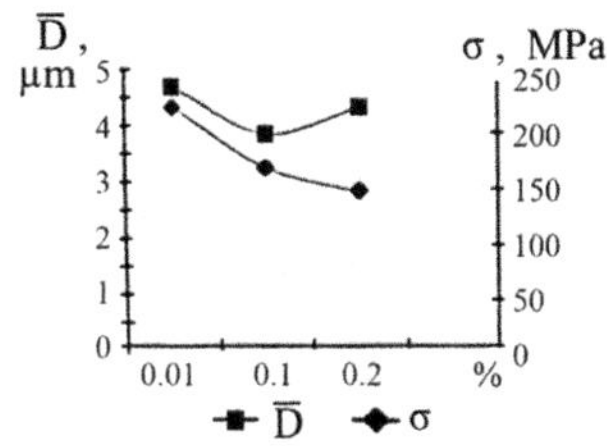

Fig. 3.18 Results of modification of SS with MnO_2: dependence of grain size and strength on modifier concentration

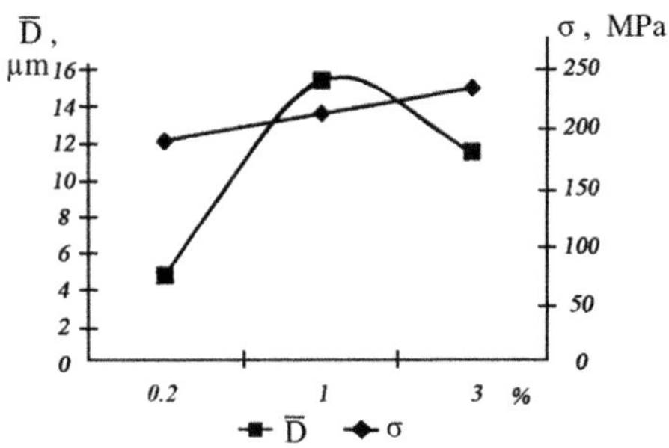

Fig. 3.19 Results of Cr_2O_3 modification of SS: dependence of grain size and strength on modifier concentration

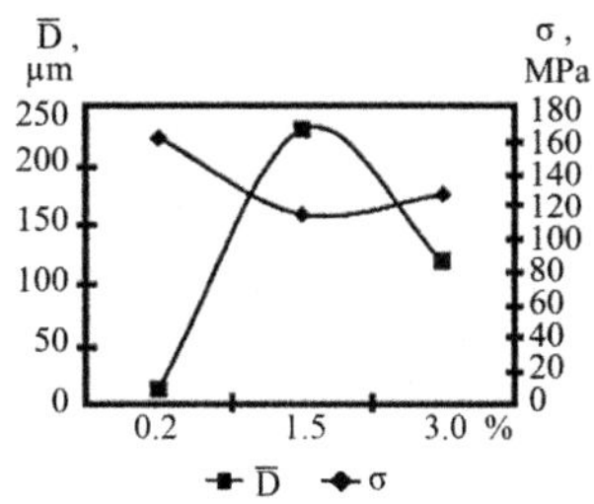

Fig. 3.20 Results of Ag_2O_3 modification of SS: dependence of grain size and strength on modifier concentration

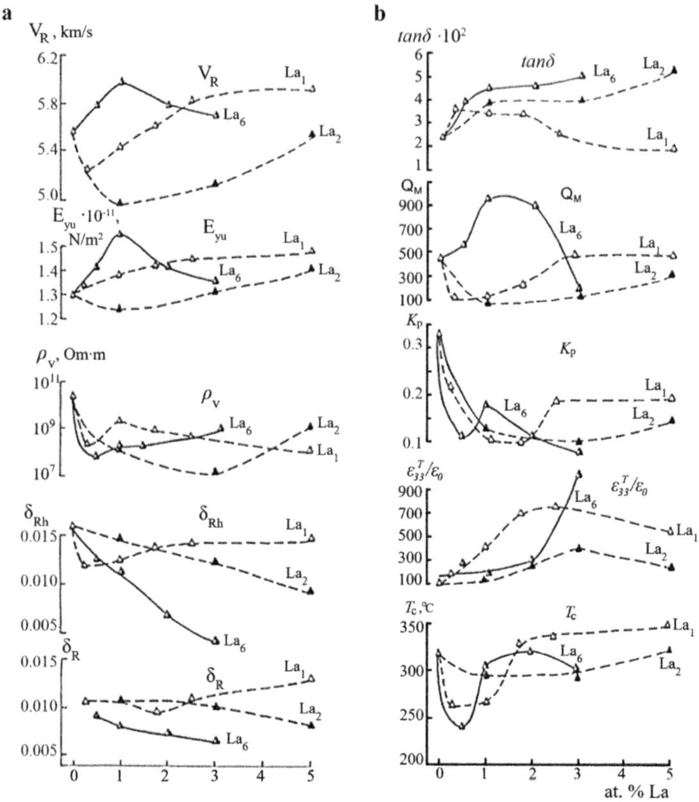

Fig. 3.21 **a** Comparison of structural and ceramic properties of SS with La^{3+}, introduced by different methods [SS-1 (La_1), SS-2 (La_2), SS-6 (La_6)]. **b** Comparison of dielectric and piezoelectric properties of SS with La^{3+}, introduced by different methods [SS-1 (La_1), SS-2 (La_2), SS-6 (La_6)]

increases, this crust disappears, the microstructure becomes more homogeneous, and $\overline{D}$ decreases (Fig. 3.20). The high strength values are observed in SS with 3 at% Cr.

The introduction of Si, U, and Ag does not result in the strengthening of ceramic. Uranium forms an extremely heterogeneous structure. The effect of Ag on the microstructure and, subsequently, the strength characteristics of SS is unusual. The largest secondary recrystallized grains are concentrated in the central region of the samples, whereas most small grains are concentrated at the sample's edges.

Similarly, as with the previous instance, as the Ag concentration rises, the consequence of disrupting the coherence of coarse-grained structures through the formation of areas comprising minutes, loosely packed crystallites become evident. The observed morphology of the latter may be attributed to the presence of residual LF in the articulating surfaces of large grains, which appear to facilitate their growth.

The occurrence of a "reverse crust effect" in SS with Ag may be attributed to the following factors. Of the analyzed modifiers, Ag is the only one place in the

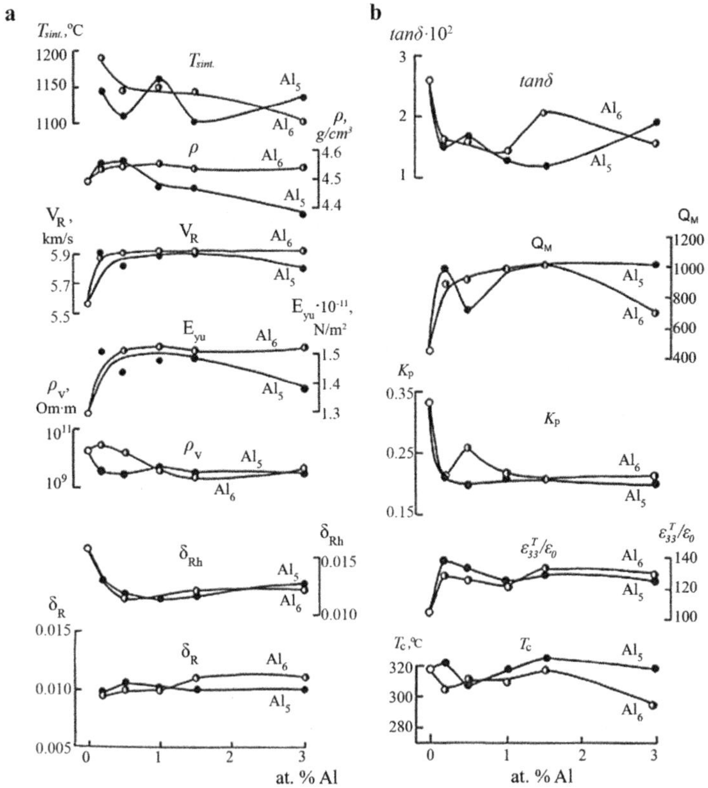

Fig. 3.22 **a** Comparison of structural and ceramic properties of SS with Al^{3+}, introduced by different methods [SS-5 (Al_5), SS-6 (Al_6)]. **b** Comparison of dielectric and piezoelectric properties of SS with Al^{3+}, introduced by different methods [SS-5 (Al_5), SS-6 (Al_6)]

A-positions, which displaces alkaline earth ions. These ions react with F-containing impurities, forming LF and promoting secondary discontinuous recrystallization.

3.1.5 A Comparison of the Properties of Niobate Ferro Piezoceramics, Modified by Different Methods

In consideration of the revealed dependence of HF on the degree of covalency of A-O bonds, a comparison is made between the properties of SS-1 and SS-2. Given the identical character of parameter changes at the introduction of La^{3+}, Sb^{3+}, and Mg^{2+}, the comparison will be carried out on the example of the lanthanum (Fig. 3.21).

At low concentrations of modifiers (up to 1.5 at%), SS-1 exhibits greater super plasticity (SF) than SS-2, as indicated by the parameters δ_R, T_c, and Q_M. However, SS-2 displays enhanced SF at concentrations exceeding 1.5 at%. The disparate degree of

HF observed in SS-1 and SS-2 can be attributed to the differing values of the vacancies created. At low concentrations of modifiers, the number of vacancies appearing is a determining factor (SS-1 with a high number of vacancies exhibits greater SF than SS-2). At the same time, at higher concentrations, the position within the structure (SS-2 with vacancies in the more covalent *B*-sublattice exhibits greater SF) becomes a significant factor. The greater SF of SS-2 likely also results in more pronounced fluctuations in the value of $\Delta Q_M / \Delta Q_{MO}$ as a function of EN within them, which is linked to the augmented flexibility of malleable systems to Q_M growth during the modification process [17].

The analysis of the data obtained permits one to conclude that *B*-vacancies are present in the system. This is evidenced by the formation of impurity-free SS-2, which exhibits distinct properties from SS-1, as well as have a more pronounced dependence of $\Delta Q_M / \Delta Q_{MO}$ on the individual characteristics of modifiers. Based on the observed characteristics of the modifiers, it was anticipated that there would be a more pronounced (compared to SS-1) change in dependence of Q_M on concentration in SS-2. However, this hypothesis was not substantiated by the experimental results. One possible reason for this phenomenon may be the redistribution of vacancies in SS-2 across *B*- and *A*-seats in a manner like that which was previously observed in the PZT system [17]. It is important to note that at high concentrations of modifiers, the character of change $\varepsilon_{33}^{\mathrm{T}}/\varepsilon_0$, depending on the type of SS, does not align with the character of change observed for SD, T_C and Q_M. Therefore, at concentrations of 1.5–5 at% La^{3+} in SS-1 is greater HF according to values for SD, T_C and Q_M, as well as greater SF according to values for $\varepsilon_{33}^{\mathrm{T}}/\varepsilon_0$, in comparison with SS-2. It is likely that the smaller values of $\varepsilon_{33}^{\mathrm{T}}/\varepsilon_0$, observed in SS-2, are a consequence of the preferential 90° reorientation of domains during polarization [18]. This assumption is corroborated by the fulfillment of the inequality: $\varepsilon_{33}^{\mathrm{T}}/\varepsilon_0 < \varepsilon/\varepsilon_0$ in SS-2.

Figure 3.21 presents the evolution of the parameters for SS with La introduced super stoichiometrically. At this method of modification, the structural changes are like those, observed in SS-2 (δ_{R} and δ_{Rh} decline), namely a decrease in both δ_R and δ_{Rh}. However, the changes in SS-6 (2) are more pronounced. This results in a similarity between the $\varepsilon_{33}^{\mathrm{T}}/\varepsilon_0$ behavior in SS-2 and SS-6 (2).

At this method of modification, the structural changes are like those, observed in SS-2 (δ_{R} and δ_{Rh} decrease), namely a more pronounced decrease in both δ_{R} and δ_{Rh} at a higher rate. This leads to a similarity between the $\varepsilon_{33}^{\mathrm{T}}/\varepsilon_0$ behavior in SS-2 and SS-6 (2). In the latter, $\varepsilon_{33}^{\mathrm{T}}/\varepsilon_0$ grows rapidly at $x > 0.02$. The K_p, tanδ, ρ_V changes are close in nature. All these may indicate the incorporation of super stoichiometric La^{3+} into the SS-2 lattice with the appearance of vacancies in the *B*-positions. An exception is the Q_M (x) dependence. Despite the lowest values of δ_{R} and δ_{Rh} in SS with super stoichiometric lanthanum, Q_M increases and at $x \sim 0.02$ is almost double the initial value. This Q_M behavior may be a consequence of the decrease in grain size of the ceramics. The observed increase in Q_M with reduction $\overline{\mathrm{D}}$ can be attributed to the reduction in twinning, which results in a corresponding weakening of internal stresses and mechanical losses [18]. A comparison of the character of changes in the properties of SS with stoichiometric and super stoichiometric Al^{3+} (Fig. 3.22) demonstrated their practical identity at low concentrations of x. This indicates, as previously noted, the incorporation of Al^{3+} into the lattice at low x. As x increases, it is probable that Al^{3+} will occupy irregular positions or form liquid-phase eutectics.

A comparative analysis of the properties of SS with Ti^{4+}, introduced by different methods. revealed that the most pronounced dependencies of parameters on x were observed in SS with stoichiometric Ti (see Fig. 3.23). Furthermore, the same SS exhibit heightened HF. It is evident that the observed phenomenon is attributable to the super stoichiometric introduction of Ti^{4+}, which masks the underlying effects of its incorporation and embedding into the lattice.

A modification of the W^{6+} was conducted in three ways and revealed that T_c, $\varepsilon_{33}^T/\varepsilon_0$, K_p, tanδ, ρ behaved almost identically. The changes in E_{yu} and V_R are most evident in SS-4, while the greatest alteration in Q_M is observed in SS with super stoichiometric W (Fig. 3.24). Across the entire range of W^{6+} concentrations, Q_M is consistently higher than in the initial SS. This phenomenon may be associated with the finer-grain structure of these materials (Table 3.5).

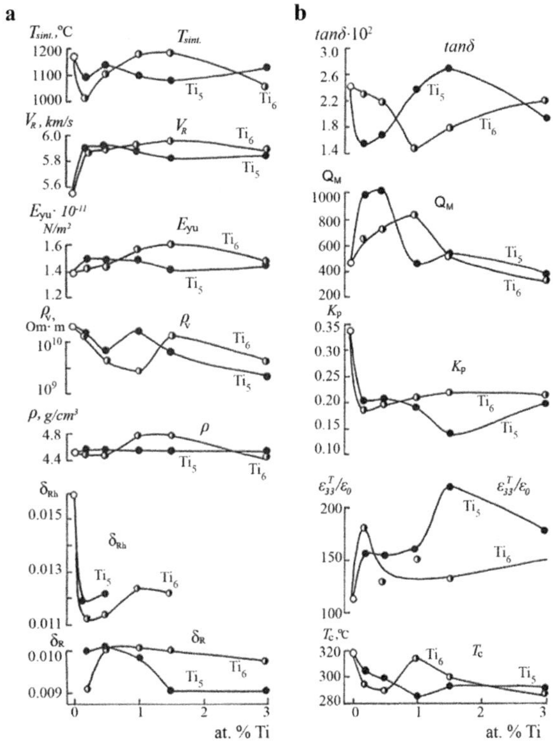

Fig. 3.23 **a** Comparison of structural and ceramic properties of SS with Ti^{4+}, introduced by different methods [SS-5 (Ti_5), SS-6 (Ti_6)]. **b** Comparison of dielectric and piezoelectric properties of SS with Ti^{4+}, introduced by different methods [SS-5 (Ti_5), SS-6 (Ti_6)]

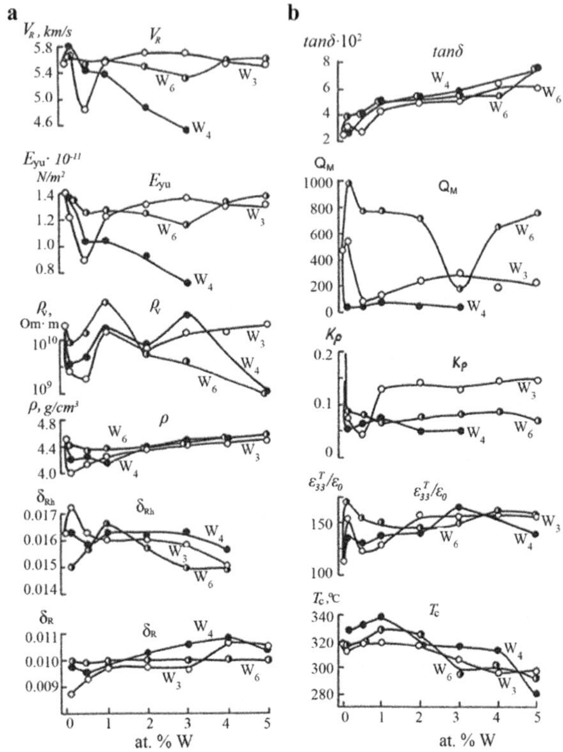

Fig. 3.24 **a** Comparison of structural and ceramic properties of SS with W^{6+} introduced by different methods [SS-3 (W_3), SS-4 (W_4), SS-6 (W_6)]. **b** Comparison of dielectric and piezoelectric properties of SS with W^{6+}, introduced by different methods [SS-3 (W_3), SS-4 (W_4), SS-6 (W_6)]

3.1.6 Dependence of the Properties of Modified Solid Solutions Based on Sodium Niobate on the Nature of Chemical Bonding and Defectivity of Crystal Structure

In this section, we will examine the relationships between the properties of modified NN-based SS and the chemical bonding (EN) of the elements involved, as well as the impact of vacancies that may arise during the modification process.

The objects of study were SS of the compositions (Na, Li) NbO_3 and (Na, $K)_{0.5}NbO_3$. For comparison, SS $Pb(Ti_{0.425}Zr_{0.407}Nb_{0.112}Mg_{0.0146}Zn_{0.0144})O_3$, which belongs to the MR of the $PbTiO_3$–$PbZrO_3$–$PbNb_{2/3}Zn_{1/3}O_3$–$PbNb_{2/3}Mg_{1/3}O_3$ system, was synthesized. The modification was carried out according to the formulae:

$$A_{1-\frac{n}{n'}x}M_x\square_{\left(\frac{n-n'}{n'}\right)x}BO_3 \tag{3.1}$$

$$A_{1-x}M_xB_{1-\left(\frac{n-n'}{6-n'}\right)x}\square_{\left(\frac{n-n'}{6-n'}\right)x}O_3, \tag{3.2}$$

$$A_{1-x}M_xB_{1-x}\square_x O_{3-x(3-\frac{n}{2})}\square_{x(3-\frac{n}{2})}, \tag{3.3}$$

$$A_{1-x}\square_x Nb_{1-x}M_x O_{3-x(3-\frac{n}{2})}\square_{x(3-\frac{n}{2})} \tag{3.4}$$

where M are the modifying ions Sr^{2+}, Cd^{2+}, La^{3+}, Cr^{3+}, Fe^{3+}, Ti^{4+}, introduced in the form of monoxides at the stage of mixing components in the process of SS synthesis; x is the M content ($0 < \boldsymbol{x} \leq 0.05$); $\boldsymbol{n'}$ and $\boldsymbol{n}$ are the valences of the substituted $A(B)$ and modifying (M) cations; □ are the vacancies arising in $A(1)$, $B(2)$, B and O(3), A and O(4)—positions.

In the initial two instances, modifiers were incorporated while maintaining the stoichiometry of the composition. In the subsequent cases, they were introduced above the stoichiometry of $(\mathbf{1}-\boldsymbol{x})ABO_3 + xMO_{n/2}$, if they were incorporated into the lattice. Formulae (3.1)–(3.3) reflect the occurrence of replacement within the A-sublattice, while formula (3.4) reflects the occurrence of replacement within the B-sublattice of the initial SS. The mean electronegativity of all A- or B-elements $\overline{EN}_A$, $\overline{EN}_B$ is selected as a characteristic of the chemical bond, that as demonstrated earlier can be considered a measure of the degree of covalence of the A–O and B–O bonds. Figures 3.25, 3.26, 3.27 and 3.28 demonstrate the relationships between the homogeneous strain parameter δ (in the case of two-phase SS, the average values of δ ($\delta_{\text{aver.}}$) account for the concentration of each coexisting phase) and the relative dielectric permittivity of the polarized $\varepsilon_{33}^{T}/\varepsilon_0$, samples as a function of x. The numbers at the curves correspond to the numbers of the modification formulae.

Furthermore, Figs. 3.25, 3.26, 3.27 and 3.28 demonstrate additional axes with values of $\overline{EN}_A$. In calculating the latter, the influence of modifiers was also considered. The number of modification formulae and the type of modifiers are provided near the axes as well as in the curves.

Let us consider the character of dependencies for SS based on PZT (Fig. 3.25). Here, we observe a decrease in δ_T (T phase) with increasing x, accompanied by an increase in $\varepsilon_{33}^{T}/\varepsilon_0$ for all cases of modification. Similarly, the EN, that is, the degree of covalency of the A-O bond, is observed to decrease. This indicates a reduction in the degree of HF [19] of the SS with increasing x. The same relations between the SS parameters and their EN are observed for all the cases previously considered in the study of PZT-based systems. It is important to highlight that the observed sharper and noticeably different from linear change of parameters with increasing x at the modification by the formula (3 for La) can be associated with the increasing influence of the number of vacancies. As is well established [20], this latter scenario results in the interruption of the "polarization chains", a reduction in P_s, and consequently, a concomitant decline in δ, accompanied by an increase in $\varepsilon_{33}^{T}/\varepsilon_0$. The largest impact of the increase in the number of vacancies is especially evident as it is accompanied by the smallest change $\overline{EN}_A$, compared to the other two modification cases, considered in Fig. 3.25.

However, the impact of the number of vacancies is particularly evident in the alteration of parameters in (Na, Li) NbO_3-based SS at substitutions in the B-sublattice (Fig. 3.26). As x increases, the parameter $\delta_{\text{aver.}}$ decreases, whereas $\varepsilon_{33}^{T}/\varepsilon_0$ increases,

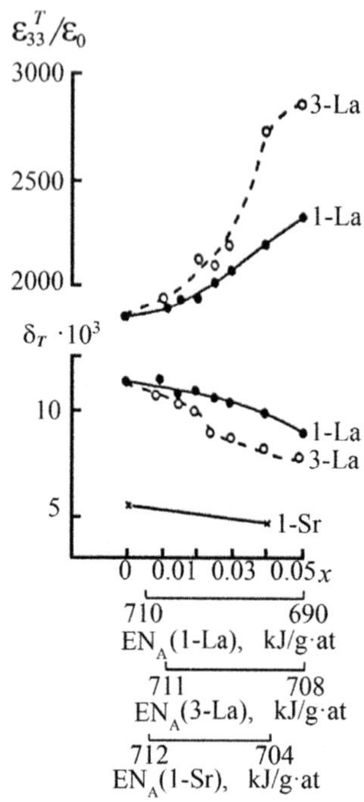

Fig. 3.25 Dependencies of $\varepsilon_{33}^{\mathrm{T}}/\varepsilon_0$ and δ_T on x and $\overline{EN}_A$ in PZT-based SS under substitution in the A-sublattice according to schemes (1) and (3)

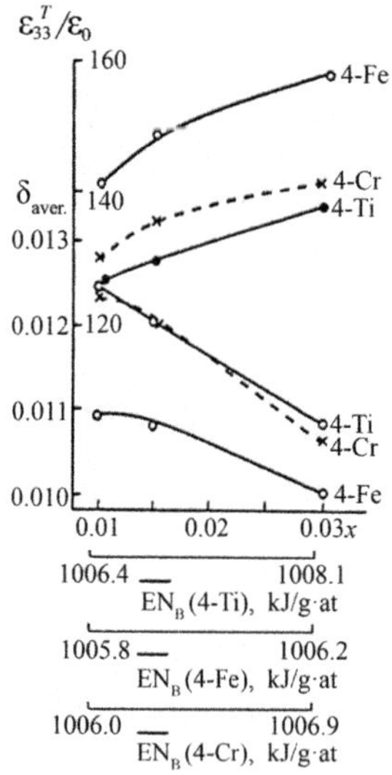

Fig. 3.26 Dependencies of $\varepsilon_{33}^{\mathrm{T}}/\varepsilon_0$ and δ_T on x and $\overline{EN}_B$ in (Na, Li) NbO_3-based SS, upon substitution in the B-sublattice according to scheme (4)

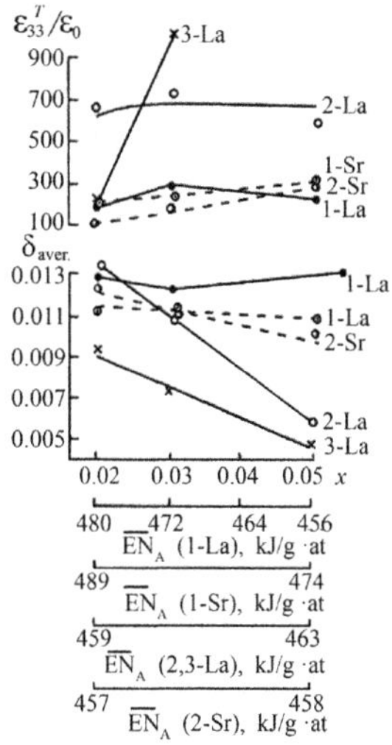

Fig. 3.27 Dependencies of $\varepsilon_{33}^{\mathrm{T}}/\varepsilon_0$ and δ_T on x and $\overline{EN}_A$ in (Na, Li) NbO_3-based SS, upon substitution in the A-sublattice according to schemes (1)–(3)

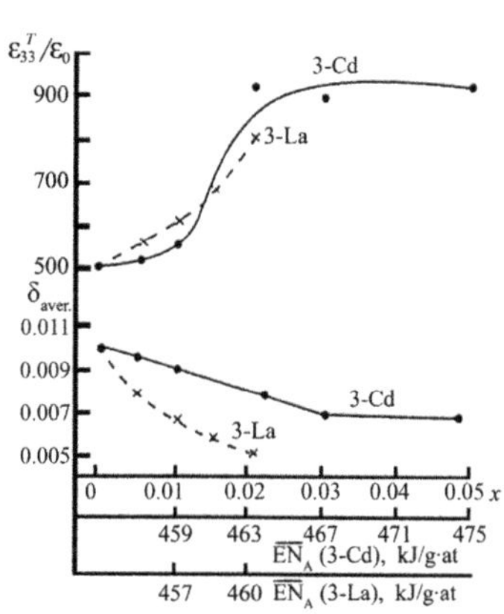

Fig. 3.28 Dependencies of $\varepsilon_{33}^{\mathrm{T}}/\varepsilon_0$ and δ_T on x and $\overline{EN}_A$ in (Na, K) NbO_3-based SS, upon substitution in the A-sublattice according to scheme (3)

despite a slight growth of $\overline{EN}_B$. The most notable alteration in the parameters occurs at the mildest change in $\overline{EN}_B$ (formula 3.4 for Fe). In our estimation, the findings of the modification, which involves the substitution in the A-sublattices for SS, based on (Na, Li)NbO_3 and (Na, K)NbO_3 (Figs. 3.27 and 3.28, respectively), are of the greatest interest.

In these two cases, new, previously unidentified, regularities are established. The main distinction is the significant reduction in the values of $\overline{EN}_A$ in comparison to $\overline{EN}_A$ in Fig. 3.27 and $\overline{EN}_B$ in Fig. 3.26.

As shown in Fig. 3.27, the correlation between changes in properties and $\overline{EN}_A$ is evident only for formula (1 for Sr). As $\overline{EN}_A$ decreases from 489 to 474 kJ/g at, $\delta_{\text{aver.}}$ exhibits a slight decline, leading to an increase in $\varepsilon_{33}^{\mathrm{T}}/\varepsilon_0$ The same set of parameters for the initial change in $\overline{EN}_A$ from 480 to 472 kJ/g at, is maintained for the formula (1 for La). A further decrease in $\overline{EN}_A$ from 472 to 456 kJ/g at results in a reversal of the observed properties. In the case of the separate dependencies (formulae 2, 3 for La and 2 for Sr), the correlation between the parameters and $\overline{EN}_A$ is reversed for all values of $\overline{EN}_A$, ranging from 459 to 463 kJ/g at in the first instance and from 457 to 458 kJ/g at in the second example. As shown in Fig. 3.28, the same character of change of mentioned properties is observed: the increase of $\overline{EN}_A$ from 457 to 467 kJ/g at is accompanied by a sharp decrease of $\delta_{aver.}$ and an increase of $\varepsilon_{33}^{T}/\varepsilon_0$ (at the range of $\overline{EN}_A = 471$–475 kJ/g at, the mentioned behavior of parameters slows down sharply). It can thus be concluded that the values of $\overline{EN}_A = 472$–474 kJ/g at represent a boundary, above which the dependencies are normal, as previously observed, and below which they are directly opposite. It is interesting to note that when the dependence relationships of the SS parameters of niobate systems (unmodified) were investigated, the conventional behavior of the parameters was observed at $\overline{EN}_A \geq 473$ kJ/g at. The behavior of δ at reduced values of EN can be explained by reference to the formula for calculating the chemical bond directional parameter (v), as set forth in Ref. [8]. According to [8], v is related to the link inequality parameter Ω, which is a linear combination of the link directivity parameters. Concurrently, the formation of unequal, mutually uncompensated bonds results in the appearance of uncompensated dipole moments, which can be defined as spontaneous polarization P_s.

P_s can be derived from the fundamental thermodynamic relations [21], which demonstrate a quadratic dependence with the spontaneous deformation ($\boldsymbol{\delta} \sim P_s^2$). Accordingly, it is plausible to posit that δ changes in a manner analogous to v. In the formula: $v = (\zeta \cdot q)/\tau$, where ζ is a measure of the anisotropy of the electron density distribution, q is the overlapping charge of the valence orbitals of bonding atoms, and τ is a measure of the perturbation (destruction) of atomic orbitals by the field of the surrounding atoms (specifically oxygen atoms, in the case of oxides with a perovskite-type structure) in one way or another, related to the EN. As stated by [8], $\zeta = |\psi|_{max.} - |\psi|_{min.}$, where $|\psi|_{max.}$ is the absolute value of the wave function of valence electrons at the distance of the radial electron density maximum in the case of its highest concentration, and $|\psi|_{min}$ is the same for the lowest concentration. For s-electrons, $|\psi|_{max} = |\psi|_{min,}$, and the degree of anisotropy $\zeta = 0$; for p- and d-elements, $|\psi|_{min} = 0$ and, therefore, $\zeta = |\psi|_{max}$. The value of q, approximately proportional to the energy of covalent interaction, is equal to the product of $n \cdot s$, where n is the number of electrons involved in the bond, s is the range of overlapping of the valence orbital of this atom with the valence orbital of the surrounding atom. The expression $\boldsymbol{\tau} = 1 + (f_0/f)$, is used for τ, where f is the field strength of the atom at the point of maximum radial electron density for the valence orbital, whose degree of perturbation is being considered, f_0 is the field strength, produced by the surrounding atoms at the same point. Considering the electron density distribution in the atom as spherically symmetric, the following specific expressions are adopted in estimating the field strength: $f = Z_{eff}/r_{max}^2, f_0 = Z_0/r_0^2$, where r_{max} is the radius of the maximum of the radial electron density distribution for the valence orbital of the atom under consideration, Z_{eff} is the effective charge for the orbital calculated by Slater's rules. In [8], the absolute values of the wave functions $|\psi|_{max}$ of the electrons of the three outer electron layers at the distances of the radial density maxima r_{max}, the values of the integrals of the overlapping between the *p*-orbitals of oxygen and the *d*-orbitals of *ds*-atoms or the *p*-orbitals of *sp*-atoms, and the values of the function $f_0(r_0)$ for different distances r_0 are given. Using the aforementioned crystallochemical characteristics, in conjunction with the values of Z_{eff}, approximated by the oxidation degree of the atom in crystals with predominant ionic bonding [8], we obtained the functions $f_0(r_{max})$ and $f(r_{max})$ and their ratios f_0/f for the *p*-orbitals of *sp*-atoms, which we have taken as examples. The changes of f_0/f, $|\psi|_{max}$ and *s*, with increasing EN are shown in Fig. 3.29. As the EN value increases, the f_0/f ratio tends towards one, indicating a shift towards a predominance of covalent bonds. Additionally, other characteristics exhibit growth with increasing EN. In the region of elements with high EN, the rate of observed change decreases. Therefore, the experimentally obtained dependence δ (EN) is the consequence of the mutual influence of all the quantities, included in the formula, with the prevailing contribution of one of them by changing EN. Thus, in the region of large EN values (the region of more covalent bonds) the decisive role, apparently, is played by the energy of covalent interaction and the anisotropy of valence orbitals, the increase of which leads to the observed increase in δ. In the low EN region, the contribution of τ is predominant, and its increase results in a decline in δ. In our case, this region of small EN extends up to 472–474 kJ/g at, below which we observe an "anomalous" decrease of δ with increasing EN [22].

The following *M*, introduced as oxides in excess of the stoichiometric composition, were selected for investigation of the properties of NL under cationic composition variation: divalent Ca (0.2–2.0) at%; Zn, Mg (0.5–5.0) at%; trivalent La (0.2–5.0) at%; Sc, Fe, Cr, Al (0.5–5.0) at%; four valent Zr, Ti, Sn (0.5–2.0) at%; hexavalent U (0.2–2.0) at%; W (0.5–5.0) at%.

A notable characteristic of nearly all compositions (except for Sc and Mg) is the incomplete incorporation of *M* into the lattice within the specified concentration limits (*x*). At all *x* values (Ca, La, Fe, U, W), or at their limiting values (Zn, Al, Cr, Sn, Zr, Ti), there are extraneous phases that are enriched with *M* and/or Nb^{5+} ions. The number of these phases increases with increasing *x* values (Ca, Zn, Fe, Al, Ti, Sn, U), or remains relatively constant (for the remaining *M*). Most often such phases

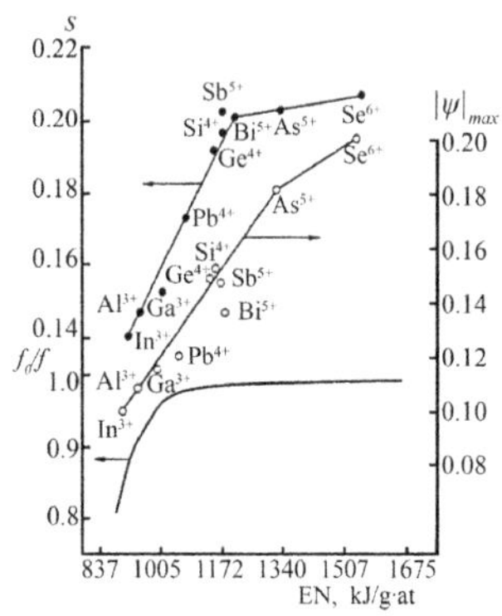

Fig. 3.29 Dependencies of *s*, $|\psi|_{max}$, f_0/f for *p*-orbitals of *sp*-atoms on EN (calculated from data [11])

are statistically distributed on the surface of samples. In the case of Fe, Cr, Zr they are localized mainly in pores.

The different occurrence of *M* in the lattice has a different effect on the structure, microstructure and electrophysical properties of NL. In accordance with the character of influence on the latter, three groups of *M* have been distinguished: donors (La, Ca, Zn, Mg), which increase ρ_V and $\varepsilon_{33}^T/\varepsilon_0$ (it should be noted that the degree of ρ_V increase decreases in the sequence La → Ca → Mg → Zn); acceptors (Sc, Fe, Cr, U), which decrease ρ_V. The behavior of the SS parameters at the introduction of the specified *M* corroborates the *p*-type conductivity of NL. The third group includes Al, Zr, Sn, Ti, and W, which gives rise to complex dependencies ρ_V (x) due to alterations in the change of valence of *M*(W), their restricted solubility (Ti, Zr), and microstructure characteristics (Zr, Ti, Sn, W).

All modified samples exhibit a coefficient of resistance (PTCR), which is observed across the entire range of *M* concentrations (except for compositions, containing Ca and Fe, where the PTCR effect is only discernible at low *M* contents). Two groups of modified SS were identified based on the nature of changes observed in the main posistor parameters, specifically the temperature at the end of the PTCR section and the width of the PTCR region. The first group encompasses PLs with elements Zn, La, Sc, Cr, Al, Zr, Sn, Ti, U, and W, which result in nonmonotonic changes in the main posistor parameters. The SS of the second group (with Fe, Mg) are characterised either by a monotonic decrease of these parameters (in the case of SS with Mg) or by the disappearance of the PTCR effect at relatively low concentrations of modifiers (in the case of Fe). The observed phenomena are shown to depend on the localization of *M* atoms and the peculiarities of phase formation in NL+*M* systems. The posistor properties undergo a decrease and complete suppression because of the enrichment of crystallite surfaces with *M* atoms and the precipitation of side phases.

In the modified NL ceramics, four distinct regions of temperature have been identified, which are characterised by distinct conduction activation energies: I—W_1 = 0.65 eV at 400 K < T < 650 K; II—W_2 = 0.92 eV at 560 K < T < 680 K; III—W_3 = 0.96 eV at 680 K < T < 860 K; IV—W_4 = 0.79 eV at 860 K < T < 1110 K, they are characterised by different forms of electrical conductivity: I is a consequence of intrinsic atomic defects ($\boldsymbol{V}_{\mathbf{Li}}$ and $\boldsymbol{V}_{\mathbf{0}}$); II is a result of impurity ionic effects (attributable to *M* ions); III is a consequence of impurity ionic effects (attributable to impurities present in the raw material); IV is a consequence of electronic defects that emerge concurrently with the introduction of heterovalent *M*.

The non-monotonic dependencies of conduction activation energies (W_i), and temperatures of fracture (T_i) on x were demonstrated. This is associated with the transition from an electronic to an atomic mechanism of heterovalent *M* charge compensation. The impact of NL modification is thoroughly examined in [22].

3.2 Features of Defect Formation in Alkali Metal Niobates

3.2.1 *Own Disorder in Alkali Metal Niobates*

For accurate interpretation of the experimental outcomes and identification of the influence of modifiers on the structure and electrophysical characteristics of NL-based SS, it is essential to understand the nature of the disorder, both intrinsic and induced by the incorporation of cations, resulting from the formation of both point and extended defects. To describe real crystalline objects, the Kreger-Wink nomenclature is employed. This nomenclature is based on the concept of structural elements, which include regular nodes and internodes of the lattice, as well as free or occupied nodes by atoms of a certain nature. Additionally, it incorporates the concept of structural defects, which are energetically excited states of the crystal lattice associated with a change in the way the lattice nodes are filled. As stated in Ref. [23], in binary and more complex stoichiometric compounds, point atomic defects occur as combinations forming six basic types of intrinsic disorder. In the case of the binary compound $M_n X_m$, these are:

(i) *symmetric types:*

S—Schottky disorder, associated with the occurrence of vacancies (unoccupied lattice nodes), in both sublattices: V_M and V_X;

I—interdomain disorder due to the formation of interdomain atoms M_i and X_i;

A—antistructural disorder, i.e. exchange of dissimilar atoms by crystallographic positions M_n and X_m.

(ii) *asymmetric types* (they can be considered as combinations of symmetric types):

S-I—disorder (according to Frenkel), namely vacancies and interdomain atoms of the same type, e.g., $V_M + M_i$; $V_X + X_i$;

S-A—disorder, namely vacancies and atoms, occupying foreign nodes, e.g., $V_M + M_X$; $V_X + X_M$;

I-A—disorder, namely interdomain atoms of one species and atoms of another species, occupying foreign nodes, e.g., $M_i + X_M$; $X_i + M_X$.

For PTS of ABO_3 type: *S*-type includes V_A, V_B, V_O; *I*-type includes A_i, B_i, O_i; and *A*-type includes A_B, A_O, B_A, B_O, O_A, O_B.

In the case of non-molecular crystals, it can be stated that all kinds of point defects are present simultaneously within the lattice. However, due to the difference in disorder energy and the nature of the basic matrix, as well as the concentration of modifying elements, the defects formed can be either dominant or subordinate.

In the general case of PTS and, in particular of PZT systems, the possibilities for the placement of ions in the internodes are constrained by the close packing of lead and oxygen ions, which approaches the densest possible packing. The significant difference in electronegativity between oxygen and cations, as well as the variation in the sizes and valences of the latter, makes it challenging to place them in "foreign" sublattice nodes. Consequently, only the Schottky S-type disorder is practically realized [24]. The occurrence of Frenkel defects (V_O and O_i, V_A and A_i, V_B and B_i) can be expected only in limited concentrations and only in locations, where the potential barrier is compromised, such as the crystal surface or an intergranular layer in a polycrystal. Among the "remaining" possible defects, the V_A, V_B, and V_O vacancies in B-positions are energetically unfavorable. This is due to two factors: the structure-forming character of BO_6 octahedrons, and the large formal charge and small size of B cations. The relative disadvantage of V_B formation, compared to V_A and V_O in [25], is explained on the basis of the mutual arrangement of the energy electronic levels of elements A and B. Specifically, the arrival of electrons from the valence band at the levels corresponding to the d-states of Ti, Zr, Hf, Fe, and 5s-states of Sn does not lead to the transition of these ions to the atomic state (when the particle breaks chemical bonds with the crystal) because there are corresponding to these atoms unfilled s- and p-levels located energetically higher. It is unlikely that electrons will fall into these levels. B-ions are only capable of reducing their charge without transitioning to the atomic state and without exiting the crystal structure. Therefore, the most prevalent type of defects in PTS are vacancies of A-cations and oxygen, which respectively form acceptor and donor centers. Additionally, B-V_O-B complexes are formed by these vacancies with captured 1 or 2 electrons, as well as the state of V_A and twelve adjacent oxygen ions with localized 1 or 2 holes [25].

Information regarding the character of dominant defects can be derived from the analysis of electrical conductivity, Hall effect, electron paramagnetic resonance (EPR), optical absorption, Mössbauer effect, and self-diffusion coefficients [26] The following works offer empirical validation of the conclusions:

(i) at high temperatures in vacuum in equilibrium with $Pb(Ti_{1-\alpha}Zr_{\alpha})O_3$ ($0 \leq \alpha \leq 1$) phases there is a vapor consisting of lead oxide molecules [27–29];
(ii) the micromechanism of perovskite phases formation from pelletized oxides consists in the one-way transfer of lead and oxygen through the crystal lattice of the reaction product, and the transfer rate is limited by the diffusion of lead cations by the vacancy mechanism [24];
(iii) the compaction mechanism at the initial stage of sintering of SS PZT consists in bulk diffusion (vacancy source—contact isthmus of particles, effluent—intergranular boundary) with the limiting role of oxygen vacancies [24];
(iv) low values of enthalpies of formation of Pb and O vacancies, with lower values characteristic of V_{Pb}.

Defect formation can be described by the following quasi-chemical equations [30], which observe the law of conservation of mass and charge (including effective charge), while maintaining the fixed ratio between the number of different lattice nodes inherent to a given crystal structure (Fig. 3.30).

$$\begin{array}{l} \mathrm{Pb}_{\mathrm{Pb}}^{x} \overset{(a)}{\longleftrightarrow} V_{\mathrm{Pb}}^{x} + \mathrm{Pb}_{(gas)} \overset{(c)}{\longleftrightarrow} \underset{(gas)}{V_{\mathrm{Pb}}''} + \mathrm{Pb}_{(gas)} + 2h^{\cdot} \\ \mathrm{O}_{x}^{o} \overset{(b)}{\longleftrightarrow} V_{o}^{x} + 1/2\mathrm{O}_2 \overset{(d)}{\longleftrightarrow} V_{o}^{\cdot\cdot} + \begin{array}{c} 1/2 \\ (e) \end{array} \begin{array}{c} \mathrm{O}_{2(gas)} \\ \Updownarrow \\ \mathrm{PbO}_{(gas)} \end{array} + \begin{array}{c} 2e'(f) \\ \Updownarrow \\ (zero) \end{array} \end{array} \tag{3.5}$$

Since the partial pressure of oxygen in the atmosphere is typically many times higher than the equilibrium dissociation pressure of gaseous Pb oxide [reaction (e)], the equilibrium positions of (d) and (b) are shifted to the left, while those of (a) and (c) are shifted to the right.

Consequently, the concentration of $[V''_{\mathrm{Pb}}]^{\cdot} >> [V_{\mathrm{o}}^{\cdot\cdot}]$ and holes, which can be localized on lead ions capable of oxidation to +4, exceeds that of free electrons, resulting in the experimentally observed *p*-type conductivity. In this instance, the electroneutrality condition is approximated by the equality:

$$2\left[V_{\mathrm{Pb}}''\right] = \left[h^{\cdot}\right] \tag{3.6}$$

By applying the mass action law to (3.5), the equilibrium conditions of defect formation in PZT can be expressed in the form of the following equations [29]:

$$\left[V_{\mathrm{Pb}}''\right] \cdot P_{\mathrm{Pb}}^{**} \cdot \left[h^{*}\right]^{2} = K_{l}^{***} \tag{3.7}$$

$$\left[V_{o}^{\cdot\cdot}\right] \cdot P_{\mathrm{O}_2}^{1/2} \cdot \left[e'\right]^{2} = K_2 \tag{3.8}$$

$$\frac{P_{\mathrm{PbO}}}{P_{\mathrm{Pb}} \cdot P_{\mathrm{O}_2}^{1/2}} = K_3; \quad \left[e'\right] \cdot \left[h^{*}\right] = K_4 \tag{3.9}$$

Upon combining these equations, the result is:

$$\left[V_{\mathrm{Pb}}''\right] \cdot \left[V_{o}^{\cdot\cdot}\right] = K \cdot P_{\mathrm{PbO}}^{-1} \tag{3.10}$$

Therefore, at a fixed temperature and vapor elasticity, the product of the lead ion and oxygen ion vacancy concentration remains constant. The formation of holes, which result in the *p*-type conductivity, is attributed to the ionization of lead vacancies.

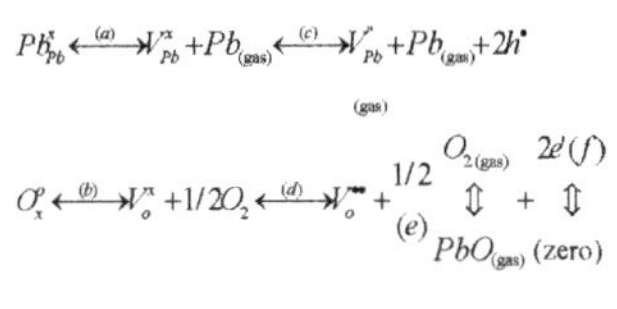

Fig. 3.30 Quasi chemical equations: (dashes indicate effective negative charge relative to the lattice normal component; dots indicate effective positive charge; *x*—indicates neutrality; *h*—hole; *e*—electron; *v*—vacancy.

The reaction: $V^{x}{}_{Pb} \leftrightarrow V''{}_{Pb} + 2h^{\cdot}$ is connected to the formation of lead vacancies, which attract electrons to complete the electron shells of oxygen ions, surrounding the vacancies, thereby acting as acceptors.

At thermal equilibrium, a certain part of the energetically excited states, associated with local lattice periodicity violations (point defects) are realized in a crystal. Accordingly, the partially disordered crystal structure represents an equilibrium lattice, and the degree of its defectiveness is a function of the thermodynamic parameters of the medium and the properties of the objects under study [24].

The introduction of modifiers may result in the number of defects, exceeding the equilibrium number. In such instances, these defects are classified as "biographical", that may also emerge due to kinetic factors stemming from the realization of specific synthesis conditions [31].

In addition to point defects, as previously stated, extended defects may also occur in PTS. These are classified by the sign of dimensionality into two categories: linear (1D) and non-linear (2D). Linear defects include a linear sequence of lattice nodes along which the periodic or nearly periodic lattice structure is interrupted. Examples of such defects include edge dislocations and helical dislocations. Surface (2D) defects include errors in the superposition of layers and boundaries separating regions of ideal or near-ideal periodic crystal structure. Examples of these boundaries include coherent scattering regions, crystallographic shift planes, and grain boundaries at their junctions in micro- and macromosaics. Volumetric (3D) defects include ordered point defects, associates, clusters, open and closed pores, cracks, and phase inclusions.

Such defects are not observed in thermodynamically equilibrium systems at any temperature. The occurrence of these defects is dependent on the thermal history of the crystal, the conditions under which it was produced, and is manifested when the disordering state is "quenched," which represents the equilibrium state for higher temperatures. While atomic point defectivity, accompanied by electronic disorder, creates conditions for the development of electrical transport and polarization processes, the role of extended defects, which are not necessarily electrically charged, in changing the concentration of free charge carriers may be small, although they are able to change the conditions for the movement of carriers, released by point defects [31]. In response to changes in external conditions, the equilibrium concentration of point defects undergoes a corresponding alteration, which in turn gives rise to a change in electrical properties. The formation of extended defects at the temperatures of object formation results in their "freezing" at lower temperatures. Therefore, these defects cannot arise as equilibrium defects and, thus, do not affect the electrical properties. However, they can significantly impact the dielectric and piezoelectric characteristics of PTS.

The structural defects, observed in oxygen-octahedral substances, are of a similar nature, allowing one to extend the results described above to $A^{I}B^{V}O_3$ compounds (A^{I} = K, Na, Li; B^{V} = Nb). However, their crystallochemical characteristics impart a degree of specificity to the processes of defect formation and disorder. Consequently, the small size of *A*-cations and the resulting inconsistency in their crystallochemical characteristics with the structure-forming octahedrons of NbO_6 lead to the formation of large cubo-octahedral voids and the unrestricted placement of alkali elements. The low atomic weight of these cations results in increased mobility and a greater tendency to interchange with other ions, leading to the accumulation of vacancies, the formation of microfluctuations, and the emergence of phases with variable compositions. The proximity in size of *A*- and *B*-cations ($LiNbO_3$) opens the possibility to "exchange of their places". All this, as well as the presence of vacant octahedrons of $LiNbO_3$ with favorable conditions for the introduction of foreign elements (the radii of spheres that "fit" into octahedrons without distorting them are 0.67, 0.81 and 0.83 Å for Nb, Li, and the vacant octahedron, respectively, with the ionic radius of oxygen equal to 1.32 Å [27]), contributes to the realization of a high degree of own disorder in alkali metal niobates [24]. As noted in [6], in general, the following point defects are possible in the structure of lithium metaniobate (LMN): vacancies and interstitial ions of lithium (${V^-}_{Li}$, ${Li^+}_i$), niobium (${V^{5-}}_{Nb}$, ${Nb^{5+}}_i$), oxygen (${V^{m+}}_O$, ${O^{2-}}_i$), niobium ions replacing lithium (${Nb^{4+}}_{Li}$), lithium ions at niobium sites (${Li^{4-}}_{Nb}$), as well as quasi-free electrons (e^-) formed by the equation ${O^x}_o \leftrightarrow {V^{\cdot}}_o + e + 1/2O_2$. At high temperatures, when defects are fully ionized, the electroneutrality equation has the following form [32]: $m[{V^{m+}}_o] + [Li^+] + 5[{Nb^{5+}}_i] + 4[{Nb^{4+}}_{Li}] = 5[{V^{5-}}_{Nb}] + [{V^+}_{Li}] + 4[{Li^{4+}}_{Nb}] + 2[{O^{2-}}_i] + [e^-]$, where $m = 1$ or 2 and the upper indices are equal to the effective charge of the defect. The defect concentration, as determined by this equation, is dependent upon two variables: the [Li]/[Nb] ratio within the crystal and the oxygen partial pressure. In the presence of low oxygen partial pressure, the concentration of oxygen vacancies is expected to exceed that of cationic defects. At elevated oxygen pressures, cationic defects are expected to prevail in the LMN, resulting in ionic conductivity [32]. The concentration of specific point defects in the LMN under varying external conditions is calculated in Ref. [33].

The crystallochemical analysis allowed us to construct specific models of the SS of alkali metal niobates. For LMN, the most appropriate model at this time is the "model with packing disorder" proposed in [34], which is described by the following formulae:

$$\mathrm{Li}^{+}_{1-5x}\mathrm{Nb}_{1+x}\left(V^{-}_{Li}\right)_{4\mathrm{x}}O^{2}_{3} \quad \text{at} \quad [\mathrm{Li}]/[\mathrm{Nb}] < 1,$$

$$\mathrm{Li}^{+}_{1+z}\mathrm{Nb}_{1-z}\mathrm{O}^{2-}_{3-2z}\left(V^{2+}_{o}\right)_{2z} \quad \text{at} \quad [\mathrm{Li}]/[\mathrm{Nb}] > 1.$$

The name of the model underscores the fact that when lithium is replaced by niobium (with a ratio of [Li]/[Nb] less than 1), the latter does not assume the Li position but instead occupies a vacant oxygen-octahedral position. This results in a disruption in the alternation of layers in the cation sublattice, as well as a change in the Nb–Nb and Li–Li distances, as evidenced in Ref. [31]. The lithium (or, in the case of excess lithium, oxygen) vacancies resulting from the introduction of Nb are distributed throughout the crystal volume rather than being localized at the cation introduction sites. In the oxygen-octahedral structure of LMN, the lithium ions are located at the common face of two octahedrons and exhibit slight displacement towards the center of one of them. If we accept the "model with packing disorder," which is in better agreement with the experimental data than the other models, we can consider that the structure of LMN allows for the introduction of Nb into the free octahedron. Concurrently, lithium vacancies are formed to maintain electroneutrality. Since the embedded Nb has the characteristics of transition metals and is surrounded by anions in an octahedral configuration, its position within the octahedron will be stable.

The application of the model to SS containing an excess of Li results in the disruption of the oxygen-niobium octahedron and the formation of two oxygen vacancies for each additional lithium atom. In actual conditions, a substantial excess of lithium oxide in the melt is necessary to create a solution with a stoichiometric composition according to [35], this is 8 mol%. It is proposed that the LMN structure may exhibit several potential defects, including the presence of free electrons, single- and double-charged oxygen vacancies, and Nb^{4+} ions in various crystallographic positions [32].

In order fully to understand the contribution of each type of defect to the real structure of the alkali metal niobates, it is necessary to conduct an experimental study of diffusion processes, electrical conductivity, spectral characteristics, and so forth.

In the case of alkali metal niobates, there are significantly greater opportunities for ion substitution and defect formation than in PZT systems. Indeed, the occurrence of vacancies in *B*-positions is a very real possibility when small *A* cations are displaced and introduced into the interstices of the crystal lattice. It should be noted that the intercalation of a small-sized Li ion into the lattice of complex oxides of various structural types is widely known [9].

The proclivity of Nb to easily alter the oxidation degree ($Nb^{5+} \leftrightarrow Nb^{4+}$) and the ordering of oxygen vacancies that arise in Nb_2O_5 result in the elimination of point defects through the formation of extended crystallographic shift planes (CSPs) [9, 24, 36, 37]. As demonstrated in [38–40], these characteristics are inherited by complex Nb-containing oxides, influencing their formation kinetics and properties.

3.2.2 *Effect of Modifiers on the Equilibrium State of Intrinsic Defects of the Matrix Backbone*

During the modification process, the introduced ions occupy vacancies in the matrix basis within the nodes of the crystal lattice, or alternatively, they embed themselves in the inter-nodes. Furthermore, they form associations with vacancies and influence the character of intrinsic disorder. Concurrently, variations in the dimensions of the matrix and incorporated atoms can induce stress within the lattice. Furthermore, discrepancies in their polarizing actions, electronic structures, and stereo chemistry can markedly influence the characteristics of the base. Furthermore, an inconsistency in the ionization energies will result in a modification of the electroneutrality conditions.

Thus, the dissolution of the Me_2O_3 impurity in the MO monoxide can be described by the following quasi-chemical reaction, which occurs by the substitution type (metal and oxygen atoms occupy regular nodes in the matrix):

$$2Me^x_{Me} + 3O^x_O \leftrightarrow 2Me^x_{Me} + V^x_M + 3O^x_O,$$

and when considering the ionization of atomic defects:

$$2Me^x_{Me} + 3O^x_O \leftrightarrow 2Me^{\cdot}_{Me} + V''_M + 3O^x_O.$$

The equation demonstrates that the dissolution of two metal atoms of the Me_2O_3 impurity, in accordance with the principle of conservation of the ratio of regular sites in the MO crystal, will result in the formation of one cationic vacancy. The general condition of electroneutrality for the impurity crystal in question is given by the expression: $[Me^{\cdot}_M] + 2[V''_O] = 2[V''_M]$, which differs from the condition of electroneutrality of the modified monoxide, which is given by the expression: *zero* $\leftrightarrow V''_O + V''_M$ (Schottky defects).

The general condition of electroneutrality for such an impurity crystal is written by the expression:

$$[Me^{\cdot}_M] + 2[V''o] = 2[V''_M] \tag{3.11}$$

that can be transformed to the form:

$$[V''_M]([V''_M] - 1/2[Me^{\cdot}_M]) = K_s \quad \text{or} \quad [V''_M]([V''_M] + 1/2[Me^{\cdot}_M]) = K_s \tag{3.12}$$

The functional dependence of the concentrations of intrinsic defects on the amount of impurity introduced into the crystal demonstrates that at an impurity concentration that is commensurate with the concentration of intrinsic defects, a notable alteration in the latter occurs.

In the SS of PZT, heterovalent modification results in a change of electroneutrality, as illustrated in the following scheme (with the introduction of the donor modifier $n_{MA} = 3$ or $n_{MB} = 5$):

$$M^X_{don} \leftrightarrow M^{\cdot}_{don} + e' \tag{3.13}$$

(charge is given relative to the normal component of the lattice). Considering (3.13), we should change the electroneutrality condition (3.6) to:

$$[e'] + 2[V''_{Pb}] = [h^{\cdot}] + [M^{\cdot}_{don}] \quad (3.14)$$

It follows that the charge of the introduced $[M^{\cdot}{}_{don}]$ donors is compensated by the charge of the $(n - p)$ $([e'] = n; [h^{\cdot}] = p)$ electrons and $[V''{}_{Pb}]$ vacancies.

A small quantity of $[M^{\cdot}{}_{don}]$ in the system is compensated by $V''{}_{Pb}$, even when $n - p \approx [M^{\cdot}{}_{don}]$. Conversely, when $2[V''{}_{Pb}] \approx [M^{\cdot}{}_{don}]$ a minimal proportion of $M^{\cdot}{}_{don}$ is compensated by electrons.

Thus, the charge compensation of the introduced modifier $(n_{MA,B} > n_{A,B})$ relative to the normal lattice component can occur by two mechanisms: the formation of both atomic and electronic defects.

In the case of atomic compensation, the hole conductivity decreases according to the expression:

$$[h^{\cdot}] = k^{-1/2} \cdot P_{O_2}^{1/4} \cdot [V''_{Pb}]^{-1/2} \quad (3.15)$$

The above considerations regarding the two mechanisms of charge compensation are also applicable to the introduction of an acceptor modifier.

As observed in [41], the intrinsic atomic disorder constant in the PZT system is marginally higher than the intrinsic electronic disorder constant. This can be expressed as follows:

$$[V''_{Pb}] \cdot [V^{\cdot\cdot}_{O}] >> [e'] \cdot [h^{\cdot}] \quad (3.16)$$

This finding is consistent with the experimental evidence that the introduction of donor modifiers in an amount exceeding 1 at% does not result in a further reduction in conductivity or a shift in the conductivity type. Additionally, the extent of the conductivity increase upon the introduction of acceptor modifiers is constrained to a single order of magnitude [42, 43].

3.2.3 Controlled Deviation from Stoichiometry as a Special Case of Heterovalent Modification

A special case of non-isovalent modification is a controlled deviation from stoichiometry leading to the creation of intrinsic defects: ion A (V_A) and oxygen V_O vacancies [44].

The simplest approach to non-stoichiometry, originally developed by Schottky and Wagner [45–47], is based on the absence of equilibrium of intrinsic point defects in the crystal. In a system, consisting of a binary MO oxide, and its equilibrium gas phase, there is always a mass transfer, expressed by the following quasi-chemical equations:

$$\frac{1}{2}O_2 \leftrightarrow O_O^x + V_M^x \leftrightarrow O_O^x + V_{M'} + h^{\cdot} \leftrightarrow O_O^x + V_M^{''} + 2h^{\cdot}$$

$$\frac{1}{2}O_2 \leftrightarrow O_i^x \leftrightarrow O_{i'} + h^{\cdot} \leftrightarrow O_i^{''} + 2h^{\cdot}$$

$$O_O^x \leftrightarrow V_O^x + \frac{1}{2}O_2 \leftrightarrow V_O^{\cdot} + e' + \frac{1}{2}O_2 \leftrightarrow V_O^{\cdot\cdot} + 2e' + \frac{1}{2}O_2$$

$$M_M^x \leftrightarrow O_O^x \leftrightarrow M_i^x + \frac{1}{2}O_2 \leftrightarrow M_i^{\cdot} + e' + \frac{1}{2}O_2 \leftrightarrow M_i^{\cdot\cdot} + 2e' + \frac{1}{2}O_2$$

The prevalence of the initial two processes results in the generation of an oxide with an excess of oxygen, whereas the predominance of the subsequent two processes is associated with the formation of an oxygen-deficient phase. The energy of quasi-chemical reactions is comprised of two distinct summands: one associated with a change in the number of point defects and the other with a change in the binding energy within the crystal, which can be conceptualized as a change in effective valence [48]. The relationship between the non-stoichiometry of a binary crystal, the concentration of point defects in it, and the partial pressure of the volatile component in the equilibrium gas phase was initially established by Schottky and Wagner [46].

In ABO_3 oxides, a variety of deviations from the stoichiometric composition are possible. These are the oxygen non-stoichiometry $ABO_{3-\square}$, which is common for oxides, and non-stoichiometry by cation ratio ($A_{1-\square}BO_{3-\square}$). The types of non-stoichiometry, which coexist within the same crystalline substance, exhibit significant differences in their character, structural organization, and impact on material properties. A significant characteristic of oxide multicomponent phases, including perovskite phases, is the potential for compositional alteration in multiple directions. In principle, such a possibility is always present. To unambiguously specify the composition of the N-component phase, it is necessary to fix the $(N-1)$ parameter. This can be achieved most simply by determining the concentration of the $(N-1)$ independent component of the phase. In the case of ABO_3 perovskites, which are three-component compounds with variable composition, it is convenient to select a simple stoichiometric oxide (AO) and oxygen as the independent components. The highest and most conveniently determined equilibrium pressures are observed in the gas phase. In this case, the composition of the non-stoichiometric perovskite can be represented by the formula $A_{1-\gamma}BO_{3-\delta}$.

The parameter δ is used to describe the change in oxygen non-stoichiometry along the line extending from the top of the phase triangle for the A–B–O system. Moreover, the ratio of cationic components A/B in the perovskite phase remains unchanged. The parameter γ reflects additional non-stoichiometry in the ratio of cationic components, which varies independently along the line passing through point AO on the phase triangle.

For a given ratio of A/B cations, the value of oxygen non-stoichiometry$\square$ under equilibrium conditions is a single-valued function of temperature T and oxygen partial pressure P_{O_2}. Therefore, a variety of chemical analysis techniques, including

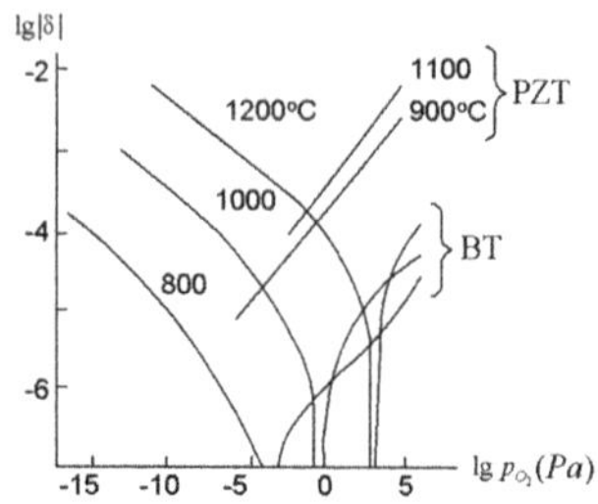

Fig. 3.31 Oxygen non-stoichiometry of $BaTiO_3$ and $Pb(Ti_{0.5}Zr_{0.5})O_3$

gravimetry (mass measurements), electrochemical property measurements to determine the concentration of excess electrons (holes), and diffusional mobility of oxygen and cations for the samples, obtained at different values of T and P_{O_2}, are employed to study oxygen non-stoichiometry.

Figure 3.31 shows the experimentally determined values of δ in barium titanate (BT) and PZT. In the latter, the interval of permissible values of P_{O_2} is limited from below by the so-called low-oxygen boundary, below which the perovskite phase decays as the result of its reduction to metallic lead and other products. In the entire permissible range of oxygen pressures, the value of δ is negative, indicating that PZT contains a non-stoichiometric excess of oxygen. For BT, the right branches of the curves correspond to negative values of δ, while the left branches correspond to positive values; thus, both oxygen excess and oxygen deficiency are possible in BT. The precise stoichiometry $\delta = 0$ in Fig. 3.33 is equivalent to the dips ($\lg|\delta| \to \infty$ at $\delta \to 0$) [49]. The maximum value of oxygen non-stoichiometry does not exceed the value of approximately $\delta \approx 10^{-2}$.

The study of electrical conductivity at elevated temperatures allows for the identification of thermodynamic conditions of oxygen stoichiometry ($\delta = 0$), which in oxide phases with a small fraction of ionic conductivity exactly correspond to the transition from electron to hole type of conductivity. Furthermore, it allows for the specification of the numerical values of small deviations from stoichiometry by the condition of the balance of electric discharges. The dependencies $\delta = f\ (P_{O_2})$ for several temperatures (see Fig. 3.31) provide a summary of the results of gravimetric, analytical, and electrokinetic studies.

BT is a phase with two-sided oxygen non-stoichiometry at higher temperatures there is an excess of oxygen and at lower temperatures there is a deficiency of oxygen. As the temperature decreases, the stoichiometric point shifts towards a lower value.

The cationic non-stoichiometry of compounds, comprising volatile cationic components, is established at a specified temperature and partial pressure of the volatile component.

For example, for PZT containing volatile lead oxide PbO, $\gamma = f\ (T,\ P_{PbO})$, perovskite phases $A^{II}B^{IV}O_3$ are characterized by significant non-stoichiometry, namely the absence of AO ($\gamma > 0$). In lead titanate and zirconate, the non-stoichiometry is maximal and reaches $\gamma \approx 0.1$, whereas in the region $x = 0.3$–0.4 it is minimal (Fig. 3.32) [24].

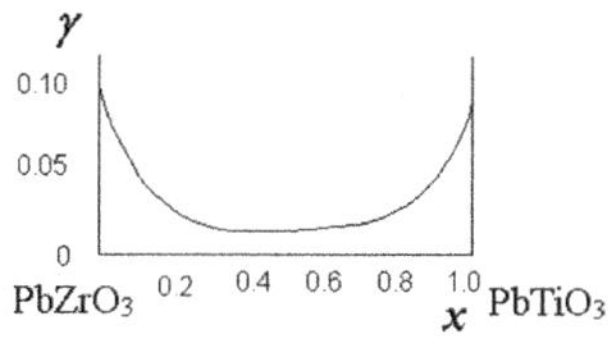

Fig. 3.32 Limit of γ-non-stoichiometry in $Pb(Ti_xZr_{1-x})O_3$ solid solutions

The homogeneity region in $NaNbO_3$ also extends to $\gamma \approx 0.1$ [50]. The characteristics of other PTS of $A_{1-\gamma}BO_3$ and $ABO_{3-\delta}$ types, as well as their solid solutions, are discussed in detail in [51]. In addition, we present data on the non-stoichiometry of complex oxides of mixed structure, containing fragments of perovskite-type structure, as well as monoxides (e.g., TiO_2, ZrO_2, Nb_2O_5, etc.).

3.2.4 *Change of Properties of Solid Solutions Based on Sodium Niobate at Introduction of Modifying Elements*

As previously stated, the modification capabilities of alkali metal niobate-based SS, particularly NN, but also $PbTiO_3$ (PT) and $PbZrO_3$ (PZ), have been insufficiently explored in the literature. In the first case, the (Na, K) NbO_3 system was the subject of many studies [52–59]. Table 2.1, located in Sect. 2.1, presents a list of classifying modifiers for the system, accompanied by an indication of the method of introduction of additives and the presumed position of the modifiers within the crystal lattice. The data presented in Table 2.1 also includes the impact of modifiers on the ceramic ($T_{sint.}$, ρ), dielectric ($\varepsilon_{33}^T/\varepsilon_0$,$T_c$), piezoelectric ($d_{31}$, K_p), and elastic (V_R, E_{yu}) characteristics of the SS system. The ranges of modifier concentrations (in mol.%) in which the parameters undergo change are indicated.

Immediately following the publications devoted to the modification of the (Na, K) NbO_3 system, the first reports on isomorphic substitutions of ions in another system, based on (Na, Li) NbO_3, appeared (since 1975). The works were produced by employees of the Research Institute of Physics of the Russian State University, who elected to modify the system to search for and develop new promising ferropiezoelectric materials for various purposes.

The selected object of study was the SS of composition $Na_{0.875}Li_{0.125}NbO_3$, situated at the rhombohedral (Rh) boundary of the MR of the $NaNbO_3 - LiNbO_3$ system [1, 2, 60]. This material exhibits exceptional values for piezoelectric parameters [2], making it an ideal candidate for investigation. A synthesis of the findings from the conducted studies and a comparison of the obtained results with the data on the modification of the PZT system in [5] led to the conclusion that the mechanism of heterovalent modification is the same in both PZT-type systems and in NN-based SS.

The formation of the complex of physical properties of modified SS is the mobility of domain walls, which is determined by SD and their interaction with vacancies, formed by the introduction of appropriate modifiers. In this case, the extent of parameter alteration is contingent upon the concentration of vacancies and the character of their distribution within the crystal structure. The relationship between the HF, stability of the domain structure to external influences [43], and the nature of chemical bonding in modified SS of the (Na, Li) NbO_3 system is like that observed in PZT-based SS systems [61]. This is due to a change in the degree of covalency of bonds, their orientation, and inequality. In Ref. [62], research was conducted on the modification processes of this system, with an emphasis on the introduction of elements more than stoichiometry. The modification of this system by specially prepared glasses of varying compositions is the subject of several works, including [7, 17, 63–66]. In [62, 67–70], the focus shifted to another component of the system, namely $LiNbO_3$. A comparative analysis of the potential for isomorphic substitution of atoms in alkali metal niobates and PZT-type SS revealed that cations exhibit a more extensive range of isomorphism for substitutions within sublattice *A*, whereas the PZT system displays a broader range of cation isomorphism within sublattice *B*. Furthermore, in alkali metal niobates, small-sized cations (Mg, Mn, Ni, Co, Zn, Sb, Sc, In) occupy *A*-positions in contrast to the PZT system. The wide isomorphism in alkali metal niobates is due to the (i) favorable size factor ($\Delta r/r_{\min} \leq 72–85\%$) for alkali cations, which favors isomorphic substitutions of atoms; (ii) favorable electronegativity factor, namely the large difference in EN for alkali cations (ΔEN = 77–140 kcal/g at) is a tolerably; (iii) crystallochemical features of their structure (presence of vacant octahedrons, high concentration of intrinsic defects, cation deficit non-stoichiometricity of alkali cations, etc.). Modified SS based on alkali metal niobates have found their application in a number of specific fields: high frequency piezo techniques [71–78] and devices, operating in the low and medium frequency range [79]; pyroelectric receivers [80]; electrets [81], frequency selective devices [82], including those operating on surface acoustic waves; highly sensitive piezoelectric transducers [83, 84], high temperature piezo techniques [85–90]. Vacancy modification of cations (controlled deviation from stoichiometry) was considered in [44, 91, 92]. In [44] it was found that with increasing deficiency of the alkali component, the T_C of alkali metal niobate-based materials decreases and the AFE increases, compared to stoichiometric compositions.

3.3 Conclusions

In this section, the modification schemes have been developed with discussion of the construction of quasichemical reactions of defect formation, and the creation of modification formulae for the solid solutions of (Na, Li) NbO_3 and $LiNbO_3$ systems. The stoichiometric modification of the material by the addition of donor and acceptor modifiers was evaluated to ascertain their influence on the structural, dielectric, ferroelectric, and mechanical characteristics of the material. The regularities of changes in the above properties at super stoichiometric modification of one (Ag)-, two (Cd)-, three (Al, Cr, Fe, La)-, four (Ti, Si, Mn)-, five (V)-, and six (U, W)-valence elements of sodium niobate-based solid solutions were investigated. The analysis of defect formation and types of order-disorder led to the proposal of a "model with packing disorder" that described the crystallochemical situation in solid solutions, based on (Na, Li) NbO_3. Finally, the influence of non-stoichiometry on the electrical conductivity of the obtained materials was studied.

References

1. Reznichenko, L.A., Shilkina, L.A.: Study of morphotropic regions in the system of solid solutions $NaNbO_3$–$LiNbO_3$. Ser. Phys. **39**(5), 1118–1121 (1975)
2. Reznichenko, L.A., Shilkina, L.A.: Extreme properties of niobate ceramics in the region of morphotropic transition. JTF **47**(5), 453–455 (1977)
3. Harate, K.H., Hennings, D.: Distribution of A-site and B-site vacancies in (Pb, La)(Ti, Zr)O_3 ceramics. J. Am. Am. Ceram. Soc. **55**(5), 230–234 (1972)
4. Titov, S.B., et al.: Influence of stoichiometry violations in the PZT system on the processes of phase formation, structure and properties of ceramics. In: Proc. Int. Sci. and Pract. Conf. Fundamental Problems of Piezoelectronics (Piezotechniques-95), vol. 2, pp. 5–12. Rostov-on-Don, Azov (1995) (in Russian)
5. Reznichenko, L.A.: Phase transitions and physical properties of solid solutions of 77-component systems based on sodium niobate. PhD Thesis (Physical and Mathematical Sciences), Rostov State University Press, Rostov-on-Don (1980) (in Russian)
6. Kuzminov, Yu.S.: Electro-optical and nonlinear-optical crystal of lithium niobate. Science, Moscow (1987). (in Russian)
7. Grineva, L.D.: Study of the patterns of changes in physical properties when modifying ferroelectric solid solutions of complex oxides with a perovskite-type structure. Ph.D. Thesis (Physical and Mathematical Sciences), Rostov State University Press, Rostov-on-Don (1975) (in Russian)
8. Fesenko, E.G.: The Perovskite Family and Ferroelectricity. Atomizdat, Moscow (1972). (in Russian)
9. Rao, C.N.R., Gopalakrishnan, J.: New Directions in Solid State Chemistry, vol. 520. Cambridge University Press, Cambridge (1990)
10. Chen, J., Feng, D.: *In situ* TEM studies of para-ferroelectric phase transitions in $NaNbO_3$. Phys. Stat. Sol (A) **109**, 427–434 (1988)
11. Akdogan, E.K., et al.: Origin of high piezoelectric activity in ferroelectric $(K_{0.44}Na_{0.52}Li_{0.04})(Nb_{0.84}Ta_{0.1}Sb_{0.06})O_3$ ceramics. Appl. Phys. Lett. **92**(11) (2008)
12. Kingery, W.D.: Introduction to Ceramics, 2nd edn. Stroyizdat, Moscow (1970). (in Russian)
13. Orowan, E.: Fracture and strength of solids. Rep. Prog. Phys. **12**, 185–232 (1949)
14. Krasovsky, A.Y.: Physical Foundations of Strength. Naukova Dumka, Kiev (1977). (in Russian)
15. Budnikov, P.P., Gistling, A.M.: Reactions in Mixtures of Solids, p. 488. Stroyizdat, Moscow (1971) (in Russian)
16. Reznichenko, L.A., Razumovskaya, O.N., Klevtsov, A.N.: On the manufacturability of manganese-containing ferropiezoceramics. In: Proc. Int. Sci. Pract. Conf. Fundamental Problems of Piezoelectric Instrumentation (Piezotekhnika-99), Rostov-on-Don, pp. 268-275 (1999) (in Russian)
17. Chernyshkov, V.A., et al.: Modification of lithium metaniobate ceramics. In: Proc. XII All-Union Conf. Physics of Ferroelectrics, vol. 3, p. 85. Rostov-on-Don (1989) (in Russian)
18. Fesenko, E.G., Danziger, A.Ya., Razumovskaya O.N. Dielectric and piezoelectric properties of solid solutions of multicomponent systems of complex oxides. Izv. USSR Ser. Inorg. Mater. **14**(5), 928–931 (1978) (in Russian)

19. Fesenko, E.G., et al.: New Piezoceramic Materials. Rostov State University Press, Rostov-on-Don (1983). (In Russian)
20. Jhomann, H.: Piezoelectric mechanism in bleizirconat-titanat. J. Angew. Phys. **20**(6), 554–557 (1966)
21. Devonshire, A.F.: Theory of barium titanate. Part. I. Phil. Mag. **40**(309), 1040–1063 (1949)
22. Reznichenko, L.A., et al.: Dependence of properties of modified solid solutions based on sodium niobate on the nature of chemical bonding and defectivity of the crystal structure. Inorg. Mater. **37**(10), 1254–1258 (2001)
23. Kröger, F.A., Vin, K.H.J.: Solid State Physics, vol. 3, p. 307. Academic Press, New York (1956)
24. Petrenko, A.G., Prisedsky, V.V.: Structure Defects in Ferroetoelectrics. UMK VO, Kiev (1989) (in Russian)
25. Prokopalo, O.I., Raevskiy, I.P.: Electrophysical Properties of Oxides of Perovskite Family, p. 103. Rostov State University Press, Rostov-on-Don (1985). (in Russian)
26. Trofimov, A.I.: Ultrasonic Systems for Monitoring Curvatures of Technological Channels of Nuclear Reactors. Energoatomizdat, Moscow (1994). (in Russian)
27. Härdtl, K.H., Rau, H.: PbO vapour pressure in the $Pb(Ti_{1-x})O_3$ system. Solid State Commun. **7**(1), 41–45 (1969)
28. Northrop, D.A.: Vaporization of lead zirconate–lead titanate materials. J. Am. Ceram. Soc. **50**(9), 441–445 (1967)
29. Northrop, D.A.: Vaporization of lead zirconate–lead titanate materials: II, Hot-pressed compositions at near theoretical density. J. Am. Amer. Ceram. Soc. **51**(7), 358 (1968)
30. Klimov, V.V.: Development of physicochemical bases for creation of new piezoelectric materials and methods of their preparation. Ph.D. Thesis (Physical and Mathematical Sciences), Donetsk (1973) (in Russian)
31. Zhukovsky, V.M., Petrov, A.I.: Introduction to Solid State Chemistry. Ural State University Press, Sverdlovsk (1978). (in Russian)
32. Jarzebski, Z.M.: Oxide Semiconductors, vol. 4, p. 285. Pergamon Press, Oxford (1973)
33. Bollmann, W.: The origin of photoelectrons and the concentration of point defects in $LiNbO_3$ crystals. Phys. Stat. Solid (a) **40**(1), 83–91 (1977)
34. Naidich, Yu.V., Lavrinenko, I.A., Eremenko, V.N.: The role of capillary phenomena in the process of compaction during sintering in the presence of a liquid phase. Powder Metall. **1**, 5–11 (1964). (in Russian)
35. Men, A.N., Vorobyev, Y.P., Chufarov, G.I.: Physico-chemical Properties of Non-stoichiometric Oxides. Khimiya, Moscow (1973). (in Russian)
36. Tretyakov, Yu.D.: Chemistry of Non-stoichiometric Oxides. Moscow State University Press, Moscow (1974)
37. Tretyakov, Yu.D.: Solid-Phase Reactions. Khimiya, Moscow (1979). (in Russian)
38. Reznichenko, L.A., Razumovskaya, O.N., Shilkina, L.A.: Aspects of the design of functional materials for piezotechnics based on alkali metal metaniobates. In: Proc. Int. Sci. Pract. Conf. Piezotechnics-97. Obninsk, pp. 191–196 (1997) (in Russian)
39. Reznichenko, L.A., Shilkina, L.A., Razumovskaya, O.N.: Secondary structural effects in polycrystalline ferro-, antiferroelectrics. In: Proc. XV All-Russian Conf. Physics of Ferroelectrics (VKS-XV), Rostov-on-Don (1999) (in Russian)
40. Shilkina, L.A., Reznichenko, L.A., Razumovskaya, O.N.: Structure modulation in antiferro-, ferroelectric Nb-containing complex oxides. In: 8th Int. Symp. Physics of Ferroelectric Semiconductors (IMFS-8), pp. 192-193. Rostov-on-Don (1998) (in Russian)
41. Takahashi, M.: Electrical resistivity of lead zirconate titanate ceramics containing impurities. Japan. J. Appl. Phys. **10**(5), 643 (1971)
42. Gerson, R., Jaffe, H.: Electrical conductivity in lead titanate zirconate ceramics. J. Phys. Chem. Solids **24**(8), 979–984 (1963)
43. Jaffe, B., et al.: Piezoelectric Ceramics. Academic Press, London, New York (1971)
44. Dambekalne, M.Y., et al.: Effect of modification on piezoelectric properties and texture of potassium and sodium niobate solid solution ceramics. Electron Technique. Materials **14**(8), 33-35 (1970) (in Russian)

45. Grineva, L.D., et al.: Influence of isovalent and heterovalent substitutions of ions in the lattice of solid solutions of alkali metal metaniobates on their physical properties. In: Proc. I All-Union Conf. Physical and Chemical Bases of Technology of Ferroelectric and Related Materials. Zelenograd (1980) (in Russian)
46. Reznichenko, L.A., Grineva, L.D., Razumovskaya, O.N.: Effect of modification on the properties of solid solutions based on alkali metal niobates. In: Proc. V Int. Conf. State and Prospects of Development of Methods for Obtaining and Analyzing Ferrite Ferro-, Piezoelectric and Capacitor Materials and Raw Materials for Them, pp. 87–88. Donetsk (1975) (in Russian)
47. Reznichenko, L.A., Grineva, L.D., Razumovskaya, O.N.: Study of the regularities of physical properties changes during the modification of solid solutions based on alkali metal niobates. In: Proc. IV All-Union Conf. State and Prospects of Development of Methods for Obtaining and Analyzing Ferrite, Ferropiezoelectric, Capacitor and Resistive Materials and Raw Materials for Them, vol. 1, pp. 62–72 (1981) (in Russian)
48. Grineva, L.D., et al.: Influence of isomorphic ion substitutions on processability and electrical properties of lithium metaniobate piezoceramics. In: Proc. II All-Union Conf. Physical and Chemical Bases of Technology of Ferroelectric and Related Materials. Zelenograd, p. 97 (1983) (in Russian)
49. Cross, E.: Lead-free at last. Nature **432**(4), 24–25 (2004)
50. Grineva, L.D., et al.: Investigation of the influence of modifiers on the mechanical strength of niobate piezoceramics. In: Proc. III All-Union Seminar. Strength of Materials and Elements of New Structures at Sound and Ultrasonic Loading Frequencies, pp. 37–38. Kiev (1981) (in Russian)
51. Reznichenko, L.A., et al.: Modification of solid solutions of alkali metal niobates. In: Proc. 7th Int. Seminar. Physics of Ferroelectrics-Semiconductors, vol 6, pp 168–169. Rostov-on-Don (1996) (in Russian)
52. Ananyeva, A. A., et al.: Piezo- and dielectric properties of modified potassium-sodium niobate ceramics. In: Methods for Obtaining and Studying Piezo- and Ferroelectric Materials and Feedstock for Them, pp. 89–95. MDNTP, Moscow (1966) (in Russian)
53. Ananyeva, A.A., Ugryumova, M.A.: Development and research of ferroelectric ceramics for electroacoustic transducers. In: Barium Titanate, pp. 201–207. Science, Moscow (1973) (in Russian)
54. Reznichenko, L.A., et al.: Dependence of properties of modified ferroelectric solid solutions from the nature of chemical bonding and defectivity of the structure. In: Proc. XV All-Russian Conf. Physics of Ferrooelectrics (VKS-XV), p. 235. Rostov-on-Don (1999) (in Russian)
55. Ugryumova, M.A.: Research and development of new piezoceramics based on solid solutions of barium and lead niobate, potassium and sodium niobate. Ph.D. Thesis (Technical Sciences), Moscow (1968) (in Russian)
56. Ugryumova, M.A.: Piezoceramics based on (K, Na)NbO_3 solid solutions. Electronic Technology. Materials (Inorganic Dielectrics) **14**(20), 130–138 (1969) (in Russian)
57. Freidenfeld, E.Zh.: Physico-chemical principles of synthesis and electrical properties of ceramic and glass-crystalline piezoelectrics. DrSc Thesis (Technical Sciences), Moscow (1972) (in Russian)
58. Yanson, G.D., et al.: Ceramic solid solutions of potassium and sodium metaniobates modified with alkaline earth metal ions. In: Piezo- and Ferroelectric Materials and Their Application, pp. 36–38. MDNTP, Moscow (1972) (in Russian)
59. Yanson, G.D., et al.: Kinetics of sintering of ferroelectric ceramics based on alkali niobates modified with oxides of alkaline earth elements. In: Proc. IV Intersectoral Meeting, Moscow, pp. 40–45 (1972) (in Russian)
60. Chernyshkov, V.A., Reznichenko, L.A., Shilkina, L.A.: Effect of modification on the quality and physical and mechanical characteristics of ceramic lithium metaniobate. In: Proc. I All-Union Council on Solid State Chemistry, Sverdlovsk. Pt., vol. 3, p. 149 (1985) (in Russian)
61. Raevskiy, I.P., Reznichenko, L.A., Panich, A.E.: Peculiarities of obtaining segmented ceramics based on alkali metal niobates by methods of hot pressing and glass-ceramic technology. In: Proc. of IX All-Union Council on Ferroelectricity, vol. 2, p. 133. Rostov-on-Don (1979) (in Russian)

62. Chernyshkov, V.A., Reznichenko, L.A., Baranov, V.V.: Electrical resistance and piezo properties of modified lithium metaniobate in a wide temperature range. In: Proc. II All-Union Conf. Actual Problems of Obtaining and Application of Ferropiezoelectric Materials and Their Role in Acceleration of Scientific and Technical Progress, Pt. 1, p. 202. MDNTP, Moscow (1984) (in Russian)
63. Fesenko, E.G., et al.: Piezoelectric ceramic material. USSR Patent No. 697463, 20.07.79 (1979) (in Russian)
64. Fesenko, E.G., et al.: Piezoelectric ceramic material. USSR Patent No. 968011, 30.04.81 (1981) (in Russian)
65. Fesenko, E.G., et al: Piezoelectric ceramic material. USSR Patent No. 1008198, 01.12.82 (1982) (in Russian)
66. Reznichenko, L.A., Chernyshkov, V.A.: Modification of lithium methaniobate. In: Proc. I Sci. Techn. Conf. Research in the Field of Powder Technology: Transfer Technologies, Complexes and Equipment, Pt. 2, pp. 189–190. Perm State Technical University Press, Perm (1993) (in Russian)
67. Fesenko, E.G., et al.: Piezoelectric ceramic material. USSR Patent No. 974771, 14.07.82 (1981) (in Russian)
68. Fesenko, E.G., et al.: Piezoelectric ceramic material. USSR Patent No. 1073220, 15.10.83 (1983) (in Russian)
69. Fesenko, E.G., et al.: Piezoelectric ceramic material. USSR Patent No. 1073221, 13.12.83 (1983) (in Russian)
70. Fesenko, E.G., et al.: Piezoelectric ceramic material. USSR Patent No. 1294791, 08.11.86 (1986) (in Russian)
71. Fesenko, E.G., et al.: Piezoelectric ceramic material. USSR Patent 587129, Invention Bulletin No. 1 (1978) (in Russian)
72. Fesenko, E.G., et al.: Piezoelectric ceramic material. USSR Patent 642272, Invention Bulletin No. 2 (1979) (in Russian)
73. Fesenko, E.G., et al.: Piezoelectric ceramic material. USSR Patent 694478, Invention Bulletin No. 40 (1979) (in Russian)
74. Fesenko, E.G., et al.: Piezoelectric ceramic material. USSR Patent No. 975673, 21.07.82 (1981) (in Russian)
75. Fesenko, E.G., et al.: Piezoelectric ceramic material (its variants). USSR Patent No. 1000439, 02.11.82 (1982) (in Russian)
76. Fesenko, E.G., et al.: Piezoelectric ceramic material. USSR Patent No. 1089075, 03.01.84 (1983) (in Russian)
77. Fesenko, E.G., et al.: Piezoelectric ceramic material USSR Patent No. 1096251, 08.02.84 (1983) (in Russian)
78. Fesenko, E.G., et al.: Piezoelectric ceramic material. USSR Patent No. 1294790, 08.10.86 (1986) (in Russian)
79. Fesenko, E.G., et al.: Piezoelectric ceramic material. USSR Patent No. 1258035, 15.05.86 (1986) (in Russian)
80. Fesenko, E.G., et al.: Piezoelectric ceramic material. USSR Patent No. 1264538, 07.01.85 (1985) (in Russian).
81. Fesenko, E.G., et al.: Piezoelectric ceramic material. USSR Patent No. 1274248, 01.08.86 (1984) (in Russian)
82. Fesenko, E.G., et al.: Piezoelectric ceramic material. USSR Patent No. 1436440, 08.07.88 (1987) (in Russian)
83. Fesenko, E.G., et al.: Piezoelectric ceramic material. USSR Patent No. 1439946, 22.07.88 (1988) (in Russian)
84. Raevsky, I.P., et al.: Study of the defect structure of single crystals and ceramics of sodium niobate. In: Crystallization and Properties of Crystals. Novocherkassk (4), pp. 76–81 (1977) (in Russian)
85. Anderson, J.S.: Nonstoichiometric compounds. Adv. Chem. **39**, 1–22 (1963)

86. Chernyshkov, V.A., et al.: Influence of composition non-stoichiometry and conditions of production on the properties of lithium ferropiezoceramic metaniobate. In: Proc. VIII All-Union Conf. State and Prospects of Development of Methods for Obtaining and Analyzing Ferrite, Ferropiezoelectric, Capacitor and Resistive Materials and Raw Materials for Them. Pt. 1. Donetsk, p. 102 (1983) (in Russian)
87. Lisitsyna, S.O., et al.: Effect of deviation from stoichiometry on sintering temperatures and electrophysical properties of polycrystalline alkali metal niobates. Ser. Inorg. Mater. **21**(6), 997–1000 (1985)
88. Schottky, W.Z.: On the mechanism of ion movement in solid electrolytes. Phys. Chem. B **29**, 355 (1935)
89. Schottky, W.Z., Wagner, C.Z.: Theory of ordered mixed phases. Phys. Chem. B **11**, 163 (1931)
90. Wagner, C.Z.: On the nature of the electrical conductivity of α-silver sulfide. Phys. Chem. B **22**, 181 (1933)
91. Wadsley, L.D.: Non-stoichiometric compounds. In: Inorganic Non-stoichiometric Compounds, pp. 102–200 (1971)
92. Bouillaud, Y.: Contribution a l'etude du systeme mNb_2O_5–nNa_2O. II. Domain m/n > 1. Bul. Soc. Chimiquede France **1**, 112-117 (1968) (in France)

Chapter 4
Structure, Grain Structure and Electrophysical Properties of Polycrystalline Solid Solutions of (Na, Li) NbO_3 System

In this section, we will examine the characteristics of the phase composition of SS based on the system (Na, Li, Sr) (Nb, Al) O_3 modified by various oxides. X-ray phase analysis revealed that the impurity content in the samples did not exceed 5%, except in the case of SnO_2-modified SS, where the total impurity content exceeded 20%. It was observed that all the obtained ceramics exhibited high density ($\rho_{rel.} > 90\%$), which contributed to the realization of maximum electrophysical parameters.

4.1 Polycrystalline Solid Solutions of Modified System (Na, Li) NbO_3

At room temperature, polycrystalline solid solutions of the modified system (Na, Li) NbO_3 exhibit a perovskite-type structure and are a mixture of monoclinic (M_2), rhombohedral (R_2) and cubic (C_2) phases. The indices at the phase designation indicate the multiplication of the perovskite cell axes (in the *M* phase, the multiplication of the *b*-axis).

Figure 4.1 shows the dependencies of the SS cell parameters of the (Na, Li) NbO_3 system on the modifier concentration. The modification was carried out according to exceed stoichiometry for MnO_2, CuO, MnO_2 + CuO and according to stoichiometry for MnO_2. The cell parameters changes with increasing modifier concentration (x), with the cell volume either increase (see Fig. 4.1a, c) or have non-monotonic behavior (see Fig. 4.1b). The most notable changes take place at $x = 0.03$. Upon modification with a mixture of MnO_2 + CuO oxides (see Fig. 4.1c), the Rh_2 phase is no observed in SS with $x = 0.03$. The indices near the symmetry designations are the multiplication of the perovskite cell axes, and in the *M* phase, the multiplication of the *b*-axis.

In the modified MnO_2 SS, the observed increase in the cell volume with increasing x can be attributed to the partial reduction and subsequent incorporation of Mn^{+3} ($R = 0.70$ Å) into the crystal lattice. This is because, in the case of incorporation of

I. A. Verbenko et al., *Environmentally Friendly Smart Materials with Special Electrical and Magnetic Properties*, Engineering Materials,
https://doi.org/10.1007/978-3-031-90387-8_4

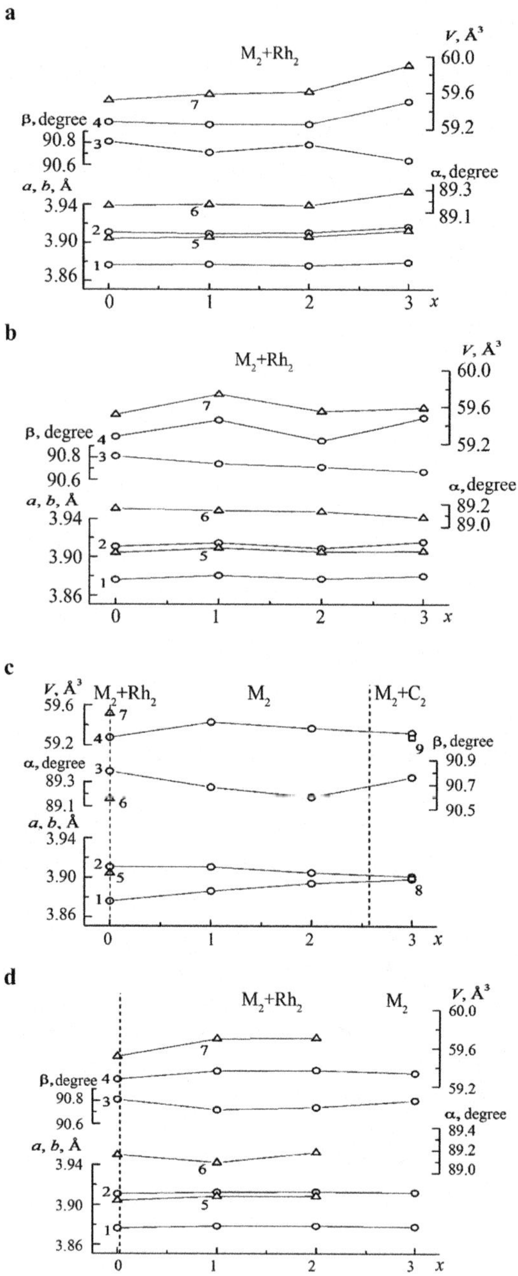

Fig. 4.1 **a** Parameters of SS based on (Na, Li) NbO_3 system: curves 1–4 correspond to b_M, $a_M = c_M$, β and V_M; curves 5–7 correspond to a_{Rh}, α and V_{Rh}, respectively; modification above stoichiometry of MnO_2. **b** Parameters of SS based on (Na, Li) NbO_3 system: curves 1–4 correspond to b_M, $a_M = c_M$, β and V_M; curves 5–7 correspond to a_{Rh}, α and V_{Rh}, respectively; modification above stoichiometry of CuO. **c** Parameters of SS based on (Na, Li) NbO_3 system: curves 1–4 correspond b_M, $a_M = c_M$, β and V_M; curves 5–7 correspond to a_{Rh}, α and V_{Rh}; curves 8, 9 correspond to a_C and V_C, respectively; modification beyond stoichiometry of MnO_2 + CuO. **d** Parameters of SS based on (Na, Li) NbO_3 system: curves 1–4 correspond b_M, $a_M = c_M$, β and V_M, curves 5–7 correspond to a_{Rh}, α and V_{Rh} respectively; stoichiometric modification of MnO_2

Mn^{+4} ($R = 0.52$ Å), which has a smaller ionic radius than Nb^{+5} ($R = 0.66$ Å), the cell volume should decrease. In CuO-modified SS (see Fig. 4.1b), the non-monotonic behavior of the cell volume is likely a consequence of the complex formation kinetics of Cu-containing alkali metal niobate-based SS.

The X-ray analysis revealed that the width of X-ray peaks exhibited a pronounced increase with the introduction of the modifier. Additionally, diffuse scattering in the region of their wings was observed to intensify, which provided indirect confirmation of the incorporation of manganese and copper into the crystal lattice of SS.

Significant number of planar defects (CSP) have been observed to facilitate the dissolution and agglomeration of dissociated ions. This is attributed to the formation of energetically non-equilibrium positions that are conducive to ion occupation. The segregation of such ions results in the appearance of diffuse scattering, concentrated near the main reflections, which is observed in the X-ray peaks. In contrast, the picture of stoichiometric modification of SS by MnO_2 oxide (see Fig. 4.1d) is different. Two morphotropic transitions are observed in the phase diagram: $M_2 + Rh_2 \rightarrow M_2 \rightarrow C_2$. At $x = 0.03$ (see Fig. 4.1d), the slight broadening of single lines indicates a very small distortion of the cubic lattice. Furthermore, superstructural reflections correspond to the doubling of the perovskite cell axes, which allows us to identify this phase as pseudocubic.

Figure 4.2 shows the dependence of cell parameters and the homogeneous strain parameter δ on NiO concentration. In the range of $0 \leq x \leq 1.0$, two phases coexist in SS-Rh and R(M_2). However, as x increases, the amount of R(M_2) decreases, and in the range of $1.0 < x < 2.0$, SS becomes practically single Rh phase. Figure 4.4 presents fragments of X-ray peaks, including diffraction lines $(111)_k$, $(200)_k$, and $(220)_k$, of the investigated SS.

Figure 4.2 shows that at the point $x = 1.5$, there is a notable decline in both the cell volume and the angle α: $\Delta V = -0.07$ Å^3, $\Delta\alpha = -0.06°$. For values of x greater than 1.5, the cell parameters exhibit minimal variation with increasing x, except for the volume, which displays a slight increase. The phase composition, however,

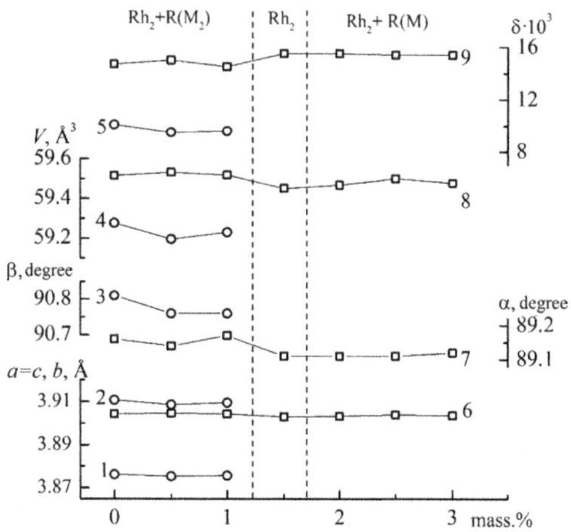

Fig. 4.2 Dependencies of SS structure parameters based on (Na, Li) NbO_3 modified by NiO on the modifier content: 1—b_M, 2—a_M, 3—angle β, 4—V_M, 5—δ_M, 6—a_{Rh}, 7—angle α, 8—V_{Rh}, 9—δ_{Rh}

changes in a gradual and smooth manner. Figure 4.2 shows that the intensity of P(M) phase lines increases gradually with increasing NiO concentration, indicating that SS again becomes biphasic. However, the amount of R(M) phase is much less than Rh. Therefore, at super stoichiometric modification of (Na, Li) NbO_3-based SS with nickel oxide, the latter is embedded in the crystal lattice of the initial SS, significantly changing its phase composition.

Figure 4.3 shows the dependencies of perovskite cell parameters and homogeneous strain parameter δ on the concentration of (Bi_2O_3 + Fe_2O_3). The X-ray data analysis indicates that within the range of $0 \leq x \leq 1.5$, the R(M_2) phase content and cell volume exhibit an increase with increasing x. Within the range of $1.5 \leq x \leq 2.0$, the transition to a single-phase R(M_2) structure is accompanied by a pronounced reduction in cell volume ($\Delta V = -0.13$ Å^3 at $x = 2.0$) and an increase in β and δ.

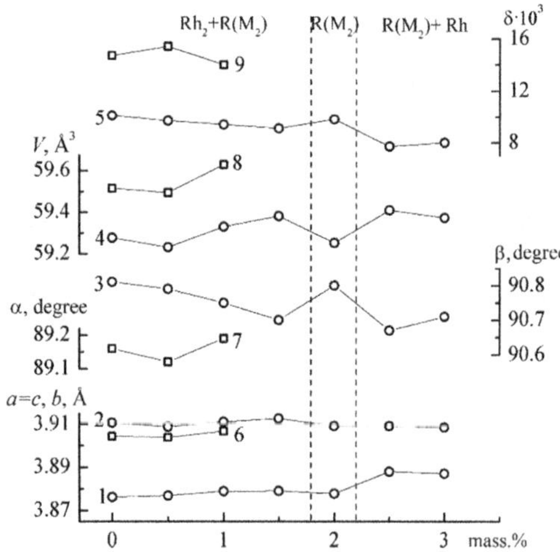

Fig. 4.3 Dependencies of SS structure parameters based on (Na, Li)NbO_3 modified by Bi_2O_3 + Fe_2O_3 on the modifier content: 1—b_M, 2—a_M, 3—angle β, 4—V_M, 5—δ_M, 6—a_{Rh}, 7—angle α, 8—V_{Rh}, 9—δ_{Rh}

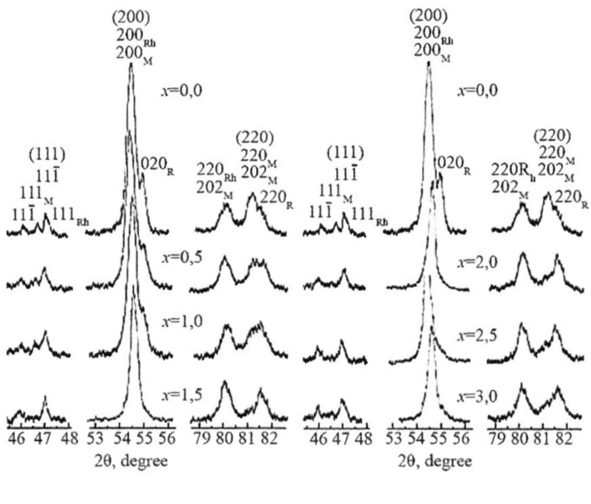

Fig. 4.4 Fragments of X-ray diffraction lines $(111)_k$, $(200)_k$, and $(220)_k$ of investigated SS based on (Na, Li)NbO_3, modified beyond the NiO stoichiometry

In the range of $2.0 \leq x \leq 2.5$, the same characteristics change dramatically again, but in the opposite direction. Specifically, the cell volume increases dramatically ($\Delta V = 0.16$ Å^3 at $x = 2.5$), while β and δ decrease. At $x \geq 2.5$, the parameters exhibit minimal change, yet pronounced diffuse maxima emerge in the wing region. These may be attributed to either disordered SS or the admixture of another phase with similar cell parameters. X-ray diffractograms (see Fig. 4.4) reveal that at $x = 3.0$, the SS once again become biphasic (the Rh phase appears).

Thus, when modifying the above stoichiometry with a $Bi_2O_3 + Fe_2O_3$ mixture of SS of the (Na, Li)NbO_3 system, the components of the latter, incorporated into the crystal lattice of the original SS, are observed at least up to 2.5 mass%.

4.2 Grain Structure of Ceramics of the Multicomponent System (Na, Li) NbO_3

Let us consider the distribution of modifiers in the volume of the obtained polycrystalline media using scanning electron microscopy. The image analysis of the spall and pore surface of the ceramic modified with 1.0 mass% CuO (Fig. 4.5) shows that this SS has a homogeneous microstructure [1]. The spall (see Fig. 4.5a) passes through the grains, which indicates the "cementing" character of intergranular interlayers. The packing of grains is dense, and the number of pores is not significant. Grain boundaries are clear, and the shape is nearly cubic (see Fig. 4.5b). The crystallite size is 2–4 μm.

As the concentration of CuO increases to 2.0 mass%, the thickness of intergranular interlayers rises, the clarity of grain boundaries decreases, and the perfection of grain faceting declines (Fig. 4.6) [1]. This phenomenon is likely attributable to the emergence of liquid phases and contact melting at the crystallite and intergrain interface. The grain size increases to a maximum of 5–7 μm.

The observed shape of grains, increase in their sizes, growth of intergranular interlayers, and traces of contact melting indicate that the studied polycrystalline material in the presence of a liquid phase is a typical phenomenon for copper-containing SS

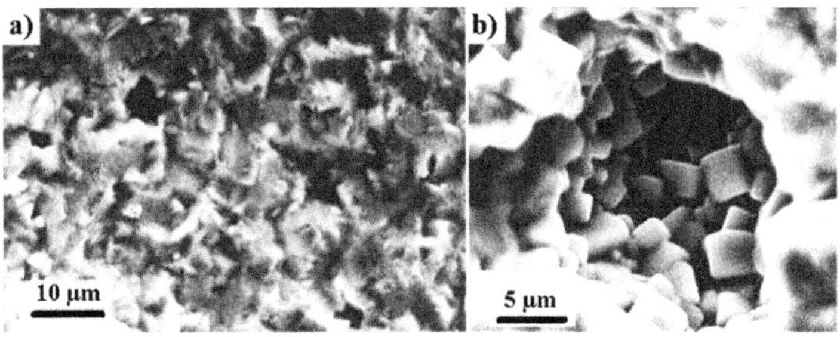

Fig. 4.5 Pictures of spall (**a**) and pore (**b**) surfaces of SS modified with 1.0 mass% CuO

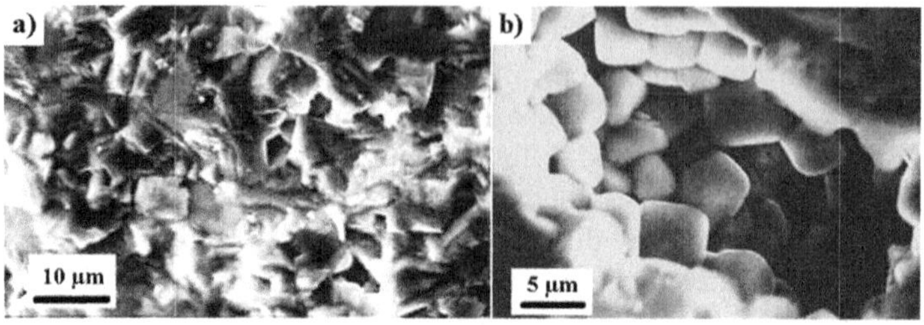

Fig. 4.6 Pictures of spall (**a**) and pore (**b**) surfaces of SS modified with 2.0 mass% CuO

based on alkali metal niobates. As the CuO content increases (Fig. 4.6), the degree of microstructure homogeneity rises, yet the character of chipping remains largely unchanged (see Fig. 4.6a). The crystallite structure of the ceramics modified with 1.0 mass% CuO, after polishing and etching, exhibited several distinctive features (Fig. 4.7) [1].

The semi-quantitative determination of Cu^{2+} ions on a chip of SS, modified with 1.0 mass% CuO (Fig. 4.7c), using a scanning electron microscope, revealed that the introduced Cu ions are predominantly located in the form of compact inclusions at the grain interface. The concentration of copper within the volume of grains is relatively low, with a maximum value of 0.05 vol%.

Therefore, the main part of copper atoms is not integrated within the crystalline lattice but rather exists as a phase that is unevenly distributed throughout the material's composition, which is in a liquid state during the sintering process within the intergranular layers, based on its observed location.

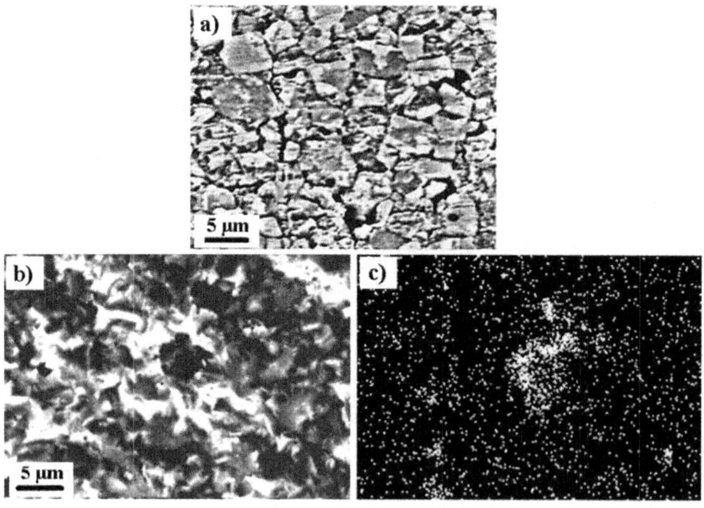

Fig. 4.7 Ceramic surface modified with 1.0 mass% CuO after polishing and etching (**a**), chip without etching (**b**) and Cu^{2+} distribution (**c**) on a SS chip, modified with 1.0 mass% CuO

As noted previously, the synthesis of Cu-containing niobate ceramics allows one to produce SS with a minimum content of impurity phases when pre-synthesized $CuNb_2O_7$ is employed. This suggests that the incorporation of copper atoms into the lattice of alkali metal niobates with perovskite-type structure is subject to significant kinetic difficulties, which are probably related to the formation of liquid phases.

The analysis of the surface of spalls and pores of ceramics, modified by 1.0 mass% in secondary electrons in a scanning electron microscope (Fig. 4.8a) reveals a homogeneous microstructure, although in some areas there is a notable dispersion of grain sizes [1]. The spall runs along the grain, exhibiting dense grain packing, indistinct grain boundaries, and non-ideomorphic shape. Textures are visually absent (see Fig. 4.8b), and grain sizes measuring between 1 and 7 microns.

As the modifier content increases to 2.0 mass% MnO_2 (Fig. 4.9), a greater number of small pores appear, the nature of spalling remains unchanged (see Fig. 4.9a), yet the width of intergranular interlayers increases, and the shape of grains remains non-ideomorphic, the grain size ranges from 1 to 15 μm [1]. The grain structure of SS modified by MnO_2 does not correspond to that formed during sintering in the presence of a liquid phase. However, the role of Mn atoms is evident in the formation of wide and stronger interlayers between crystallites, which are wider and stronger than the grains of the main phase.

Fig. 4.8 Chip surfaces (**a**) and pore surfaces (**b**) of ceramic SS modified with 1.0 mass% MnO_2

Fig. 4.9 Chip surfaces (**a**) and pore surfaces (**b**) of ceramic SS, modified with 2.0 mass% MnO_2

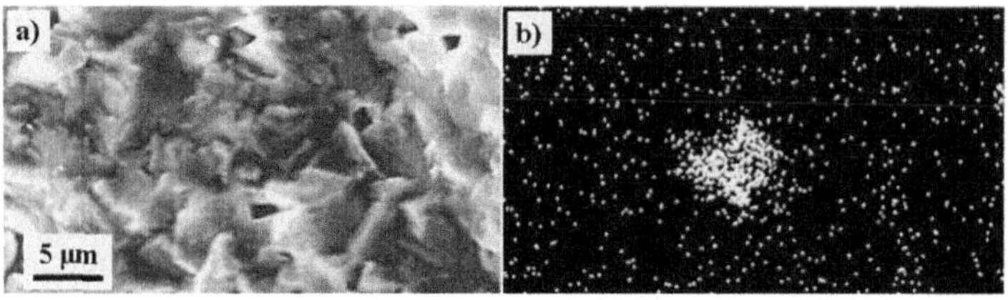

Fig. 4.10 Chip (**a**) and Mn distribution (**b**) on the chip of SS, modified with 1.0 mass% MnO_2

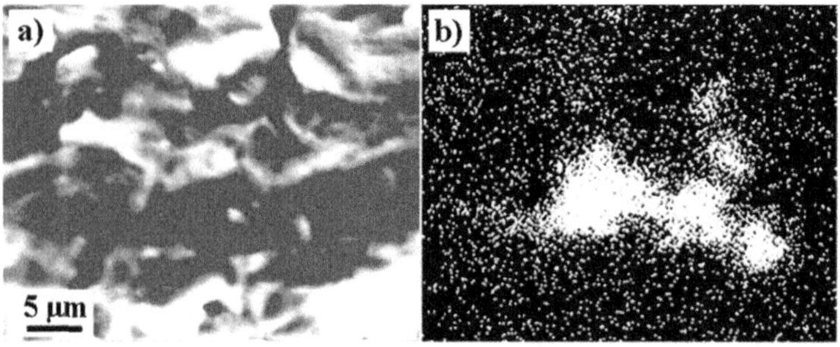

Fig. 4.11 Chip (**a**) and Mn distribution (**b**) on the chip of SS modified with 2.0 mass% MnO_2

Semi-quantitative determination of Mn on the chips of SS, modified by 1.0 mass%, when analyzing the elemental composition of grains after polishing and etching with hydrochloric acid, has established that Mn is absent on the surface of ceramic grains of these SS (Fig. 4.10), and in SS, modified by 2.0 mass% Mn content (Fig. 4.11) is 0.1 mass% [1]. As in the case of Cu, Mn ions are predominantly located in intergranular interlayers, but, unlike Cu, Mn is localized not in pores, but on dense intergranular interfaces.

The detection of Mn ions on the surface of the etched grains suggests their partial incorporation into the crystalline structure of the material. The analysis of Fig. 4.12 reveals that this ceramic exhibits an exceptionally heterogeneous microstructure, comprising a multitude of minute pores, but the observed chipping behavior remains consistent [1].

Fuzzy grain boundaries are observed. The size of crystallites does not exceed 1 micron and does not allow one to reveal the grain structure of polycrystalline objects even at maximum magnification. Increasing the modifier content up to 1.5 mass% leads to an increase in the grain size up to 2–15 µm, their faceting improves, while the grains have a non-ideomorphic shape and vary greatly in size. Further modification, up to 2.0 mass% of modifier, results in increased microstructure inhomogeneity and blurring of grain boundaries.

The analysis of fragments of polycrystalline samples, modified with 1.5 mass% (Fig. 4.12a) and 2.0 mass% (Fig. 4.12b) of the combined modifier, after grinding and etching, revealed the presence of discrete compact formations with a composition

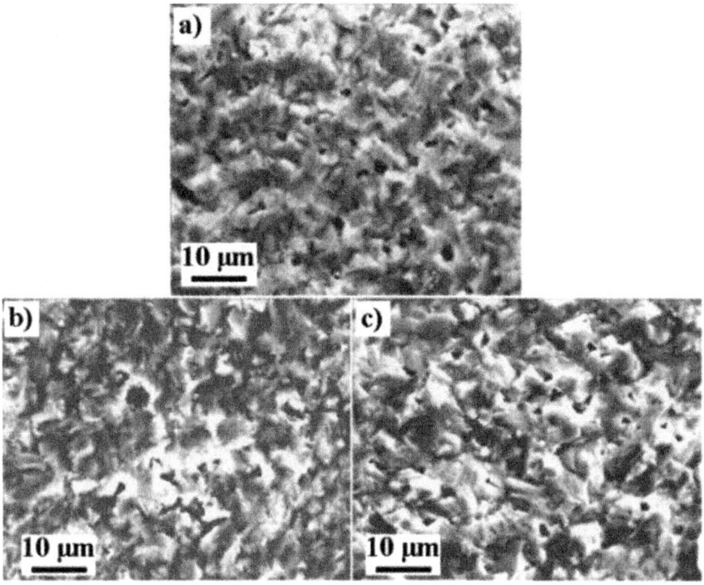

Fig. 4.12 Surface chips of SS, modified with combined modifier ($Bi_2O_3 + Fe_2O_3$) in the amount of 1.0 mass% (**a**), 1.5 mass% (**b**) and 2.0 mass% (**c**)

closely resembling that of Fe_2O_3. These observations suggest that the formation of these phases may be attributed to the generation of Bi-containing liquid phases.

Thus, polycrystalline materials based on alkali metal niobates possess a complex hierarchical structure, enabling the distribution of introduced modifiers within the volume of grains of the basic phase and at intergranular boundaries. Such a distribution of modifying additives serves to highlight the quasi-equilibrium state of the objects in question, as well as their dependence on the thermodynamic prehistory.

4.3 Electrophysical Properties of SS of the (Na, Li) NbO_3 System

The solid solutions, based on (Na, Li) NbO_3, modified with MnO_2, CuO, and NiO, exhibit the following characteristics [1]: relatively low $\varepsilon_{33}^T/\varepsilon_0$ and experimental density (Figs. 4.13, 4.14, 4.15, 4.16, 4.17, 4.18 and 4.19), which are typical of alkali metal niobate-based materials [2–5].

The relatively low values of tanδ indicate that the real part of the dielectric permittivity plays a significant role in the formation of the observed dependencies and low through conductivity.

The use of CuO, NiO and MnO_2 as modifiers leads, in the case of the corresponding SS, to an increase in Q_M: by 24%, at 1.0 mass% CuO; by 17%, at 1.0 mass% NiO;

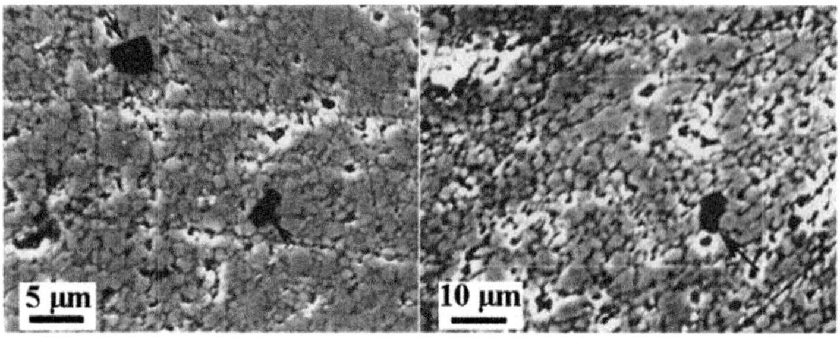

Fig. 4.13 Chip surface of SS modified with 2.0 mass% (Bi_2O_3 + Fe_2O_3)

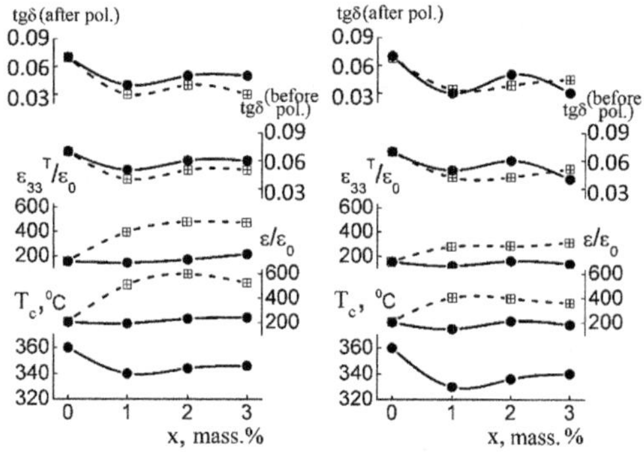

Fig. 4.14 Dependencies of dielectric characteristics of SS, based on (Na, Li) NbO_3, modified by MnO_2: T_{sint} = 1180 °C (left), T_{sint} = 1200 °C (right); solid lines correspond to the samples of Ø 10 mm, dotted lines show the samples of Ø 20 mm

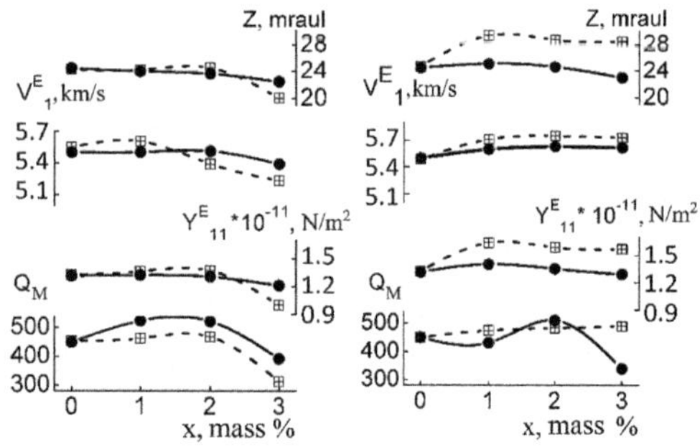

Fig. 4.15 Dependencies of elastic characteristics of SS, based on (Na, Li) NbO_3, modified with MnO_2 at sintering temperatures: T_{sint} = 1180 °C (left), T_{sint} = 1200 °C (right); solid lines correspond to the samples of Ø 10 mm, dotted lines show the samples of Ø 20 mm

by 24%, at 2.0 mass% MnO_2, and to an increase in d_{33} with a decrease in $\varepsilon_{33}^T/\varepsilon_0$, which leads, in turn, to an increase in g_{33}: by 3%, at 1.0 mass% CuO; by 45%, at 0.5 mass% NiO; by 25%, at 2.0 mass% MnO_2, and to an increase in piezoanisotropy d_{33} with a decrease in $\varepsilon_{33}^T/\varepsilon_0$, and the increase of piezoanisotropy $d_{33}/|d_{31}|$: by 18%, at 1.0 mass% CuO; by 24%, at 1.0 mass% NiO; by 46% at 1.0 mass% MnO_2 [1].

The experimental results observed can be explained by considering the following factors. The formation of electrophysical properties in polycrystalline media, based on alkali metal niobates, as previously noted, is significantly influenced by the

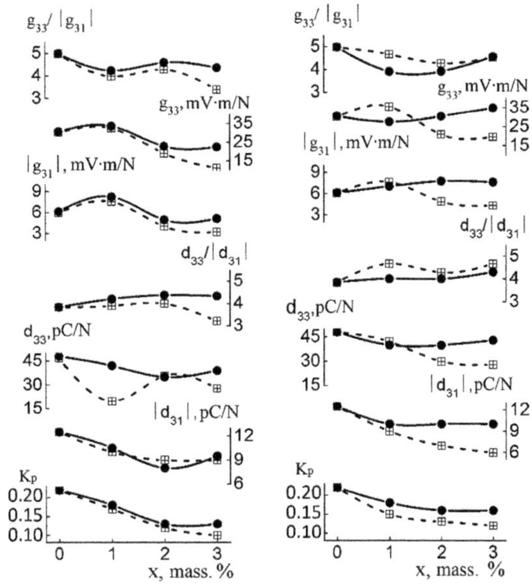

Fig. 4.16 Dependencies of piezoelectric characteristics of SS, based on (Na, Li) NbO_3, modified with MnO_2 at sintering temperatures: T_{sint} = 1180 °C (left), T_{sint} = 1200 °C (right); solid lines correspond to the samples of Ø 10 mm, dotted lines show the samples of Ø 20 mm

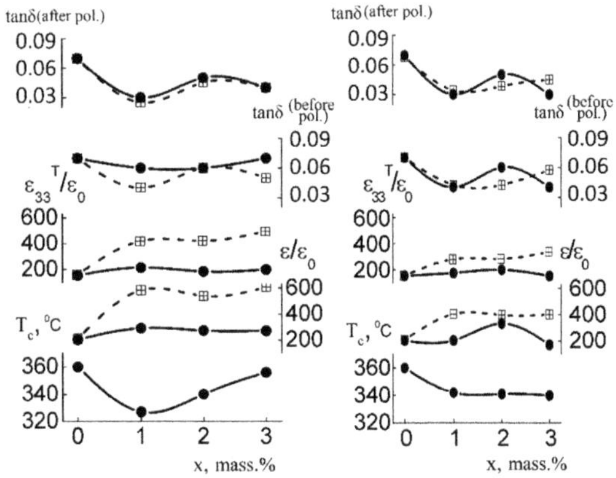

Fig. 4.17 Dependencies of dielectric characteristics of SS based on (Na, Li) NbO_3, modified by CuO: T_{sint} = 1180 °C (left), T_{sint} = 1200 °C (right); solid lines correspond to the samples of Ø 10 mm, dotted lines show the samples of Ø 20 mm

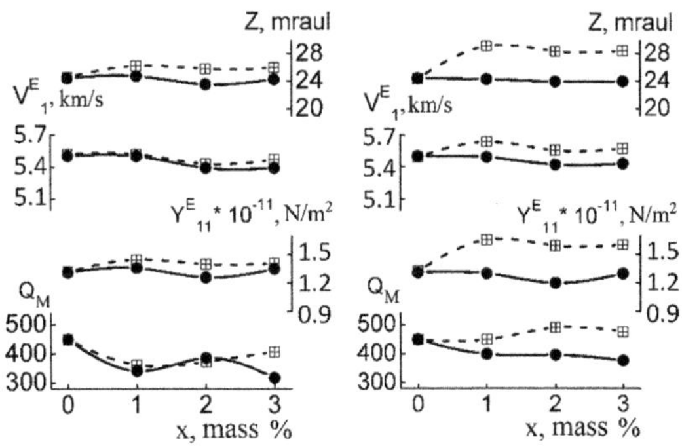

Fig. 4.18 Dependencies of elastic characteristics of SS based on (Na, Li) NbO_3, modified with CuO at sintering temperatures: T_{sint} = 1180 °C (left), T_{sint} = 1200 °C (right); solid lines correspond to the samples of Ø 10 mm, dotted lines show the samples of Ø 20 mm

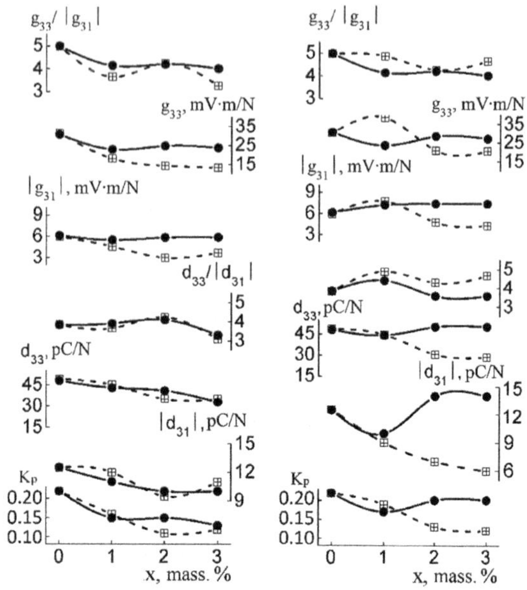

Fig. 4.19 Dependencies of piezoelectric characteristics of SS based on (Na, Li) NbO_3, modified with CuO at sintering temperatures: T_{sint} = 1180 °C (left), T_{sint} = 1200 °C (right); solid lines correspond to the samples of Ø 10 mm, dotted lines show the samples of Ø 20 mm

complex kinetics of the formation of these SS, which, typically results in an increase in the through conductivity of ceramics, a decrease in their polarization efficiency, and a rapid degradation of their electrophysical characteristics.

The electrophysical parameters of ceramics, obtained at different $T_{sint.}$ and topology (Figs. 4.14 and 4.20), show that the electrophysical characteristics of ceramics vary inconsistently with composition. This is particularly evident in the values, related directly to electrical conductivity, including $\varepsilon_{33}^T/\varepsilon_0$, tanδ, and Q_M.

Furthermore, the electrophysical parameters of FPCMs are dependent upon the compactness of the microstructure of polycrystalline samples, which are primarily

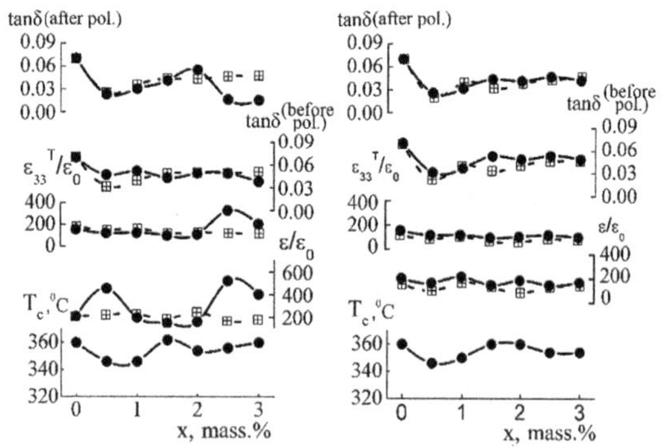

Fig. 4.20 Dependencies of dielectric characteristics of SS based on (Na, Li) NbO_3, modified by NiO: T_{sint} = 1190 °C (left), T_{sint} = 1205 °C (right); solid lines correspond to the samples of Ø 10 mm, dotted lines show the samples of Ø 20 mm

characterized by experimental density. This large density determines both the dielectric properties and the piezoelectric properties, particularly the piezomodulus d_{33} of SS due to the reduction of electromechanical losses (Q_M growth).

The presented dependencies effectively demonstrate two mechanisms of g_{33} increase: (i) due to a decrease in electrical conductivity, namely a reduction in $\varepsilon_{33}^T/\varepsilon_0$, and (ii) due to an increase in experimental density and, consequently, an increase in d_{ij}. The introduced modifiers are responsible for both mechanisms.

In this section, we will examine the impact of various cations. The geometrical conditions of existence of the perovskite-type structure [6] and the closeness of the radii of interchangeable ions allow for the incorporation of Cu^{2+} and Ni^{2+} cations into the A-positions and Mn^{4+} into the B-positions of the ABO_3 crystal lattice (Table 4.1) [7]. In accordance with the electroneutrality rule, heterovalent substitution gives rise to the realization of the following schemes: $(Na_{1-x-y}Li_xM_{2+y})NbO_{3+y/2}$ (anionic excessive SS, where M is Cu or Ni) or $(Na_{1-x}Li_x)(Nb_{1-y}\ M_{4+y})O_{3-y/2}\square_{y/2}$ (anion-deficient SS, where M is Mn; □ is a vacancy). In the case of incorporation more highly charged Cu^{2+} and Ni^{2+} cations at the A-position instead of Na^+ and Li^+ cations, the formation of anionic excessive SS occurs, in which the excessive oxygen atoms can be located either in the internodes or in positions displaced from the internodes.

In the latter case, this situation constitutes a defect subsystem, which includes, among other elements, Willis clusters of varying types. These are characterized by combinations of vacancies and different types of internodes. This results in the emergence of local ordering effects, which may be described as near order. It should be noted that excessive anions tend to accumulate mainly on specific crystallographic planes. In the case of alkali metal niobates, these accumulations can be localized on the CSP. The presence of O^{2-} anions within the internodes has the potential to generate a dipole moment, which may result in further distortion of atomic groups.

Considering the documented proclivity of niobate ceramics to self-organize defect structures [1] through ordering and microdomain formation, it is reasonable to infer that these structures exert an influence on the total dipole moment of the system, the degree of HF, and, consequently, the values of the coercive force, dielectric permittivity, piezosensitivity, and mechanical goodness of fit. The findings of the experimental investigation indicate that the degree of HF is elevated in the SS specimens of CuO- and NiO modifications. The analysis of Figs. 4.18, 4.19, 4.20, 4.21 and 4.22 also demonstrates that the maximum values of d_{33} and QM are reached at the modifier content of 0.5–1.0 mass% NiO and then exhibit a continuous decline with increasing NiO concentration. Then at CuO modification the corresponding values exceed those at the NiO modification maximums at 2.0–3.0 mass%. It is evident that

Table 4.1 Ionic radius of some ions, included in the composition of SS bases and modifiers

Cation	Mn^{4+}	Fe^{3+}	Nb^{5+}	Li^+	Ni^{2+}	Cu^{2+}	*$(Na_{0.86725}\ Li_{0.12475})^+$	Na^+
Ionic radius, Å	0.52	0.57	0.66	0.68	0.74	0.80	0.93	0.98

this behavior can be attributed to a shift in sintering kinetics, namely, the transition from solid-phase sintering to sintering in the presence of LP [8].

The introduction of liquid phases into intercrystalline layers can occur at the formation of these phases. The formation of insulating layers, which occurs because

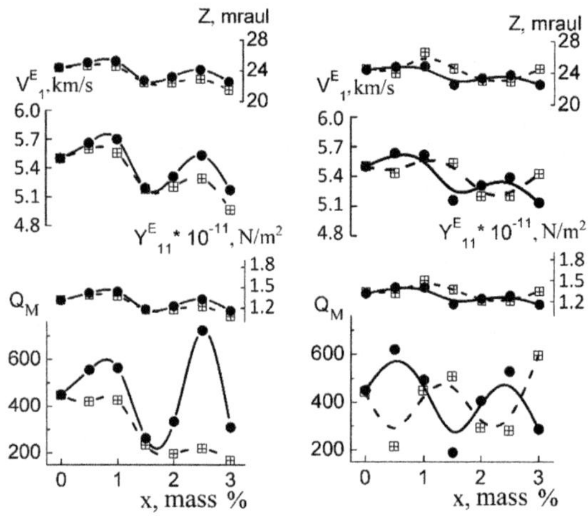

Fig. 4.21 Dependencies of elastic characteristics of SS based on (Na, Li) NbO_3, modified with NiO at sintering temperatures: T_{sint} = 1190 °C (left), T_{sint} = 1205 °C (right); solid lines correspond to the samples of Ø 10 mm, dotted lines show the samples of Ø 20 mm

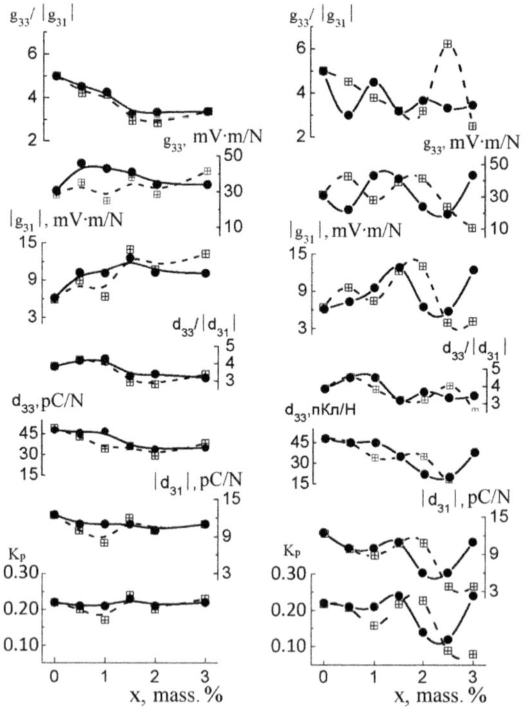

Fig. 4.22 Dependencies of piezoelectric characteristics of SS based on (Na, Li) NbO_3, modified with NiO at sintering temperatures: T_{sint} = 1190 °C (left), T_{sint} = 1205 °C (right); solid lines correspond to the samples of Ø 10 mm, dotted lines show the samples of Ø 20 mm

of this introduction, leads to a reduction in electrical conductivity. This mechanism is also confirmed by the increase in the $d_{33}/|d_{31}|$ ratio, in which the attenuation of piezoelectric oscillations in the plane perpendicular to the texture in the result of the formation of inelastic intergranular interlayers plays a significant role. Among the considered modifiers, manganese is the most highly charged and small-sized (Table 4.1) [7]. It has the potential to compete with niobium for the location in the *B*-position of the perovskite structure.

Thus, upon introduction of MnO_2, as illustrated in the above scheme, we can postulate the formation of anion-deficient structures, which is accompanied by the formation of vacancies in the oxygen sublattice and disruption of the symmetrical anionic environment of the *B* cation. This results in an increase in the dipole moment and, consequently, an increase in the segmental rigidity of ceramics, Q_M and g_{33}, as well as a decrease in $\varepsilon_{33}^{\mathrm{T}}/\varepsilon_0$.

The charge instability, associated with the manganese valence states, represents an additional source of vacancies, with the potential for reduction to Mn^{3+}. In this instance, the substitution scheme will assume the following form $(Na_{1-x}Li_x)(Nb_{1-y}M_{3+y})O_{3-y}\square_y$ (anion-deficient SS, $M = \mathrm{Mn}$; $\square$ is a vacancy).

A reduction in the oxidation degree of Mn to 3 results in the occurrence of the Jahn–Teller effect, which also contributes to the growth of the HF of the modified materials [9]. The introduction of Mn-containing phases into intergranular interlayers results in the formation of strong bonds between grains, effectively inhibiting their movement. Additionally, the development of textures in these phases promotes the growth of anisotropy in piezoelectric coefficients.

Furthermore, the crystallization of fine-grained structures with a reduced number of domains results in a decrease in the mobility of the latter. This is due to the action of a large field of volume charge, which shields the ferroelectric polarization, leading to an increase in HF. Moreover, the presence of oxygen vacancies enhances the sinter ability of ceramics, increases the density of polycrystalline materials, and, because of the reduction in electromechanical losses, increases the piezomodulus d_{33}.

The main electrophysical characteristics of materials based on the (Na, Li) NbO_3 system, modified with combined modifiers, are presented in Figs. 4.23, 4.24, 4.25, 4.26, 4.27 and 4.28. It is notable that the effects, observed when modifying SS with simple MnO_2 and CuO oxides, are also manifested, when they are jointly introduced into SS based on (Na, Li) NbO_3. Furthermore, the maximum piezoanisotropy ($d_{33}/|d_{31}| = 5.1$) can be achieved precisely when using the combined modifier ($0.02MnO_2 + 0.01CuO$). The effects resulting from the utilization of such modifiers are, in numerous aspects, analogous to those that emerge in the design of multicomponent systems. The introduction of the combined modifier ($Bi_2O_3 + Fe_2O_3$) results in nonlinear changes to most of the material characteristics. The region between $x = 1.5$ and $x = 2.0$ mass% is characterized by the presence of extreme values for the observed properties. The observed anomalies are likely associated with either the morphotropic phase boundary or the characteristics of the modifier incorporation into the matrix structure. The values of electrophysical characteristics: $Q_M > 700$, $K_p(\sim 0.2)$, $\varepsilon_{33}^{T}/\varepsilon_0 \sim 160$, and ~ 5.6 km/s at $x = 1.5$ mass% indicate that these SS are

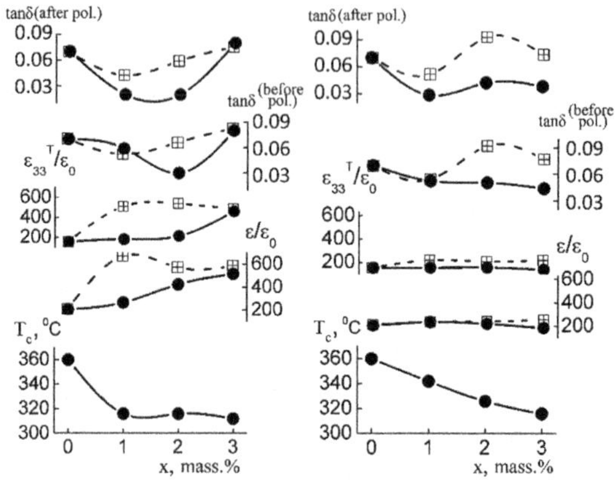

Fig. 4.23 Dependencies of dielectric characteristics of SS based on (Na, Li) NbO_3, modified by MnO_2 + CuO: T_{sint} = 1180 °C (left), T_{sint} = 1200 °C (right); solid lines correspond to the samples of Ø 10 mm, dotted lines show the samples of Ø 20 mm

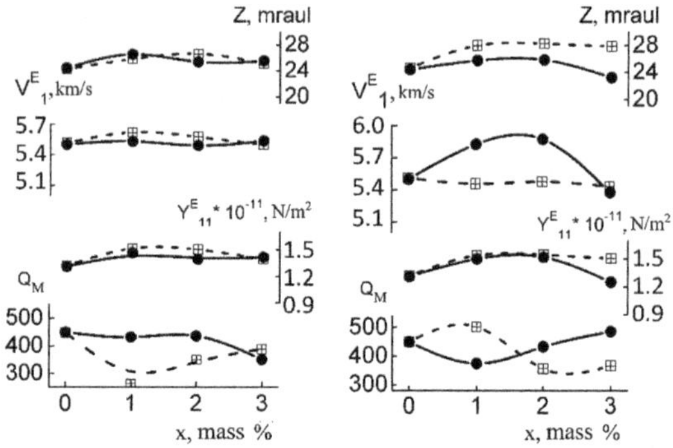

Fig. 4.24 Dependencies of elastic characteristics of SS based on (Na, Li) NbO_3, modified with MnO_2 + CuO at sintering temperatures: T_{sint} = 1180 °C (left), T_{sint} = 1200 °C (right); solid lines correspond to the samples of Ø 10 mm, dotted lines show the samples of Ø 20 mm

promising for applications in power high-frequency electromechanical transducers, such as those, used in ultrasonic logging of oil wells [2].

During the work, it was discovered that the potential for reducing the electrical conductivity of SS (reduction of tanδ) exists not only in the modification of such objects, but also in the change of the technological regulations for their production. Thus, in SS based on (Na, Li) NbO_3 with 2 mass% MnO_2, $\varepsilon/\varepsilon_0$ and $\varepsilon_{33}^T/\varepsilon_0$ decrease by one and a half times, at increasing $T_{sint.}$ from 1180 ºC to 1200 ºC, indicating a smaller contribution of electrical conductivity to these values. Modification with other modifiers (CuO, NiO) further enhances this effect.

The analysis of the dependencies of electrophysical parameters on the content of modifiers (Figs. 4.14, 4.15, 4.16, 4.17, 4.18, 4.19, 4.20, 4.21, 4.22, 4.23, 4.24, 4.25, 4.26, 4.27 and 4.28) also demonstrates that, although the general shape of the obtained curves by changing $T_{sint.}$ remains consistent, the sintering mode has a significant impact on the values of electrophysical parameters for individual compositions. Usually, the attainment of elevated piezoelectric characteristics is contingent upon an increase in $T_{sint.}$ The impact of elevated T_{sint} can be manifested in the establishment of conditions that are more closely aligned with equilibrium for phase formation,

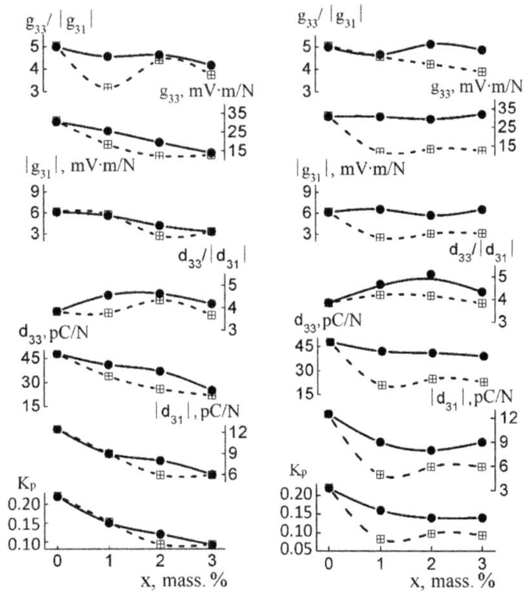

Fig. 4.25 Dependencies of piezoelectric characteristics of SS based on (Na, Li) NbO_3, modified with MnO_2 + CuO at sintering temperatures: T_{sint} = 1180 °C (left), T_{sint} = 1200 °C (right); solid lines correspond to the samples of Ø 10 mm, dotted lines show the samples of Ø 20 mm

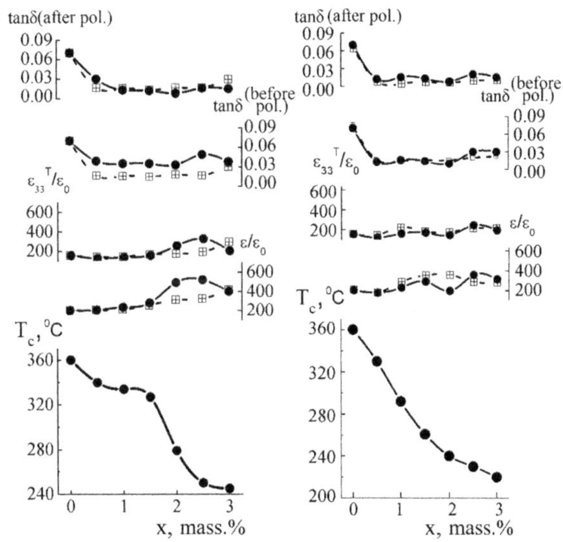

Fig. 4.26 Dependencies of dielectric characteristics of SS based on (Na, Li) NbO_3, modified by Bi_2O_3 + Fe_2O_3: T_{sint} = 1190 °C (left), T_{sint} = 1205 °C (right); solid lines correspond to the samples of Ø 10 mm, dotted lines show the samples of Ø 20 mm

the acceleration of diffusion processes, an expansion in grain size, the formation of bimodal structures, and, therefore, an enhancement in the piezoelectric response.

In the case of most samples, an increase in diameter results in an increase in tanδ and $\varepsilon/\varepsilon_0$, as well as a decrease in piezoelectric activity. This phenomenon is associated with the following factors. Niobate ceramics are characterised by the phenomenon

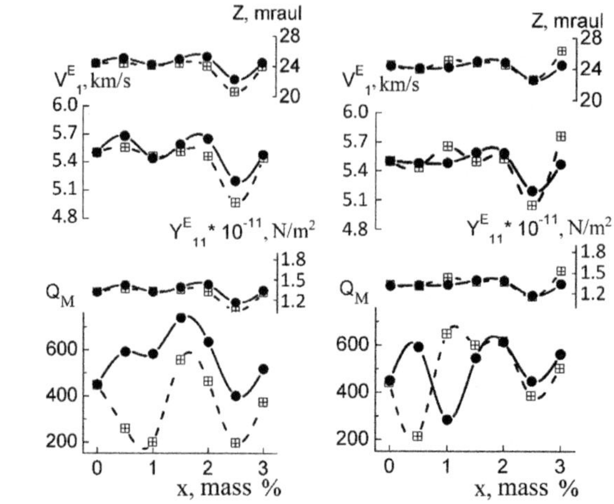

Fig. 4.27 Dependencies of elastic characteristics of SS based on (Na, Li) NbO_3, modified by $Bi_2O_3 + Fe_2O_3$ at sintering temperatures: T_{sint} = 1190 °C (left), T_{sint} = 1205 °C (right); solid lines correspond to the samples of Ø 10 mm, dotted lines show the samples of Ø 20 mm

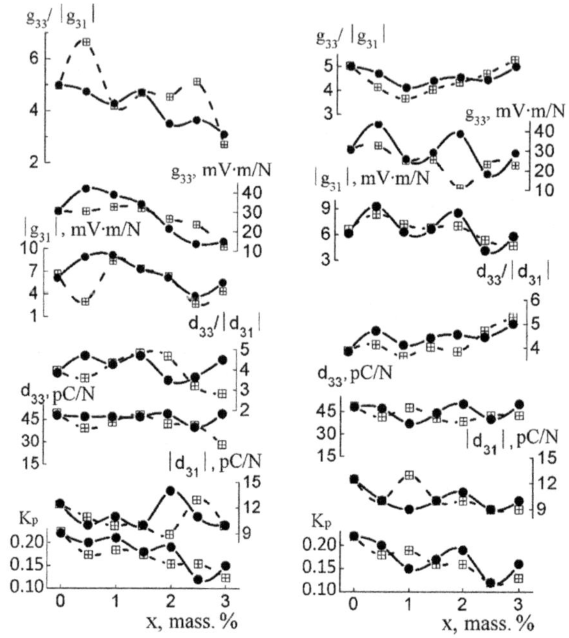

Fig. 4.28 Dependencies of piezoelectric characteristics of SS based on (Na, Li) NbO_3 modified by $Bi_2O_3 + Fe_2O_3$ at sintering temperatures: T_{sint} = 1190 °C (left), T_{sint} = 1205 °C (right); solid lines correspond to the samples of Ø 10 mm, dotted lines show the samples of Ø 20 mm

of sublimation of volatile compounds of alkali metals, which results in a violation of stoichiometry and a complex defect situation in the objects. Given that sublimation is a heterogeneous process, and its speed is directly related to the interface, it is evident that the greatest number of defects will accumulate on the top and side surfaces of the billet. Consequently, an increase in the volume of the sample will result in a proportional increase in the degree of defects characteristic of the near-surface layer.

Therefore, in the case of larger diameter samples, the observed increase in overall electrical conductivity is likely associated with the formation of a more defective and wider near-surface layer, which contributes to the high through-conductivity. In consideration of the aforementioned factors, it becomes evident that the impact of sample morphology on the depth and degree of defectivity of the near-surface layer should be considered during the manufacturing process, particularly in the case of mechanical processing of piezoelectric elements based on niobate ceramics.

4.4 The Dielectric Spectra of Solid Solutions of the (Na, Li) NbO_3 System

The modified samples by MnO_2 and CuO are shown in Fig. 4.29. The analysis of the above dependencies allowed us to identify the qualitative and quantitative compositions of the modifiers that decrease electrical conductivity and increase temperature stability of electrophysical characteristics [10].

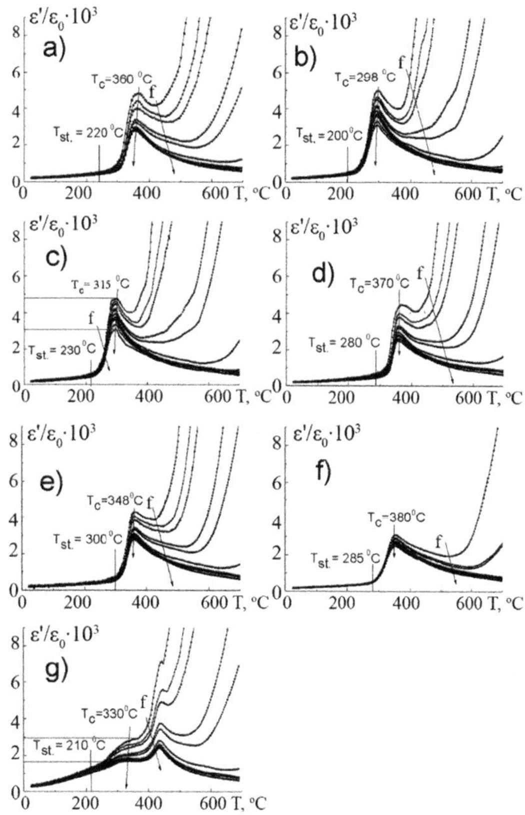

Fig. 4.29 Dependencies of $\varepsilon'/\varepsilon_0(T)$ of ceramics based on (Na, Li) NbO_3 system, not modified (**a**) and modified by 1 (**b**), 2 (**c**), 3 (**d**) mass% MnO_2 and modified by 1 (**e**), 2 (**f**), 3 (**g**) mass% CuO; the arrows indicate the directions of growth *f* and change in T_C

As the concentration of MnO_2 is increased, a notable decline in T_C is observed from 360 to 298 °C (at $x = 1$ mass%), followed by a gradual rise to 370 °C, at $x = 3$ mass%. This is accompanied by a decrease in the dependence of the T_C position on f, which shifts to the right with increasing frequency in SS with low modifier content. A non-monotonic change in T_C with a minimum in the region of small concentrations of introduced additives was previously observed, for instance, in Ref. [2]. This phenomenon was explained by the emergence of crystallochemical disorder in SS, which is associated with the difficulty of uniform dissolution (and entry into the crystal lattice of the "host") of introduced modifying additives.

In reference [11], the thermal-frequency behavior of SS (Na, Li) NbO_3 was examined, and it was discovered that the observed anomalous relaxation of T_C (with a shift to the low-temperature region) is most frequently manifested in alkali metal niobates and is a consequence of the increased low-frequency conductivity of objects at high temperatures. The independence of T_C from f in the region of large concentrations of MnO_2 indicates a decrease in the thorough conductivity in these compositions, as evidenced by the decrease in the dispersion depth. The increase of the temperature stability ($T_{st.}$) region $\varepsilon/\varepsilon_0$ is noticeable: the upper limit of this region is $T_{st.}$ ~ 210 °C, at $x = 0$ mass%; and $T_{st.}$ 290 °C, at $x = 3$ mass% (Fig. 4.30).

The behavior of CuO-modified SS based on (Na, Li) NbO_3 (Fig. 4.29e–g) differs from that observed in other cases. For these samples, T_C remains constant ($T_C = 360$ °C) throughout the modification interval, while $\Delta\varepsilon$ decreases within the range of x from 0 to 2 mass%. Furthermore, the stability region $\varepsilon/\varepsilon_0$ expands. The addition of a higher CuO content ($x = 3$ mass%) results in a deterioration of all characteristics (Fig. 4.29g) [12].

We have found that the reduction of the through conductivity of SS can be achieved not only by introducing modifiers, but also by changing the conditions of their phase formation. Thus, in SS based on (Na, Li) NbO_3 with 2 mass% MnO_2, $\varepsilon/\varepsilon_0$ and $\varepsilon_{33}^T/\varepsilon_0$ decrease in one and a half times, at increase of $T_{sint.}$ from 1180 to 1200 °C, that testifies about less contribution of defect subsystem to these values. CuO modification further enhances this effect (Fig. 4.31) [12].

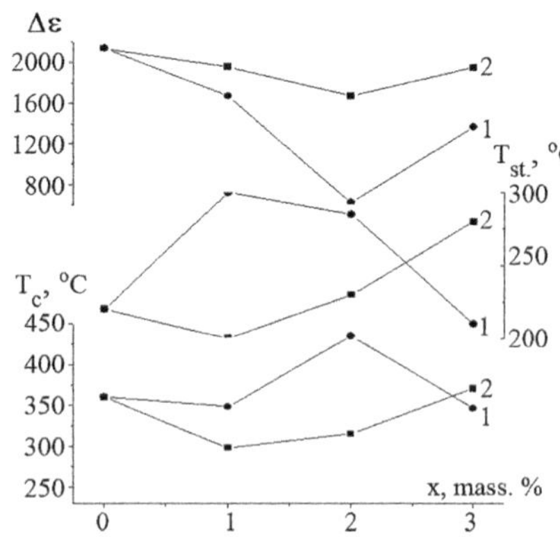

Fig. 4.30 Dependencies of T_C, upper limit, $T_{st.}$, stability region $\varepsilon/\varepsilon_0$ and dispersion depth $\Delta\varepsilon$ on x: 1—CuO, 2—MnO_2

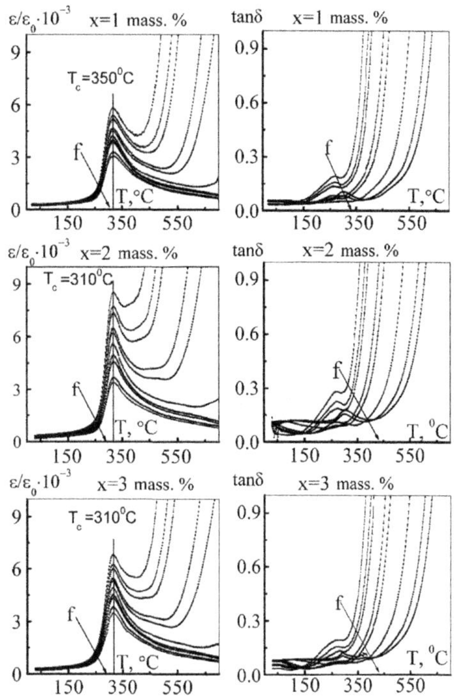

Fig. 4.31 Dependencies of $\varepsilon/\varepsilon_0$ and tanδ on temperature (T, °C) at 20–1000 Hz of (Na, Li) NbO_3-based solid solutions, modified by Mn + CuO ($T_{\mathrm{sint.}}$ = 1180 °C)

The thermal-frequency dependencies of the dielectric permittivity, $\varepsilon/\varepsilon_0$, of ceramic SS based on the (Na, Li) NbO_3 system, both unmodified and modified with a combined modifier (Bi_2O_3 + Fe_2O_3), demonstrate that the SS under study are distinguished by dielectric spectra exhibiting a prominent dispersive maximum ($\varepsilon/\varepsilon_0$) of non-relaxation character within the ferro-paraelectric (FE-PE) PT range of 250 – 350 °C. At temperatures above PT, an increase in the dielectric dispersion $\varepsilon/\varepsilon_0$ and a pronounced increase in $\varepsilon/\varepsilon_0$, particularly evident at low frequencies of the measuring field, are observed, accompanied by an increase in electrical conductivity resulting from the loss of oxygen and partial reduction of niobium [5]. In the temperature range of 20–200 °C, the addition of a modifier (x) in a concentration of 0–1.5 mass% to SS results in a high degree of temperature stability of $\varepsilon/\varepsilon_0$. As the modifier content is increased up to 2 mass%, a second maximum $\varepsilon/\varepsilon_0$ appears in the high-temperature region. This maximum increases with further modification and shifts to the high-temperature region. This is accompanied by a decrease in the PT temperature of FE-PE and an increase in its blurring, as well as a non-monotonic (with a minimum at x = 1.5 mass%) change in the dispersion depth $\varepsilon/\varepsilon_0$. The appearance of an additional maximum $\varepsilon/\varepsilon_0$ can be a consequence of both the PT between two nonpolar phases and the rearrangement of the defect structure in the polycrystalline material.

In Ref. [11], the temperature dependence of the relative dielectric permittivity () characteristic of ferroelectrics with a pronounced clear maximum at the Curie temperature (T_C) was observed in the SS of the (Na, Li) NbO_3 system (composition $Na_{0.875}Li_{0.125}NbO_3$).

In the frequency range from 25 Hz to 1 kHz, following a precipitous decline above T_C, the $\varepsilon/\varepsilon_0$ exhibits a rapid increase. As the frequency (f) of the electric measuring field increases, the starting temperature (T_I) of the rapid increase of the $\varepsilon/\varepsilon_0$ also increases. The relationship between T_I and lg$f$ is nearly linear (Fig. 4.32, inset). In the region of higher frequencies, the effect of increasing $\varepsilon/\varepsilon_0$ is not observed. In the same frequency range, first, PT maximum is observed in the dependence of tanδ(T), which gradually "diffuses" with increasing f. Outside the specified frequency range, the dependence of tanδ (T) becomes linear (Fig. 4.32, inset). As the frequency increases, the values of $\varepsilon/\varepsilon_0$ and tanδ decrease markedly at T_C. Furthermore, this change is more pronounced at higher temperatures and at low frequencies (from 25 Hz to 1 kHz). The peak values of $(\varepsilon/\varepsilon_0)_{max}$ and tanδ_{max} (at 300 °C) exhibit an inverse proportionality to lg$f$, with the dependence $(\varepsilon/\varepsilon_0)_{max}$ on lg$f$, comprising two linear sections with different rates of change in dielectric losses. At low frequencies (up to 1 kHz), the rates are an order of magnitude larger than at higher frequencies (Fig. 4.32, inset). With increasing the frequency, T_C shifts slightly (Fig. 4.32) toward lower temperatures [$\Delta T_C = T_{C(25\,Hz)} - T_{C(1MHz)} = 20$ °C].

In consideration of the findings, it can be posited that the observed effects of low-frequency dispersion of the dielectric permittivity of the investigated SS are indicative of the frequency dependence of $\varepsilon/\varepsilon_0$ is not (or is nearly not) associated with orientational polarization. Rather, this phenomenon is attributed to the influence of electrical conductivity on $\varepsilon/\varepsilon_0$ and tanδ, which may be attributed to the presence of impurities or defects [2].

Such phenomena are most often manifested in compounds and SS containing easily sublimable elements or ions of variable valence. In alkali metal niobates, there are those and others (both alkali ions and niobium), which cause their increased conductivity at high temperatures [13] despite their high resistivity at $T \approx 100$ °C (specific electrical resistance ~ 10^{10} Ω m). As observed in [14], the electrical resistivity undergoes a pronounced decrease at 450 °C in Nb_2O_5, which can be attributed

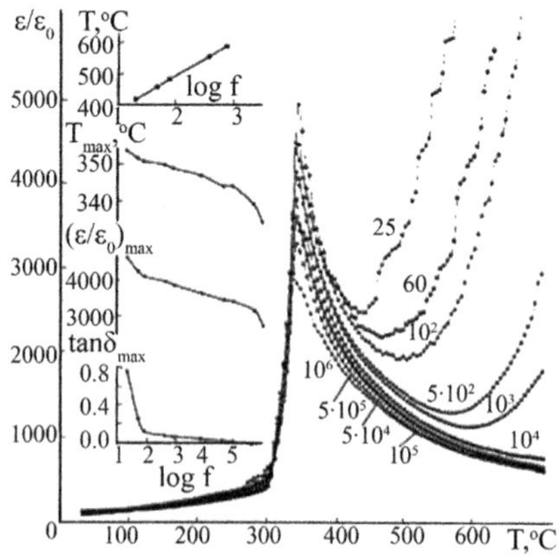

Fig. 4.32 Dependencies of SS on temperature ($\varepsilon/\varepsilon_0$) in the range of $T =$ 25 – 700 °C; the figures near the curves are frequency values in Hz; dependencies of the inflection temperature (T_I) and maximum $\varepsilon/\varepsilon_0$ (T_{max}), maximum $(\varepsilon/\varepsilon_0)_{max}$ and $(\tan\delta)_{max}$ on the logarithm frequency are shown in the inset

to the reduction of niobium to the 4-valence state. This process persists at least until 740 °C. At temperatures approaching the temperature, rapid increases in $\varepsilon/\varepsilon_0$ and tanδ at low frequencies are observed. Furthermore, niobate materials are distinguished by the presence of heterogeneous defects [15], including the non-inclusion of specific amounts of *A*-cations in the lattice. This phenomenon is prevalent in solid-phase oxide ceramics and frequently results in the formation of impurity phases [16].

The extensive polymorphism inherent in alkali metal niobates, the coexistence of phases of different nature at the same temperature, and the high mobility of different types of domain boundaries [17, 18] provide evidence of their structural instability and extreme "softness" with respect to external influences [17], including temperature and frequency of alternating electric voltage. To elucidate these effects, we have conducted a study on the thermal-frequency behavior of analogous but more intricate SS, based on the (Na, Li) NbO_3 system.

Figures 4.33 and 4.34 show the thermal-frequency dependence $\varepsilon/\varepsilon_0$ of the SS, based on the binary system (Na, Li) NbO_3, modified by MnO_2 [19]. As for the SS studied in [11], the fact that the Curie temperature shifts to the left with increasing frequency of the measuring field is noteworthy, indicating the determining role of the through conductivity in the formation of the $\varepsilon/\varepsilon_0$ dispersion.

A comprehensive examination of dielectric spectra above T_C revealed a few discrepancies from the thermal-frequency behavior of $\varepsilon/\varepsilon_0$ in the PE region as previously described in [11].

In contrast to the observations made in [11], which noted an anomalous change in $\varepsilon/\varepsilon_0$ in the high-temperature region far from the PT (FE-PE) and attributed this change to the reduction of pentavalent niobium, our findings indicate that the primary growth of $\varepsilon/\varepsilon_0$ begins in the vicinity of the PE, the second anomaly of $\varepsilon/\varepsilon_0$ (Fig. 4.33) coincides with the one described in [11] and corresponds to the temperatures at which the valence of Nb changes. Additionally, higher temperature anomalies are evident in some of the spectra under examination (see Fig. 4.34).

Given that the initial anomaly is invariably present at PT, it is reasonable to posit that it (PT) is responsible for the growth of $\varepsilon/\varepsilon_0$, which is a consequence of the

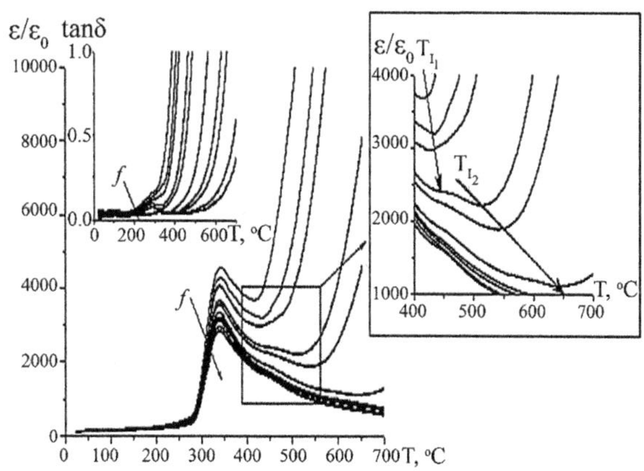

Fig. 4.33 Dependencies of $\varepsilon/\varepsilon_0$ (T) and tanδ (insets on the left) of ceramic SS based on the (Na, Li) NbO_3 system, modified with 1 mass% MnO_2; the inset on the right shows an enlarged fragment of the spectra

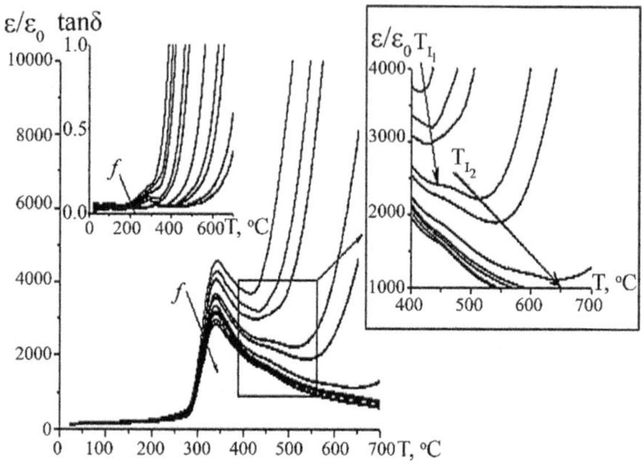

Fig. 4.34 Dependencies of $\varepsilon/\varepsilon_0$ (T) and tanδ (insets on the left) of ceramic SS based on the (Na, Li) NbO_3 system, modified with 2 mass% MnO_2; the inset on the right shows an enlarged fragment of the spectra

formation of defects at the point of structural rearrangement (breaking "old" chemical bonds and formation of "new" ones). These defects may take the form of clusters of the newly formed phase, coherent scattering regions, vacancies, and so forth. In consideration of the extensive polymorphism, observed in Nb oxide compounds [20, 21], and given that our analysis encompasses not only simple oxides but also complex polyatomic compositions derived from them (the polyoctahedral structure $(NbO_6)_{3\infty}$ in alkali metal niobates serves as the structural foundation), it can be postulated that the reversible reduction of Nb_2O_5 may occur across a considerably broader range than previously indicated.

Furthermore, the intricacy of the oxidation–reduction processes within the Nb–O system [13] indicates the multistage nature of these transformations, with which the occurrence of multiple T_i in the high-temperature region is associated. As previously noted in reference [2], the formation of such defects is not exclusive to polymorphic PT, but is also observed in morphotropic (concentration) PT. In the latter case, this phenomenon is manifested in the minimization of sintering temperatures $T_{\text{sint.}}$ [22], specific electrical resistance ρ_{v} [22, 23], activation energy of formation reactions E_{a} [24], forbidden zone width E_{g} [25], and reproducibility of electrophysical properties [26, 27]. The disparate rates of change exhibited by T_i and its change with increasing f merit particular scrutiny. The subsequent changes of T_i are even less rapid, by a factor of nearly threefold (Fig. 4.34). The maximum values, observed in the vicinity of PT, can be attributed, by analogy with the Hedwall effect (an increased reactivity of solids during or because of polymorphic transformations), to the intensification of diffusion processes, including the motion of defects, initiated by the high mobility of the lattice constituents during its rearrangement. The observation that the rate of change of $T_i(f)$ diminishes as one moves away from the PT suggests that the structural rearrangements caused by it play a progressively lesser role in the processes of motion of defects and their dislocations, formed at various stages of permanent reduction–oxidation of Nb.

4.5 Conclusions

This section has presented the results of the study of the (Na, Li)NbO_3 system, modified by various oxides and their combinations. The following oxides were considered: MnO_2, CuO, NiO, $MnO_2 + CuO$, and $Bi_2O_3 + Fe_2O_3$. Phase diagrams of solid solutions of the modified system (Na, Li)NbO_3 were constructed, and the dependencies of perovskite cell parameters and homogeneous deformation parameter δ on the modifier concentration were discussed. The impact of the quantity of the modifier additive on the grain structure of ceramics and the uniformity of the modifier distribution were elucidated. The conditions of formation of electrophysical characteristics were analyzed for modified solid solutions. The behavior of ceramics in high electric fields was investigated. Consequently, the analysis of the dependencies of the relative reversible dielectric constant on the electric field strength for unpolarized and polarized ceramics of the studied compositions permits the classification of the studied ceramics, based on (Na, Li)NbO_3 as the group of ferro-rigid FPCMs.

References

1. Sadykov, Kh.A., et al.: Effects of simple and combined modification of niobate materials that do not contain toxic elements. Des. Compos. Mater. **3**, 45–55 (2013). (in Russian)
2. Danciger, A.Ya., et al.: Multicomponent Systems of Ferroelectric Complex Oxides: Physics, Crystal Chemistry, Technology. Design Aspects of Ferro-piezoelectric Materials, vol. 1 (2001), vol. 2 (2002). Rostov State University Press, Rostov-on-Don (in Russian)
3. Guinier, A.: X-Ray Analysis of Crystals, pp. 604–611. Fizmatlit, Moscow (1961) (in Russian)
4. Trofimov, A.I.: Ultrasonic Systems for Monitoring Curvatures of Technological Channels of Nuclear Reactors. Energoatomizdat, Moscow (1994). (in Russian)
5. Verbenko, I.A., et al.: Preparation and dielectric properties of lead-free ceramics of the composition $[(Na_{0.5}K_{0.5})_{1-x}Li_x](Nb_{1-y-z}Ta_ySb_z)O_3$. Inorg. Mater. **45**(6), 762–768 (2009)
6. Zhang, B., et al.: Rhombohedral—orthorhombic phase coexistence and electrical properties of Ta and $BaZrO_3$ co-modified (K, Na)NbO_3 lead-free ceramics. Curr. Appl. Phys. **13**, 1647–1650 (2013)
7. Bokiy, G. B.: Crystal Chemistry. Textbook for Universities, p. 357. Moscow State University Press, Moscow (1960) (in Russian)
8. Jonker, S.Kh.: Grain size of sintered barium titanate. In: Ceramics. Proc. I and II Conf. of the British and Dutch Ceramic Societies, pp. 60–62. Metallurgy, Moscow (1967) (in Russian)
9. Samara, G.A.: The relaxational properties of compositionally disordered ABO_3 perovskites. J. Phys. Condens. Matter. **15**, 367–411 (2003)
10. Sadykov, Kh.A., et al.: Features of dielectric spectra of niobate materials modified with manganese and copper oxides. Izv. RAS. Ser. Phys. **78**(1), 108–109 (2014)
11. Kravchenko, O.Yu., et al.: Properties of $Na_{0.875}Li_{0.125}NbO_3$ ceramics. Inorgan. Mater. **44**(10), 1265–1280 (2008)
12. Sadykov, Kh.A., et al.: Dependence of dielectric properties of lead-free ferropiezoceramic materials on modification with 3d-metal oxides. In: Phase Transitions, Ordered States and New Materials, pp. 32–34 (2018) (in Russian)
13. Fesenko, E.G., et al.: Study of hot-pressed lithium metaniobate ceramics in a wide temperature range. J. Tech. Phys. **54**(2), 412–415 (1984)
14. Dellis, J.-L., et al.: Cole–Cole analysis of a new lead free $NaNbO_3$–$Sr_{0.5}NbO_3$–$LiNbO_3$ ferroelectric relaxor. Ferroelectrics **318**, 260–267 (2005)

15. Vezzoli, G.C.: Electrical properties of NbO_2 and Nb_2O_5 at elevated temperature in air and flowing argon. Phys. Rev. B **26**(7), 3957–3963 (1982)
16. Kuzminov, Yu.S.: Electro-Optical and Nonlinear-Optical Crystal of Lithium Niobate. Science, Moscow (1987). (in Russian)
17. Reznichenko, L.A., et al.: Dielectric and piezoelectric properties of solid solutions based on sodium niobate. Inorg. Mater. **39**(2), 187–199 (2003)
18. Tretyakov, Yu.D.: Solid-Phase Reactions. Khimiya, Moscow (1979). (in Russian)
19. Verbenko, I.A., et al.: Features of the dielectric spectra of niobate materials. In: Proc. XI Int. Sci. Techn. Conf. Fundamental Problems of Radioelectronic Instrument Making (INTERMATIC-2011), vol. 2, pp. 106–109. MIREA, Moscow (2011) (in Russian)
20. Gagarina, E.S., et al.: Domain structure of $Na_{1-x}Li_xNbO_3$ crystals. Crystallography **47**(6), 1057–1059 (2002)
21. Reznichenko, L.A., et al.: Structural instabilities, incommensurate modulations, P- and Q-phases in sodium niobate in the range of 300–500 K. Crystallography **48**(3), 501–505 (2003)
22. Rao, C.N.R., Gopalakrishnan, J.: New Directions in Solid State Chemistry, p. 520. Cambridge University Press, Cambridge (1990)
23. Fayrbrotter, F.: Chemistry of Niobium and Tantalum, pp. 276–280. Chemistry, Moscow (1972) (in Russian)
24. Reznichenko, L.A., et al.: Invar effect in n-Nb_2O_5, αBT-Nb_2O_5. Crystallography **54**(3), 502–509 (2009)
25. Reisman, A., et al.: Metastability in niobate systems. J. Am. Chem. Soc. **81**(6), 1292–1295 (1959)
26. Raevsky, I.P., et al.: Phase transitions and electrical properties of ferroelectric solid solutions based on sodium niobate. Inorg. Mater. **15**(5), 875–879 (1979)
27. Raevsky, I.P.: Phenomena, caused by the relationship of ferroelectric and semiconductor properties in substances of the oxygen-octahedral type, p. 520. DrSc Thesis (Physical and Mathematical Sciences). Rostov State University Press, Rostov-on-Don (1990) (in Russian)

Chapter 5
Regularities of Changes in the Structure and Physical Properties of Solid Solutions Based on Alkali and Alkaline Earth Metal Niobates upon Heterovalent Ion Substitution

In this chapter, we will sequentially consider the features of the microstructure (for comparison, we will show the role of $LiSbO_3$ in the "construction" of the grain landscape, varying its amount in the basic compound), the issues of phase formation, the formation of macroresponses (dielectric, piezoelectric, ferroelastic) in a wide range of external influences (temperature) of media based on alkali metal niobates, modified with various elements, including their combinations. Let us establish the regularities of the occurrence of correlations: composition—grain structure—crystal structure (phase state)—electrophysical properties—areas of application of the analyzed SS, considering their crystallochemical specificity and thermodynamic prehistory.

The possibilities of modifying SS based on alkali metal niobates are studied in detail in [1–4]. Several works described the effect of various modifiers, including the ions Pb^{2+} [5–8], Cd^{2+} [5, 6, 8], Mg^{2+} [9–12], Ca^{2+} [9–12], Sr^{2+} [9–12], Ba^{2+} [9–11], Bi^{3+} [7], La^{3+} [7], Ti^{4+} [5, 6, 8], Ge^{4+} [5, 6, 8] on the properties of the binary system $(Na_{1-x}K_x)NbO_3$.

The effect of some of them (Mg^{2+}, Sr^{2+}, La^{3+}, Sb^{3+}) and modifying glasses on ceramic and electrophysical properties is also considered in the SS of the binary system $(Na_{1-x}Li_x)$ NbO_3 [4, 13, 14], which established the main regularities of changes in their characteristics during iso- and heterovalent substitutions of cations. It is shown that after modification, the determining factors in the formation of the physical properties of alkali metal niobates, as well as in PZT materials, are the mobility of domain walls, determined by spontaneous deformation and interaction with the resulting vacancies, as well as the crystallochemical features of the structure (in the case of alkali metal niobates, they are the discrepancy between the dimensional characteristics of alkaline cations and the parameters of structure-forming octahedrons, block structure). It is revealed that the unequal degree of ferro-hardness (HF), modified SS based on alkali metal niobates is due to the unequal nature of the vacancies created. At small concentrations of modifiers, their number is decisive (SS with many vacancies are more ferro-soft, SF, and have large values of piezoelectric coefficients). At high concentrations, it is the vacancies position in the structure (SS with

I. A. Verbenko et al., *Environmentally Friendly Smart Materials with Special Electrical and Magnetic Properties*, Engineering Materials,
https://doi.org/10.1007/978-3-031-90387-8_5

vacancies in a more covalent *B*-sublattice become more SF). The basic principles of the design of SS with a greater or lesser degree of HF, which determines the level of piezoelectric and elastic characteristics in objects, were determined.

In this chapter, we describe the results of doping SS based on (Na, K) NbO_3 as the most similar in properties to PZT systems.

5.1 Influence of Thermodynamic Prehistory on the Microstructure of Ferroactive Niobate Materials

In addition to compaction, phase changes and polymorphic transformations, the key points in the sintering of dispersed-crystalline powders from which ferroactive ceramics are obtained are recrystallization processes, because of which the microstructure (grain landscape) of finished products is formed [15]. In fact, sintered polycrystalline bodies are nothing more than a conglomerate of grains (crystallites) [16], the boundaries of which form a kind of network of intergrain surfaces. The latter determines the main difference between ceramics and a single crystal [16] and significantly influences, along with the elemental composition and the nature of the crystal structure, the macroscopic properties of substances. The ability to control the grain structure (the size and habit of grains and pores, the type of their packaging, the state of the intergrain boundaries and other details of the microstructure) makes it possible to change the performance characteristics of ceramics within a wide range up to the formation of fundamentally new combinations. Both the regulations of the technological process (primarily the firing temperature) and variations in the elemental composition of the sintered compositions can act as levers for the directional organization of a given set of properties of ceramic ferropiezomaterials. In this section, we will consider the effect of thermodynamic prehistory (effects of inoculation and sintering temperature, T_{sint}) on the microstructure of ferroelectric materials based on complex niobium oxides of alkali metals, which is a continuation and development of earlier work on this topic [17].

The objects of the study were multielement compositions based on the binary SS system of sodium–potassium niobates of the compositions $(Na_{0.46}K_{0.54})_{1-x}Li_x(Nb_{0.9}Ta_{0.1})_{1-x}Sb_xO_3$ ($x = 0.02$—base 1; 0.04—base 2; 0.06—base 3), modified super stoichiometrically and (in some cases) stoichiometrically with oxides of manganese, copper, nickel, bismuth, iron, as well as combined additives is due to the demand [18–23] for ferroelectric ceramics, based on non-toxic elements of alkali metal niobates, namely alternative media to lead-containing functional materials, industrially produced on a global scale with the basic participation of PZT-SS [23–25]. The formulation of the proposed studies is due to the lack of a detailed analysis in the literature of the effects of the additives introduced into these bases on the microstructure of the resulting ceramics and, ultimately, on the macroproperties of objects that may be promising for use in sensor equipment for various purposes. The information available in the literature is very fragmentary, not systematized and, in many respects, contradictory [1, 18–31].

Figures 5.1, 5.2 and 5.3 show fragments of the microstructure of base 1–3 ceramics, sintered at different temperatures. In all cases, there is a heterogeneity of the grain landscape with crystallites, the habit of which is either parallelepipeds ("plates") (larger grains) or forms close to cubic (small grains). In each case, the size of large and small grains differs several times (sometimes by an order of magnitude). The packing of grains is chaotic, loose in places, with areas of denser structures in the form of ring areas (marked with dashed lines in the figures), sometimes consisting of several (more often two) layers.

Introduction of 2 (Base 1), 4 (Base 2) and 6 (Base 3) mol% $LiSbO_3$ to the original composition $(Na_{0.54}K_{0.46})(Nb_{0.9}Ta_{0.1})O_3$ leads, on the one hand, to a decrease

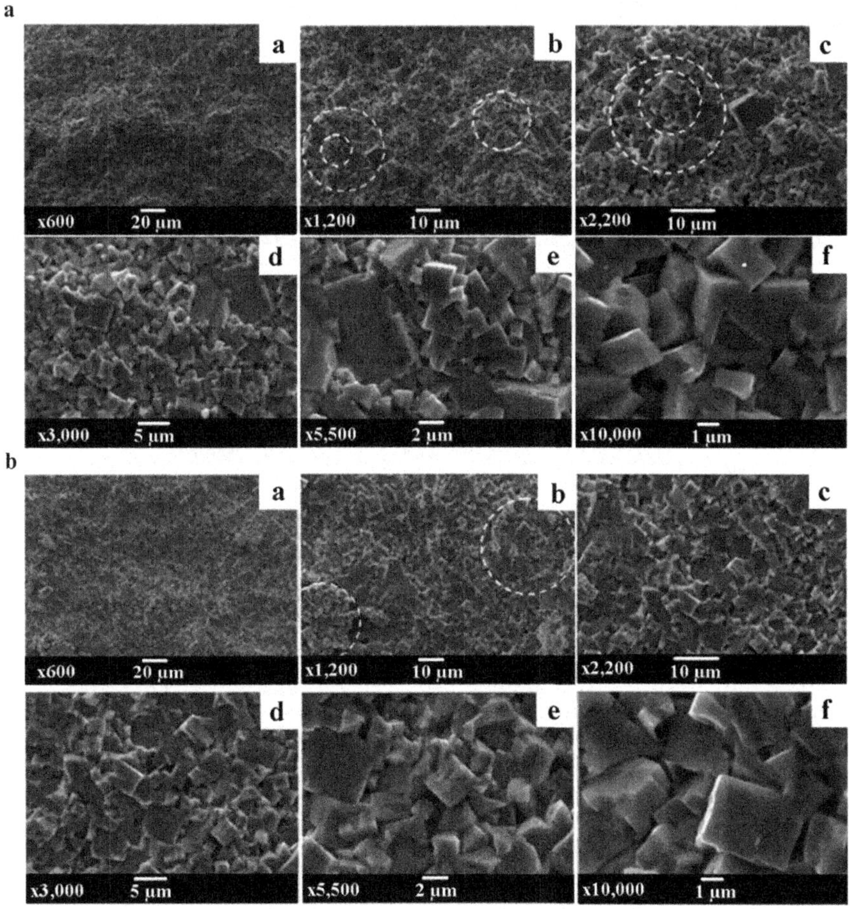

Fig. 5.1 **a** Fragments of microstructure of base 1 ceramics, sintered at 1150 °C. **b** Fragments of microstructure of base 1 ceramics, sintered at 1160 °C

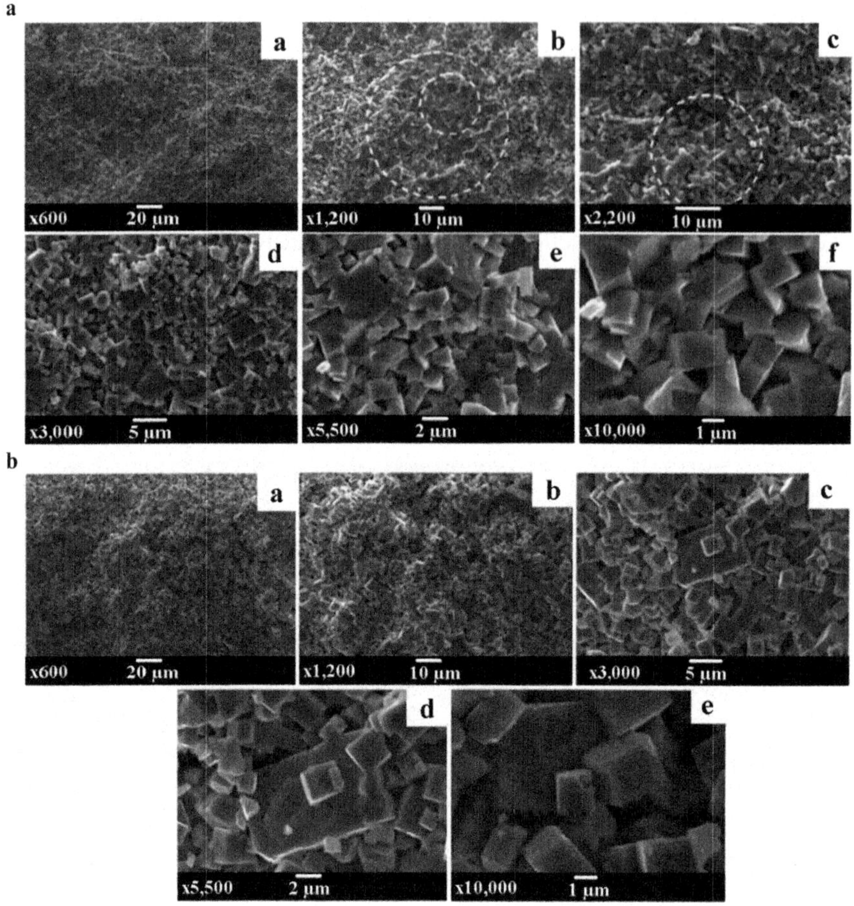

Fig. 5.2 **a** Fragments of microstructure of base 2 ceramics, sintered at a temperature of 1150 °C. Fragments of microstructure of ceramics of base 2, sintered at a temperature of 1160 °C

in the average size of grains, their alignment in size, the formation of clearer rectilinear boundaries, and, on the other hand, to a significant loosening of the structure, expressed in violations of the continuity of the grain pattern, the appearance (isolation) of a large number of single crystallites or their ensembles immersed in a certain dark phase, which is especially well seen in photographs of the microstructure of ceramics sintered at higher T_{sint} (1160 °C). In these cases, the effect, introduced by $LiSbO_3$, enhances the effect of the higher T_{sint} samples.

It should be noted that the change in the T_{sint} just 10 °C (upward) also leads to a sharp increase in the average grain size and an increase in the heterogeneity of the microstructure. The observed phenomena can be interpreted as follows.

The fact that in the process of recrystallization sintering in all the cases under consideration grains of almost regular geometric shape with clear edges are formed,

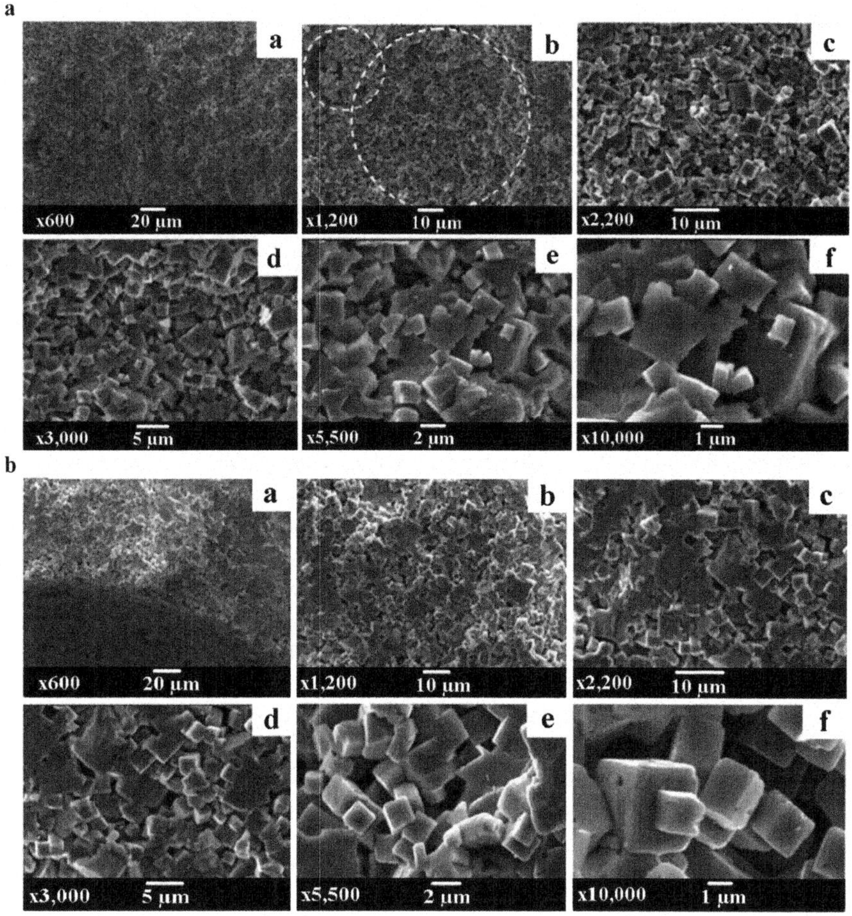

Fig. 5.3 **a** Fragments of microstructure of ceramics of base 3, sintered at a temperature of 1150 °C. **b** Fragments of microstructure of base 3 ceramics, sintered at a temperature of 1160 °C

indicates that they grow in the main volume, in which, in addition to solid grains and gaseous pores, there is a liquid phase (LF) (in Figs. 5.1, 5.2 and 5.3 we qualified it as the dark phase, with a large amount of $LiSbO_3$), in which the grains are immersed [32, 33]. Most often, such ideomorphic grains are characteristic of secondary intermittent recrystallization processes, when impurities located at the grain boundary at the recrystallization temperature form LF that actively interacts with the base material, acting not so much as a "lubricant", but as a transport medium, that is, a solvent. In this case, the selective growth of large grains occurs due to the usual dissolution and precipitation from the solution, and not to the movement of boundaries. At the same time, the liquid covering the surface of such grains with a film causes them to acquire a certain form of growth and a regular facet, identical to the habit of crystals growing from a molten solution. As a result, ideomorphic grains crystallize in the form of polyhedra with almost rectilinear boundaries [34–36].

As noted in [16], the study of the kinetics of the formation (synthesis and sintering) of ceramics based on alkali metal niobates has shown that it almost always occurs with the participation of LF [37–40]. At the same time, the latter can either be chemically inert in relation to the basic composition, or actively react with it. In the first case, we are dealing with a fluid of increased viscosity (LF1), which mainly plays the role of a "lubricant" ("binder"), in the second case, with LF2 with low viscosity (fluid), which, dissolving the solid phase, performs the function of a transport medium. At the same time, the dissolution process at the firing temperature is limited to a certain extent. Often, both types of LF coexist in the same formulation.

In each of these cases, LF has a significant impact on the elementary stages of diffusion, polymorphic transformations, sintering, recrystallization, dissociation, etc.

The source of LF is all substances that differ from the main composition, that is, impurities contained in the raw material, unreacted components, and some modifiers. For example, in alkali metal niobates, there are the products of interaction of unreacted components with impurities (complex multicomponent oxide compositions containing silicates, aluminates, calcite, ferrites, etc. of various compositions with the participation of LF1 with the participation of LF1 with a melting point of $T_{\mathrm{melting}} \leq 1000$ °C) or impurities with each other (forming low-melting eutectics), as well as artificially introduced modifiers, namely melts or glasses of various compositions with low T_{melting}. The formation of LF2 is associated with the presence of unreacted components (Na_2CO_3 with $T_{\mathrm{melting}} = 853$ °C, K_2CO_3 with $T_{\mathrm{melting}} = 891$ °C, KOH c $T_{\mathrm{melting}} = 404$ °C, LiOH c $T_{\mathrm{melting}} = 471$ °C) and low-melting eutectics in niobate charges ($NaNbO_3$ with $T_{\mathrm{melting}} = 987$ °C, $KNbO_3$ with $T_{\mathrm{melting}} = 845$ °C), as well as the presence of fluorine impurities in Nb_2O_5 of the "NbO-Pt" qualification, which actively interacts with alkali metal to form low-melting, sufficiently aggressive fluorides (LF2′). The solvent power of fluorine-enriched LF2′ is significantly higher than that of LF2. The amount of LF1 and LF2 can be small due to the small concentrations of cationic impurities in the feedstock and unreacted components in the reaction mixtures. The presence of LF2′ is much more noticeable both due to the high contamination of Nb_2O_5 with fluorine (0.2 mass%) and the aggressiveness of alkali metal fluorides. The results of the thermographic study of the sintered powders confirmed the presence of LF2 (endothermic effect at ~ 1000 °C). Analysis of possible impurities and T_{melting} showed that the basis of LF2 (in particular, in systems based on (Na, Li) NbO_3 and (Na, K) NbO_3) is sodium fluoride (NaF). By actively dissolving the main matrix of the material, alkali metal fluorides are enriched with Nb_2O_5 and, when the cured ceramics are cooled, they crystallize, most often in the form of the $Nb_2O_5 \cdot NaF$ compound. Layer-by-layer preparation of samples with simultaneous X-ray imaging showed that this impurity, for example, is "squeezed out" to the surface during hot pressing. In the center of the samples, the amount is minimal. This is due to the strong heterogeneity of the microstructure of alkali metal niobates obtained from raw materials with a high F-content, and the development of anomalous growth of giant ideomorphic grains in the near-surface layers (like the growth of single crystals from a solution in a melt—secondary intermittent recrystallization).

Electron microscopy studies have shown that LF1 is concentrated, most often, along the boundaries of grain conglomerates (less often individual grains) in the form of rather "thick bundles" (layers with double borders). Sometimes it can be observed in the form of "lakes" in which grains float (when borate glasses are introduced). LF2 is more fluid than LF1, is evenly distributed over the sample volume and surrounds the grains with a thin film, making it difficult to visualize. This is aggravated by the proximity of its composition to the base material, the possibility of spending money to compensate for stoichiometry violations, the loss of alkali metal due to evaporation, etc. Radiographically, when sintering ceramics by hot pressing in the filling, we also detected the compound $NaAl_5O_8$, the amount of which increases in the sequence: fluoride—free raw materials—F-containing raw materials—F-containing raw materials + borate glasses in the charge, that is, with an increase in the amount of LF. This is probably a consequence of facilitating the volatilization of the Na-containing component from the melts, which, in turn, leads to local violations of their stoichiometry. Thus, LF2 turns out to be very "not harmless" in relation to the base: by strongly affecting the homogeneity of the chemical composition and microstructure of ceramics, it significantly worsens the electrophysical properties.

It should be noted that in known materials based on PZT, the possibility of LF formation is mainly associated only with the interaction of impurities with each other or with unreacted components of the charge (viscous silicates, phosphates, aluminates and lead ferrites), melts of the latter ($PbO + TiO_2 + ZrO_2$), artificially introduced additives, namely melts or glasses. LF2′ is practically not formed in such materials.

Based on the above, it seems obvious that two types of LF are mainly formed in the analyzed ceramics, namely LF2 and LF2′. At the same time, the enrichment of materials with lithium stibate leads to an increase in the amount of LF2 due to the appearance of low-melting Li-containing eutectic (Li_2CO_3, LiOH, $LiSbO_3$ with $T_{\text{melting}} = 824$ °C) and a decrease in LF2′ due to a decrease in the *B*-positions of Nb, (replaced by Sb) carrier (in Nb_2O_5) of the fluoride impurity.

To prove this, we present Fig. 5.4, which shows micrographs of chips of PCR-35 ceramics (based on the $(Na_{1-x}Li_x)$ NbO_3 system with $x = 0.125$), obtained by hot pressing (a) and conventional ceramic technology (b), as well as PCR-34 material (based on the $Na_{1-x}K_xNbO_3$ system with $x = 0.50$) cooked without external pressure. In the first two cases it was not possible to identify the grain structure (according to the method used in this work), since the chip "passed" not along the boundaries of the grains, but along their surface. This indicates the sufficient strength of the intergranular layers, undoubtedly associated with the cementitious action of LF, formed due to the large amounts of Na and Li in the chemical composition of the PCR-35 material. In the absence of Li (PCR-34 material), the grain structure is visualized (Fig. 5.4c). When niobium is replaced by antimony, the phenomenon, caused by secondary intermittent recrystallization, i.e. anomalous intermittent growth of large grains, disappears (Figs. 5.1, 5.2 and 5.3), which speaks in favor of the disappearance of LF2′.

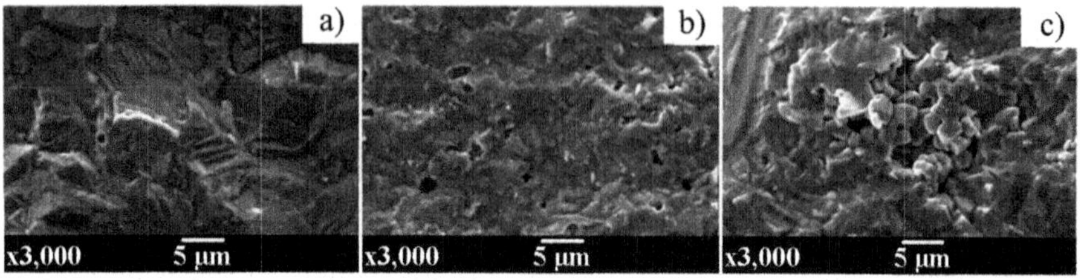

Fig. 5.4 Fragments of the microstructure of ceramics of PCR-35 (**a**, **b**) and PCR-34 (**c**), obtained by various methods

In individual fragments of microstructures presented in Figs. 5.1, 5.2 and 5.3, as noted earlier, one can see the radial-annular formation of grains previously observed in [41–44] in the study of various types of ceramics, both lead-free [45] and Pb-containing [41–43]. In these cases, large pores were framed by a layer (or layers) of densely packed, defect-free large crystallites, having a shape close to idiomorphic, and small, loosely packed fewer perfect crystallites grew away from the pores. The reason for the observed effect lies in the special role of large pores, namely sinks of vacancies and dislocations, which favor the growth of defect-free crystallites in their vicinity and the emergence of the described clusters. In our case, most likely, the mechanism of their formation is different and is associated with the heterogeneity of the structure of ceramics containing $LiSbO_3$, which we will discuss below.

On one of the micrographs (Fig. 5.1, base 1, T_{sint} = 1150 °C) at the highest magnification (×10,000) one can see an unusual shape of a single grain, presemted in the form of a rhombus. A similar effect was first observed in [29] (Fig. 5.5) and it was also explained there from the standpoint of the complex formation of nuclei of crystallite crystallization and growth [27]. Figure 5.6 shows a diagram of a possible mechanism for the formation of diamond-shaped grains.

The authors suggest that in this process there is an initial partial melting of grains and the formation of crystallization nuclei. First, Na^+ ions combine with Nb_2O_5 to form irregularly shaped crystalline $NaNbO_3$ particles. These randomly distributed $NaNbO_3$ particles then act as crystallization nuclei at the initial stage of grain growth. Further, chemically active radicals are formed due to the compaction of atoms at the

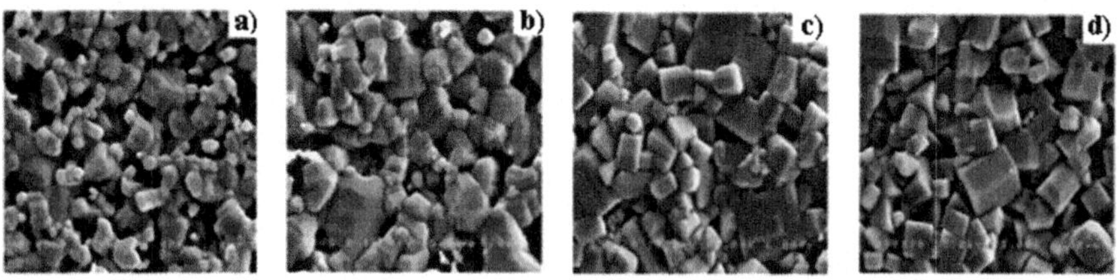

Fig. 5.5 Photographs of the microstructure taken with a scanning electron microscope of the surface of $NaNbO_3$ ceramics, sintered for 3 h at T_{sint} = 1100 °C (**a**), 1130 °C (**b**), 1170 °C (**c**), 1200 °C (**d**); magnification ×500

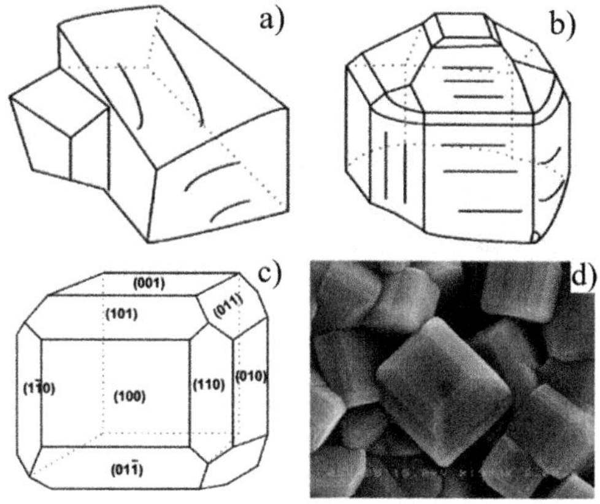

Fig. 5.6 Modeled morphology of the orthorhombic phase of $NaNbO_3$ ceramic grains: irregular form (**a**); polyhedron with stripes (**b**); ideal grain structure (**c**); a fragment of a photograph from Fig. 5.5 (**d**)

stages of the formation of geometric and functional features like helical dislocation, as noted in the description of experiments on growing crystals [46, 47].

With a certain T_{sint}, mass transfer in this process is enhanced due to higher atomic mobility and causes accelerated grain growth by merging adjacent grains. As a result, some grains form a diamond-like shape, and the mass transfer required for this process occurs by the diffusion of ions during the growth stages. In an orthorhombic crystal system, all right axes are mutually perpendicular.

Therefore, faces (1 0 0), (0 1 0), and (0 0 1) parallel to two of these axes are perpendicular to the third right axis. In this case, the growth of diamond-shaped grains with well-defined facets follows the growth model of ideal grains.

Let us consider the reasons for the decrease in the average grain size in ceramics with an increase in the concentration of $LiSbO_3$ in the composition of chemical compositions, which was observed in other works [48, 49], and by us in [50]. There are several of them: (i) polymorphism of the initial reagents, namely Sb_2O_5 (Sb_2O_3), which undergo a number of phase transformations with a change in the oxidation state of Sb in the process of heat treatment of charges: $Sb_2O_5 \xrightarrow{600\ \text{K}} Sb_6O_{13} \xrightarrow{940\ \text{K}} SbO_2$; $Sb_2O_3 \xrightarrow[air]{400\ \text{K}} SbO_2$ [50]; (ii) the development of crystal-chemical disorder due to the possibility of placing Li in the *A*- and *B*-positions of perovskite and perovskite-like structures [51]; (iii) significant disparity (from 22 to 34%) of the radii of the substituting Sb^{5+} (0.52 Å), Sb^{3+} (0.90 Å) and substituting Nb^{5+} (0.66 Å), Nb^{4+} (0.67 Å), Ta^{5+} (0.66 Å) ions, exceeding the permissible (10–15%) to maintain the stability of the crystal structure [52, 53].

The first circumstance (i) (polymorphism) is associated with the occurrence of the Hedwall effect, i.e., an increase in the reactivity of solids during or because of polymorphic transformations. This, in turn, stimulates recrystallization processes, causing the simultaneous formation of many centers of primary recrystallization of reduced mass capacity, due to which smaller crystallites grow from each embryo of the future grain picture. The second (ii) and third (iii) reasons destabilize the structure

Fig. 5.7 Fragments of the microstructure of the studied ceramics with isolated bimodal grain structures

in conditions of its increased heterogeneity, which also provokes the formation of several primary clusters and, as in the first case, leads to inhibition of grain growth.

The heterogeneity of the structure, characterized by fluctuations in the composition and density of ceramics, and the unevenness of the grain pattern are associated with another phenomenon that we observe in bases 1–3, as well as in modified environments, based on them, which we will discuss below. This phenomenon is illustrated in Fig. 5.7, which shows fragments of the microstructure of the ceramics studied with highlighted areas of bimodal grain structure.

There are columns of more densely and loosely packed grains (a); areas with a stochastic distribution of the smallest unformed grains surrounding large (b); fragments of microstructure with giant crystallites that absorb smaller grains (c); chains of "sticky" grains surrounded by randomly packed others (d); fields of ordered plate-like crystallites adjacent to single grains, probably surrounded by LF (e); intraporous structures (e). Similar bimodal structures were observed in [40, 54–56] (Fig. 5.8).

Based on the analysis of the data obtained, it can be concluded that the most promising for further research are bases 1 and 2, which we consider as basic compositions for their modification with various elements and establishing the regularities of the influence of small variations in the cationic composition on the macroproperties of the analyzed objects.

Figures 5.9, 5.10, 5.11 and 5.12 illustrate the effects of modifying bases 1 and 2 with the super stoichiometric additives MnO_2, CuO, NiO and various combinations thereof. Figures 5.13, 5.14 and 5.15 show the microstructures of materials comprising matrix 1, stoichiometrically modified with MnO_2, as well as media with a combination of dosant administration methods. Figure 5.16 shows how the characteristics of the microstructure (primarily the average grain size) change during stoichiometric doping of base 1 by simultaneous introduction of Bi_2O_3 and Fe_2O_3 oxides into the charges.

Fig. 5.8 Bimodal structures in niobate materials

The effect of Mn on the kinetics of grain formation will be considered from the standpoint of its simultaneous presence in charges in different oxidation states. Thermogravimetric, radiographic, and EPR [57, 58] have shown that, regardless of the type of Mn oxides used in the synthesis of SS, all of them provide the real presence of Mn (III) (mainly) + Mn (IV) + Mn (II) cations in SS formulations after heat treatment in an airborne atmosphere, which is usually used in the production of ferrotopies. This is a consequence of the redox properties of Mn oxides and the specifics of their behavior in the presence of oxides of other elements [57]. Due to the low T_{melting} of oxides Mn (IV) (540 °C) and Mn (III) (1080 °C), which decrease even more in complex multi-element compositions, in the considered SS, LFs of eutectic origin are formed (in addition to the already existing ones) in the process of roasting both initial reaction mixtures and synthesized powders [59]. Their presence was detected, as we can see from Figs. 5.9, 5.10, 5.11, 5.12, 5.13, 5.14 and 5.15, using microstructural analysis. Let us consider the effect of the formed LF on recrystallization processes.

At first, a small amount of it, which is "pinched" between the primary crystallites, cement them, slowing down the movement of the intergrain boundaries, which leads to a certain decrease in their average size. A further increase in the concentration of LF (with a larger amount of Mn oxide introduced) favors its filling of the gaps between the grains, facilitating diffusion and mass transfer, which causes some grain growth, while the processes of dissolution and precipitation in conglomerates of

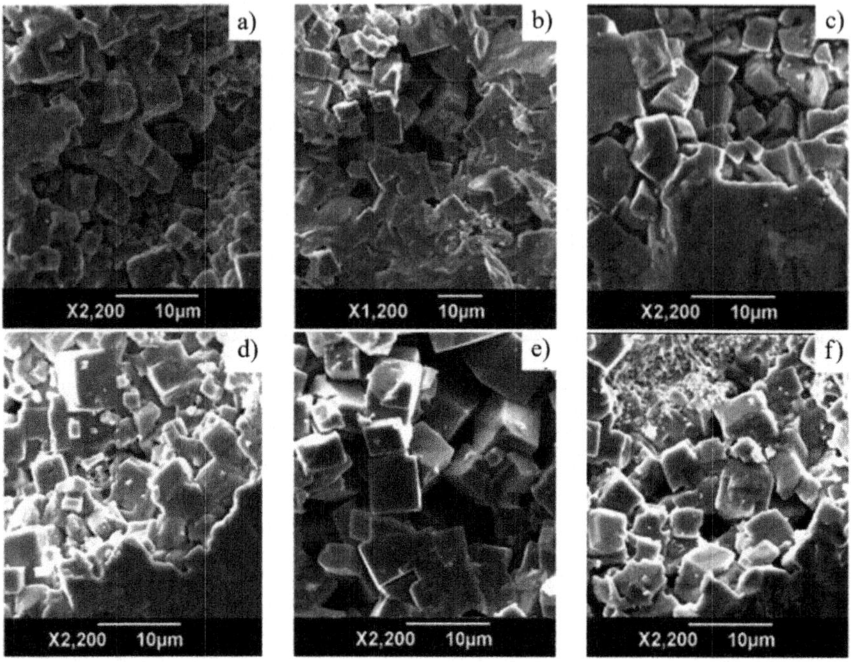

Fig. 5.9 Fragments of microstructure of base 1 ceramics modified super stoichiometrically by MnO_2: **a**, **d** 1 mass%; **b**, **e** 2 mass%; **c**, **f** 3 mass% and sintered at different temperatures

solid particles immersed in LF contribute to the transfer of matter between primary crystallites, even those not in direct contact. Increase in T_{sint} (Fig. 5.9), which leads to a decrease in the angle of LF wetting and, as a result, to deep penetration of LF into the intergranular spaces, further aggravates this process by activating grain growth [60]. In addition, at the first stage, when closed, gas-filled pores are formed, the pressure in them inhibits not only shrinkage [60], but also recrystallization processes, contributing to the grinding of the structure. At the second stage, with a larger amount of LF, the latter fills the intergrain "channels", forming some "lakes" and contributing, as a result, to easier mass transfer, which also entails the enlargement of the grain structure.

It should be noted that in the process of isothermal sintering of the analyzed materials, the role of LF is predominant only at the first two stages (liquid flow and dissolution–precipitation), which are characterized, among other things, by the processes of diffusion coalescence. At the final stage, LF can crystallize due to its enrichment with a refractory component and, as a result, the formation of a rigid solid-state framework, and then the mechanisms of diffusion transfer of matter will "lie" at the basis of recrystallization processes, obeying the laws of solid-phase sintering and the formation of grain architecture.

When analyzing the results obtained, it is necessary to consider the following circumstances. Provided that the base ceramic material is well wetted, LF contributes

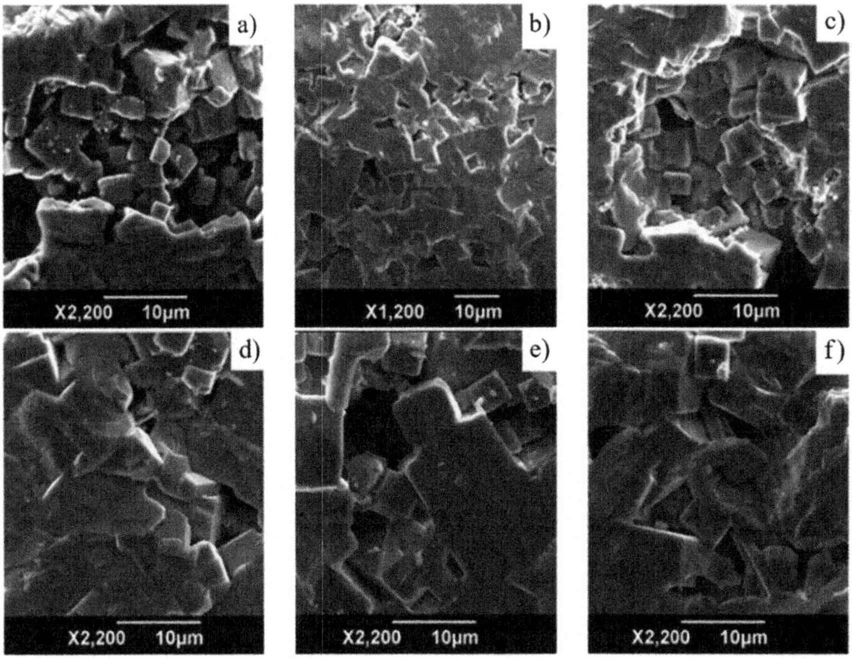

Fig. 5.10 Fragments of microstructure of base 1 ceramics, modified super stoichiometrically by CuO: **a**, **d** 1 mass%; **b**, **e** 2 mass%; **c**, **f** 3 mass% and sintered at different temperatures

Fig. 5.11 Fragments of microstructure of ceramics of base 1 (**a**, **b**), base 2 (**c**, **d**), modified super stoichiometrically by 2 mass% MnO_2 (top) and CuO (bottom), sintered at temperatures 1180 °C (top), 1200 °C (bottom)

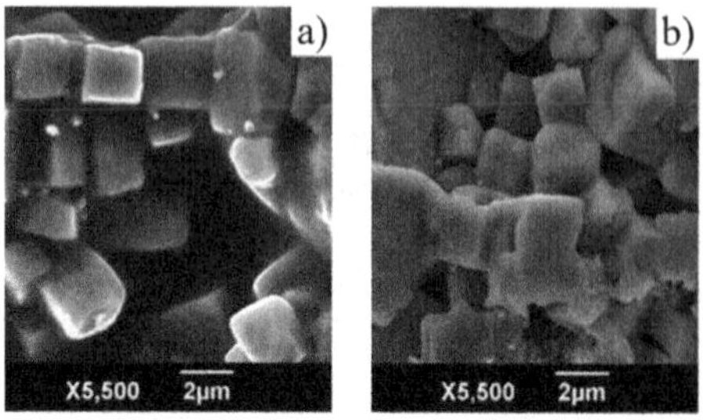

Fig. 5.12 Fragments of the microstructure of ceramics of base 1, modified superstoichiometrically by 1 mass% MnO_2 and 2 mass% NiO (**a**); 3 mass% MnO_2 and 2 mass% NiO (**b**), T_{sint} = 1155 °C in both cases

Fig. 5.13 Fragments of microstructure of ceramics of base 1, modified stoichiometrically by MnO_2: **a** $x = 0.0125$; **b** $x = 0.0250$; **c** $x = 0.0375$

Fig. 5.14 Fragments of the microstructure of the ceramics of the base 1, modified stoichiometrically by MnO_2 ($x = 0.0375$) and super stoichiometrically 2 mass% NiO, T_{sint} = 1140 °C

to the formation of a compressive capillary force at the junction of solid particles, which tightens the particles [61] (when fully wet, the "tightening force" is maximum [62, 63]), because of which the sintering process in the presence of wetting LF is significantly stimulated [61, 64, 65].

Fig. 5.15 Fragments of the microstructure of the ceramics of the base 1, modified stoichiometrically by MnO_2 ($x = 0.0375$) and super stoichiometrically by 2 mass% NiO, $T_{sint} = 1150$ °C

Fig. 5.16 Fragments of microstructure of ceramics of base 1 modified by super stoichiometric $BiFeO_3$ in quantities of 0.5 mass% (**a**); 1.0 mass% (**b**); 1.5 mass% (**c**); $T_{sint} = 1180$ °C (in all cases)

The consequence of the intensification of the sintering process of ceramics containing LF is a decrease in their T_{sint} and an increase in ρ_{exp}, which we are seeing in practice. The increase in ρ_{exp} in some cases with an increase in the concentration of LF is also the result of the "binding" of ceramic crystalline grains by the residual vitreous phase after recrystallization [66].

The decrease in the average grain size, observed in individual ss $\overline{D}$, may be associated with a slowdown in grain growth during the sintering process, since LF, filling the space between the crystallites, contributes to the recrystallization process through a certain "barrier" formed by LF.

The increase of $\overline{D}$ may be a consequence (in addition to the reasons already described) of the uneven granularity (heterogeneity of grains in size) of the microstructure, including due to the uneven distribution of LF over the sample, and secondary recrystallization, in the process of which there is an acceleration of the growth of individual grains that are in energetically more favorable conditions [67].

We have also noted the hardening of ceramics containing LF. This is due to the following reasons: (i) a decrease in stresses at the boundaries between grains and phases (crystalline and amorphous, LF) with a decrease in $\overline{D}$ [68]; (ii) an increase in the value of destructive stress, proportional to $\left(\sqrt{\overline{D}}\right)^{-1}$ [69]; (iii) barrier effect plays the strengthening role of intergranular and interfacial boundaries as powerful obstacles to moving dislocations; (iv) increased segregation of impurity atoms at the grain boundaries, fixing dislocations and increasing granular relaxation; (v) an increase in the rate of reproduction of dislocations and, in this regard, an increase in interaction between them [70]. Another factor contributing to the hardening of ceramics may be the cementation of the grains of the crystalline phase with a melt that fills the pores and intergranular voids [71, 72].

In addition, LF itself, being to some extent a relaxant of mechanical stress, reduces the sensitivity of ceramics to such surface damage that leads to stress concentration [73].

The presence of different Mn ions in materials also determines other phenomena that affect the formation of microstructure.

This is the formation of oxygen vacancies, which occurs according to the following schemes: $A'^{+}Nb^{5+}{}_{1-x}Mn^{2+}{}_{x}O_{3-3/2x}\square_{3/2x}$ (1); $A'^{+}Nb^{5+}{}_{1-x}Mn^{3+}{}_{x}O_{3-x}\square_{x}$ (2); $A'^{+}Nb^{5+}{}_{1-x}Mn^{4+}{}_{x}O_{3-x/2}\square_{x/2}$ (3) ($\square$ is a vacancy). We assume that most of the super stoichiometric Mn will be in the *B*-positions of the original structure.

It is known that the vacancies are associated with ions that modify the Coulomb interaction, in the result of which the latter are adsorbed at the crystallite boundaries [74], which slows down their growth and increases the length of the intergranular boundaries.

In addition, it is necessary to consider the slower rate of oxygen vacancies in comparison with vacancies, formed in the *A*-sublattice [75]. The compromise between these two factors determines the kinetics of grain growth. Apparently, at the first stage, the number of oxygen vacancies is decisive (and the grain is formed of a smaller size), at the second stage, namely their speed, because of which slowly moving vacancies "have time" to participate in diffusion processes, contributing to the enlargement of the grain structure.

A similar situation, associated with the occurrence of LF, which actively affects the kinetics of recrystallization processes, develops with the introduction of other modifiers, for example, by $BiFeO_3$ (Fig. 5.16). Figure 5.17 shows phase diagrams of the Cu–O, Mn–O, Bi–O, Bi_2O_3–Fe_2O_3 systems [76–85], demonstrating the possibility of LF formation in them at certain cation–anion ratios and temperatures.

Thus, all the above indicates the decisive role of the crystal-chemical specificity of alkali metal niobates of both pure and modified elements, which is associated with the formation of low-melting eutectics, which often fundamentally change the kinetics of sintering (shrinkage) from solid-phase to liquid-phase and the nature of recrystallization, as well as the occurrence of oxygen vacancies. Many research groups, dealing with the problems of creating materials based on alkali metal niobates, note this fact [39, 40, 86, 87]. Figures 5.18 and 5.19, show typical microstructures of such materials obtained under the presence of LF (according to [37, 39, 40, 75, 86, 87]).

In view of the existing relationship between the characteristics of microstructure and macroproperties (dielectric, piezoelectric, etc.) [87], it is quite obvious that the nature of the formed microstructure of functional materials, the controllability of its variations in chemical composition and external influences, which are parts of the technology, can be a powerful lever for directed changes in consumer parameters and their combinations of intelligent materials for various purposes.

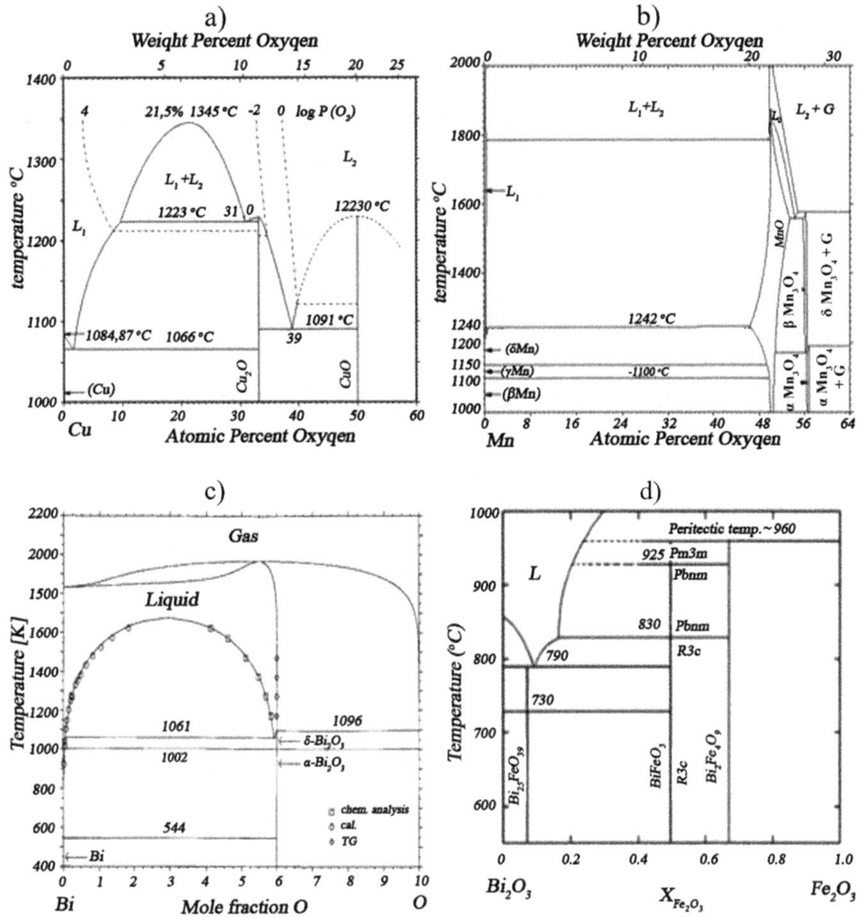

Fig. 5.17 Phase diagrams of Cu–O systems (**a**); Mn–O (**b**), Bi–O (**c**) and Bi_2O_3–Fe_2O_3 (**d**)

5.2 Phase Diagrams of States of Multielement Niobium Compounds with Low Concentrations of Dopants

It was radiographically established that base 1 has orthorhombic symmetry with a monoclinic perovskite cell R(M). The large width of the X-ray lines, the asymmetry, and the obvious splitting of lines 200 and 020 (Fig. 5.20) indicate that it is inhomogeneous and consists of several low-symmetric phases with similar cell parameters.

Considering the complex formula composition of the studied SS, it can be expected that in addition to the main R(M) phase with a small β-angle, characteristic of the SS $Na_{0.54}K_{0.46}NbO_3$, they may contain an R(M) phase with a high β-angle, corresponding to the SS $Na_{1-x}Li_xNbO_3$ at a low Li content.

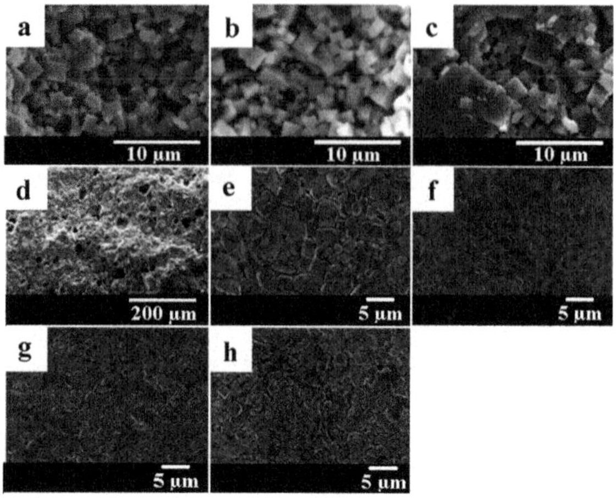

Fig. 5.18 Typical microstructures of niobate materials obtained in the presence of a liquid phase

Fig. 5.19 Typical microstructure of niobate materials based on (Na, K) NbO_3

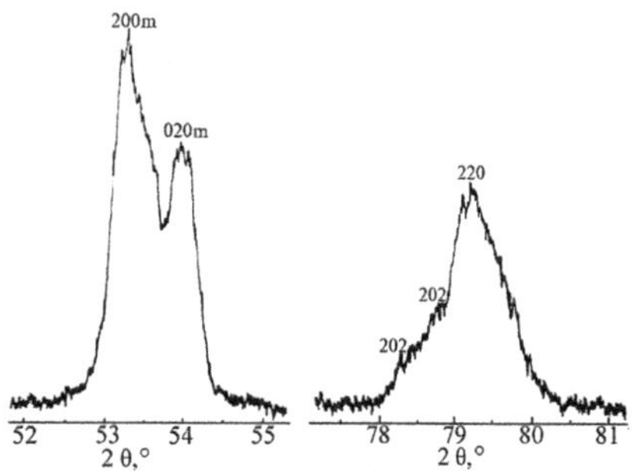

Fig. 5.20 X-ray lines of $(200)_c$ and $(220)_c$ of the initial solid solution (base 1)

5.2.1 MnO_2 Base Modification

X-ray phase analysis of the samples showed that all of them have a perovskite structure and do not contain impurities. Figure 5.20 shows the X-ray lines of $(200)_c$ and $(220)_c$ solid solutions super stoichiometrically and stoichiometrically modified by MnO_2.

Figure 5.21a shows that when MnO_2 is introduced into base 1 in the amount of 1 mass%, the intensity of line 020 corresponding to the rhombic distortion of

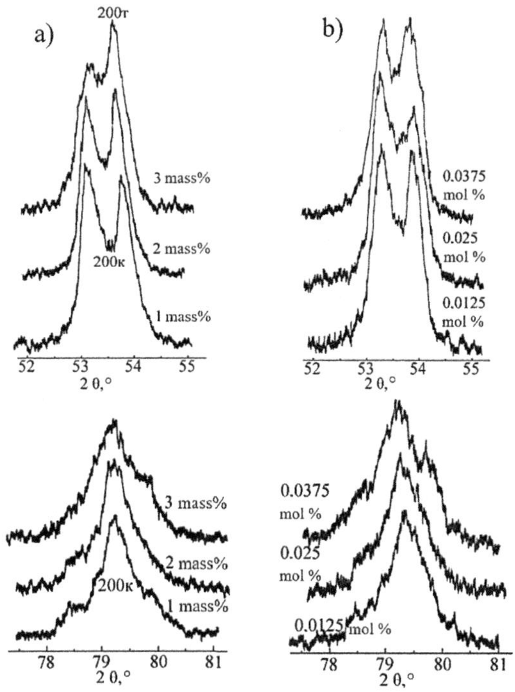

Fig. 5.21 X-ray lines of $(200)_c$ and $(220)_c$ of base 1 super stoichiometric (**a**) and stoichiometric (**b**) modified by MnO_2

the perovskite cell increases, indicating the appearance of a tetragonal (T) phase coherently coupled to the original matrix along the plane (020).

With an increase in the concentration of MnO_2, the content of the T phase increases. The symbat change in the concentrations of MnO_2 and T phases indicates that Mn^{+4} is embedded in the structure of SS. The peak next to the line 220 from the side of the large angles 2θ (≈ 79.7°) at 1 and 3 mass% of MnO_2 does not exactly correspond to the line 220 T of the phase, the parameters of which were calculated from lines 002 and 200. This can be either a different degree of distortion of the crystal lattice along the axes <100> and <110>, or another phase, modulation of the structure due to heterovalent modification by the Mn^{+4} ion, which is very different in radius and scattering capacity from the rest of the *B* cations.

In stoichiometric modification of MnO_2, there is no clear correlation between the concentration of the modifier and the phase composition of the SS (Fig. 5.21b). In samples with $x = 0.0125$ and 0.0375, there is an increase in the intensity of line 020, compared to the initial SS, possibly due to the appearance of the T phase. However, at $x = 0.025$, the intensity of line 020 decreases again, an additional peak appears near it (shown by an arrow), and a slight splitting is visible on line 200, from which we can conclude that another low-symmetric phase is distinguished. At $x = 0.0375$, traces of this phase are still visible, and an additional peak appears near line 220. Due to the uncertainty of the diffraction pattern, the phase composition of the SS

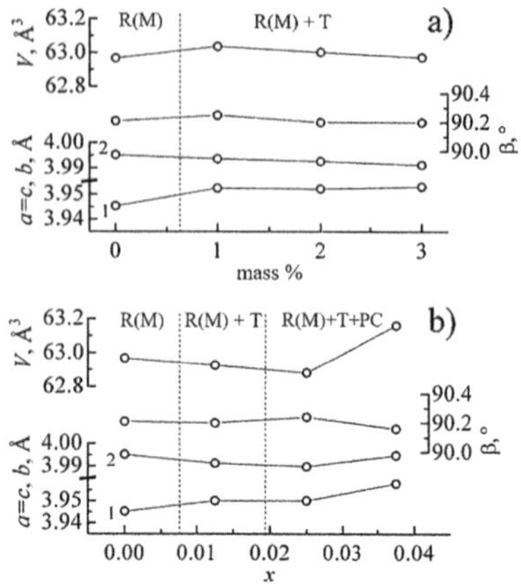

Fig. 5.22 Dependencies of the parameters of the perovskite monoclinic cell of the backbone 1, super stoichiometric (**a**) and stoichiometric (**b**) modified by MnO_2, on the content of the modifier: (1) b, (2) $a = c$

in the range of $0.0125 < x \leq 0.0375$ is reflected by us as a mixture of R(M), T and pseudocubic (Psc) phases.

Thus, based on the analysis of SS radiographs, it can be concluded that Mn^{+4} is included in the crystal lattice of backbone 1 in both modification schemes, but its placement in the structure is different.

Figure 5.22a, b shows the dependencies on x parameters a, b, angle of β and volume, V, monoclinic perovskite cell of the main P(M) phase of modified SS.

Figure 5.22a shows that with super stoichiometric modification, *a* decrease monotonically over the entire range of MnO_2 concentrations, and b, β, V increase with $0 < x \leq 1$ mass%. With further growth to 2 mass%, x and b remain practically constant, V decreases monotonically, and β decreases.

Even though the variations in the cell parameters are within the measurement error (except for b), the phased composition of the SS, which gradually changes with increasing x, allows us to consider this behavior of structural parameters reliable. Since the ionic radius Mn^{+4} has the smallest value of all B-cations ($R^6_{Mn^{+4}} = 0.52$ Å, $R^6_{Nb^{+5}} = 0.66$ Å, $R^6_{Ta^{+5}} = 0.66$ Å, $R^6_{Sb^{+5}} = 0.62$ Å along Belov–Bokiy). An increase in V may be due to the formation of a SS of Mn^{+4} insertion into the crystal lattice, and a subsequent decrease in V may be due to the formation of a SS substitution. On the other hand, the volume changes are so small that, most likely, these two processes occur simultaneously, resulting in the appearance of local areas of SS insertion and SS substitution of $Mn^{+4} \rightarrow B$-cation. Moreover, with an increase in x, the number of the former decreases, and the number of the latter increases. In fact, we have a picture of the morphotropic phase transition R(M) → T and we can expect the appearance of anomalous electrical properties here.

Figure 5.22b shows the dependencies of the parameters of the perovskite cell of the main monoclinic phase on x during stoichiometric inoculation of base 1 with manganese oxide. In the range of $0 < x \leq 0.025$, V of the cell decreases, in accordance with the inoculation formula, which implies the formation of SS substitution. A sharp increase in V at $x = 0.0375$ may be associated with either the formation of

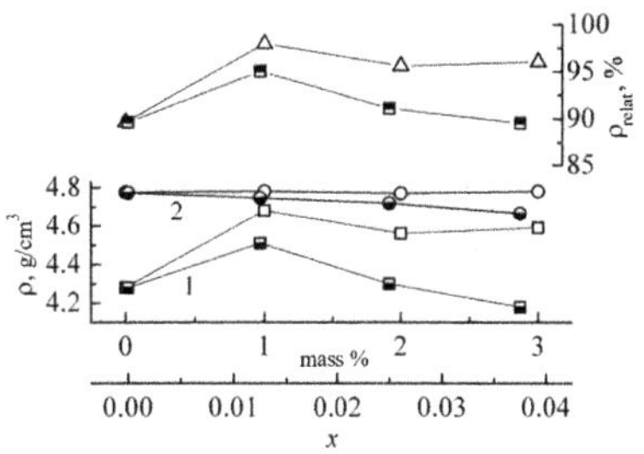

Fig. 5.23 Experimental (1), X-ray (2) and relative densities of ceramics, empty and filled icons indicate, respectively, the densities of solid solutions, super stoichiometrically modified and stoichiometrically modified by MnO_2

SS embedding, or with substitution of cations at *B*-positions with lower valence manganese ions ($R^6_{Mn^{+3}} = 0.70$ Å, $R^6_{Mn^{+2}} = 0.91$ Å).

Figure 5.23 shows the density dependencies of SS ceramics with different schemes of inoculation with manganese oxide. In both cases the inoculation significantly increases the density of base 1, however, with an increase in the concentration of the inoculant, high densities are preserved in the super-stoichiometric and fall in the case of stoichiometric inoculation.

That is, in the first case, SS are more resistant to small fluctuations in the concentration of the inoculant, which is important when using alkaline elements for the manufacture of ferroceramics. In the second case, very precise stoichiometry is required to achieve the maximum density of ceramics.

5.2.2 *Complex Modification of Base 1 by 2 Mass% NiO + MnO_2*

The addition of 2 mass% of NiO to the "base 1" makes the SS even more heterogeneous when it is modified by MnO_2. This is evidenced by the profiles of the X-ray lines $(200)_c$, $(220)_c$, shown in Fig. 5.24.

In the super stoichiometrically modified SS (Fig. 5.24a), at all MnO_2 concentrations, a mixture of two phases is observed (P(M) and T), with the largest amount of the T phase present in SS with 1 wt% MnO_2, while at 2 wt%, SS is very heterogeneous.

In stoichiometrically modified SS (Fig. 5.24b), as the concentration of MnO_2 increases, a gradual complication of the phase composition is traced from the P(M) phase in the initial SS to a mixture of two phases R(M) + T, at $x = 0.0125$, and then another phase is added to them, namely pseudocubic (Psc), the symmetry of which cannot be determined unambiguously due to strong blurring and overlapping of X-ray lines. The content of the T phase and Psc increases with an increase in the concentration of MnO_2.

Figure 5.25 shows the dependencies of the parameters of the perovskite monoclinic cell of the main R(M) phase on the concentration of MnO_2. In super stoichiometrically modified SS (Fig. 5.25a) at 1 mass% of MnO_2, a decrease in the *V* cell by

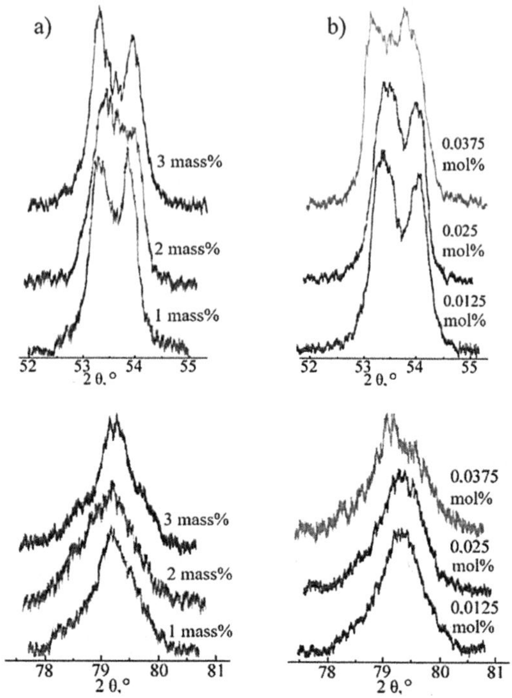

Fig. 5.24 X-ray lines $(200)_c$ and $(220)_c$ (bases 1 + 2 mass% NiO) super stoichiometrically (**a**) and stoichiometrically (**b**) modified by MnO_2

$\Delta V = -0.28$ Å^3 is observed. In the range of 1–2 mass% MnO, V remains constant, and at 2–3 mass%, it increases.

In stoichiometrically modified SS (Fig. 5.25b), at $x = 0.0125$, the cell volume, V, decreases by $\Delta V = -0.32$ Å^3. With a further increase of x, V grows monotonically.

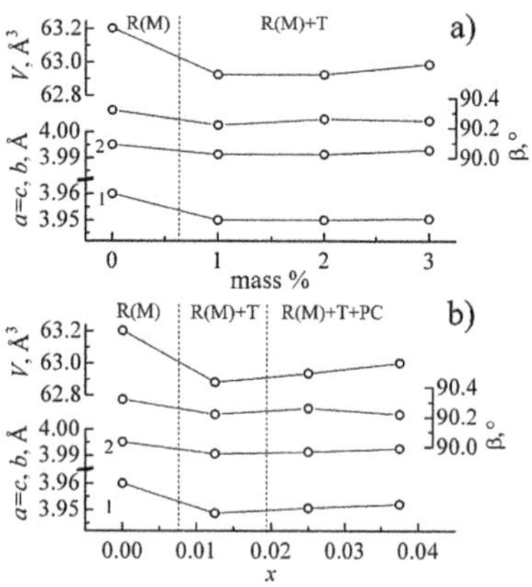

Fig. 5.25 Dependencies on x of the parameters of a perovskite monoclinic cell (base 1 + 2 mass% NiO), super stoichiometric (**a**) and stoichiometrically modified by MnO_2 (**b**): (1) b, (2) $a = c$

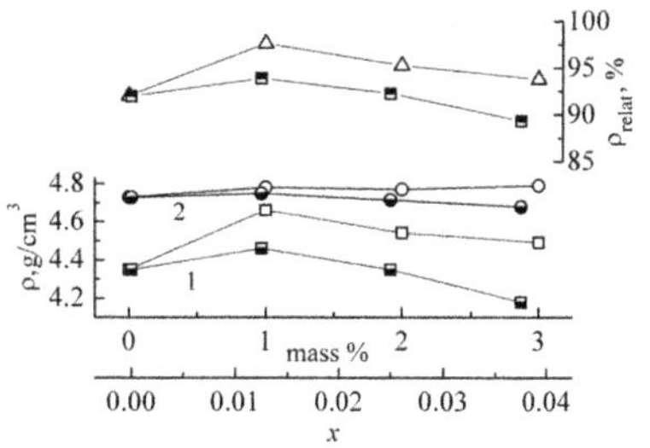

Fig. 5.26 Experimental (1), X-ray (2) and relative density of ceramics "base 1 + 2 mass% NiO", empty and filled icons indicate, respectively, the densities of solid solutions super stoichiometrically modified and stoichiometrically modified by MnO_2

This means that in both cases, with a minimum amount of MnO_2, there is a substitution of $Mn^{+4} \rightarrow B$-cation (ionic radius $R^6_{Mn^{+4}} = 0.52$ Å significantly less than the ionic radius of other cations).

An increase in V may be a consequence of either the dissolution of Mn^{+4} ions in the crystal lattice to form an insertion SS, or the formation of a substitution SS with Mn^{+3} or Mn^{+2} ions at the B-positions, since their ionic radius $R^6_{Mn^{+3}} = 0.70$ Å, $R^6_{Mn^{+2}} = 0.91$ Å greater than the ionic radius of any B-cation. Perhaps both processes are taking place.

Figure 5.26 shows the dependencies of ceramic densities on the concentration of manganese oxide. The maximum density of ceramics reaches the lowest amount of MnO_2, especially in super stoichiometrically modified ceramics.

The same pattern is observed in the monomodification of MnO_2, but in this case the densities are not so sensitive to variations in the composition of SS. This, as well as the less complex phase composition and greater homogeneity of the SS during monomodification, makes this scheme preferable.

5.2.3 *Modification of Base 1 by CuO*

Super stoichiometric inoculation of the base 1 with copper oxide, as well as MnO_2, leads to the appearance of the T phase, the content of which is maximum at 2 mass% CuO and minimum at 3 mass% CuO (Fig. 5.27).

Figure 5.28a, b shows the dependencies on x of the parameters of the monoclinic cell of the main R(M) phase (a) and the experimental, X-ray and relative densities of ceramics SS (b).

As shown in Fig. 5.28a, the changes in the linear parameters and volume of the perovskite cell with increasing concentration are non-monotonic. This indicates a complex and ambiguous mechanism for the distribution of Cu^{+2} ions in the structure.

An increase in V (2 mass%) may be associated with the formation of both SS insertion and substitution of the $Cu^{+2} \rightarrow B$ cation ($R^6_{Cu^{+3}} = 0.80$ Å, the radii of the remaining B-cations are much smaller, see the description of the modification MnO_2). The decrease in V at 3 mass% CuO may be a consequence of the decay of the supersaturated SS, because of which the amount of T phase decreases. Figure 5.28b

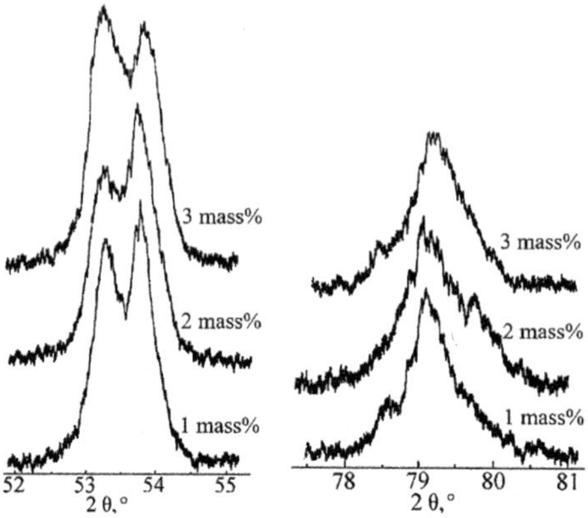

Fig. 5.27 X-ray lines of $(200)_c$ and $(220)_c$ of base 1 super stoichiometrically modified by CuO

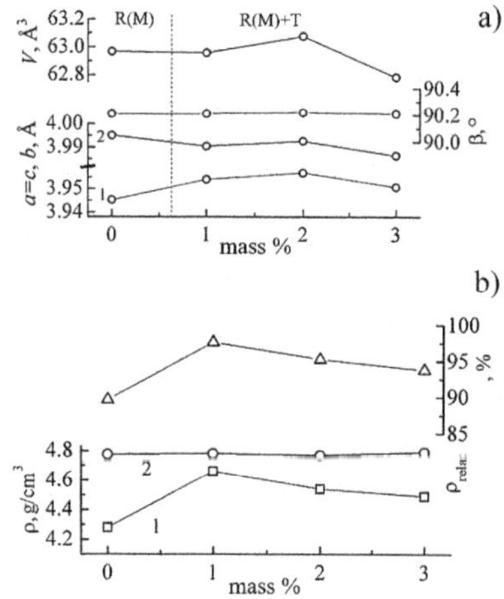

Fig. 5.28 Dependencies on x parameters: $a = c$ (2), b (1), β, V of the perovskite monoclinic cell (**a**); experimental (1), X-ray (2) and relative densities of the base 1, super stoichiometrically modified by CuO (**b**)

shows that the modification of SS with 1 mass% CuO significantly increased the density of base 1, but a further increase in the concentration of the modifier leads to its decrease.

5.2.4 *Modification of Base 1 by* $BiFeO_3$

X-ray phase analysis showed that all SS have a perovskite structure, and there are no extraneous phases.

Super stoichiometric introduction to base 1 of 0.5 mass% $BiFeO_3$ leads to the appearance, along with the R(M) phase, of the T-phase, the content of which increases with the increase in the concentration of the modifier up to 1.5 mass% $BiFeO_3$ (Fig. 5.29a).

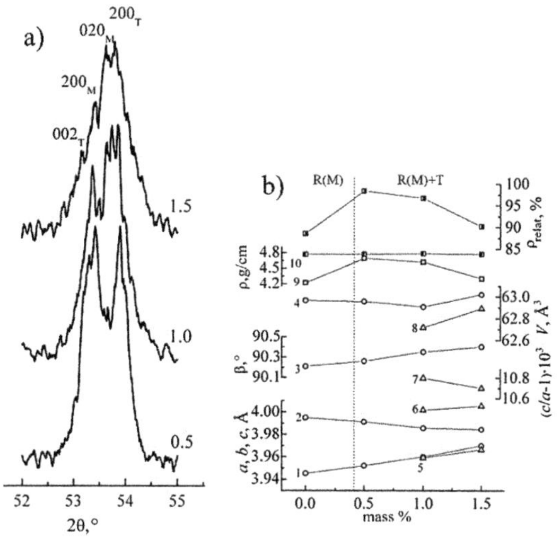

Fig. 5.29 X-ray lines $(200)_c$ (**a**) and the dependence of cell parameters and density of base 1 ceramics super stoichiometrically modified by bismuth ferrite on the content of modifier (**b**): 1—a_m, 2—b_m, 3—β, 4—V_m, 5—a_t, 6—c_t, 7—c/a, 8—V_t, 9—ρ_{meas}, 10—$\rho_{X\text{-ray}}$

Figure 5.29b shows that with an increase in the concentration of $BiFeO_3$, the linear parameters of the cell of both phases converge, the angle of monoclinity β increases, the volume of the M-cell first decreases, and above 1.0 mass% increases even at 1.5 mass% exceeds it in the base 1. Changes in the cell parameters and phase composition of the CC indicate that during its super stoichiometric inoculation with bismuth ferrite, the inoculant ions are embedded in the crystal lattice of the base 1. Given that the ionic radii of Nb^{+5}, Ta^{+5} and Fe^{+3} are the same ($R^6_{Nb^{+5}} = 0.66$ Å, $R^6_{Ta^{+5}} = 0.66$ Å, $R^6_{Fe^{+3}} = 0.67$ Å), it can be assumed that the decrease in the cell volume is associated with the formation of the SS replacement $Bi^{+3} \rightarrow K^{+1}$ ($R^6_{Bi^{+3}} = 1.20$ Å, $R^6_{K^{+1}} = 1.33$ Å). An increase in volume can be caused either by substitution $Bi^{+3} \rightarrow Na^{+1}$ ($R^6_{Na^{+1}} = 0.98$ Å), either by the introduction of Fe^{+3} ions into the crystal lattice of the base 1, or by both processes.

The density of ceramics, as in the previous cases, is maximum in composition with the smallest number of modifiers. The FE properties of the SS depend, among other things, on the degree of distortion of the perovskite cell, which is characterized by a homogeneous parameter of deformation δ, a decrease of δ makes the SS less ferrous. Table 5.1 shows the δ values for all modifiers for super stoichiometric and stoichiometric SS inoculation.

From the data in Table 5.1, it follows that, with the exception of super stoichiometric inoculation of $NiO + MnO_2$, in all other cases the introduction of the modifier reduces δ. Based on the combination of structural characteristics (the ratio of R(M) and T-phases), the homogeneity of SS, the value of δ, as well as the density of ceramics and its resistance to small fluctuations in composition, it can be concluded that the most promising modifiers are MnO_2 and $BiFeO_3$ during super stoichiometric inoculation.

Table 5.1 Homogeneous deformation parameter of the modified base 1

MnO_2		MnO_2		$NiO + MnO_2$		$NiO + MnO_2$		CuO		$BiFeO_3$	
Mass%	$\delta \times 10^3$	x	$\delta \times 10^3$	Mass%	$\delta \times 10^3$	x	$\delta \times 10^3$	Mass%	$\delta \times 10^3$	Mass%	$\delta \times 10^3$
0	5.8	0	5.8	0	5.8	0	5.8	0	5.8	0	5.8
1	5.7	0.0125	5.2	1	5.4	0.0125	5.5	1	4.9	0.5	5.6
2	5.2	0.025	5.4	2	5.8	0.025	5.7	2	4.8	1	5.2
3	5.0	0.0375	4.5	3	5.7	0.0375	5.3	3	4.8	1.5	4.7

5.3 Dependencies of Dielectric, Piezoelectric and Ferroelastic Properties of Alkaline Niobates on the Composition of Dopants, Doping Methods, Preparation Conditions and Morphology of Modified Samples

Figures 5.30, 5.31 and 5.32 show the dielectric, piezoelectric and ferroelastic characteristics of SS based on $0.98(Na_{0.54}K_{0.46})(Nb_{0.9}Ta_{0.1})$–$0.02LiSbO_3$, modified by MnO_2 (samples of different morphologies, obtained at different T_{sint}, were studied).

The fundamental difference in the behavior of macroparameters with different methods of administering supplements is clearly visible. For example, in the case of stoichiometric inoculation, the analyzed characteristics are significantly inferior

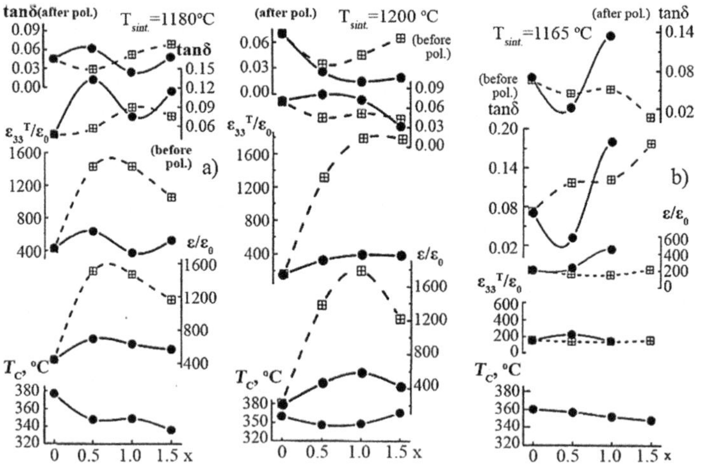

Fig. 5.30 Dependencies of dielectric characteristics of SS based on $0.98(Na_{0.54}K_{0.46})(Nb_{0.9}Ta_{0.1})$–$0.02LiSbO_3$, modified by MnO_2 (method of inoculant administration: left **a** super stoichiometric ($T_{\text{sint}} = 1180$ °C, $T_{\text{sint}} = 1200$ °C), right **b** stoichiometric ($T_{\text{sint}} = 1165$ °C)); solid lines define the samples of Ø 10 mm, dotted lines correspond to Ø 20 mm

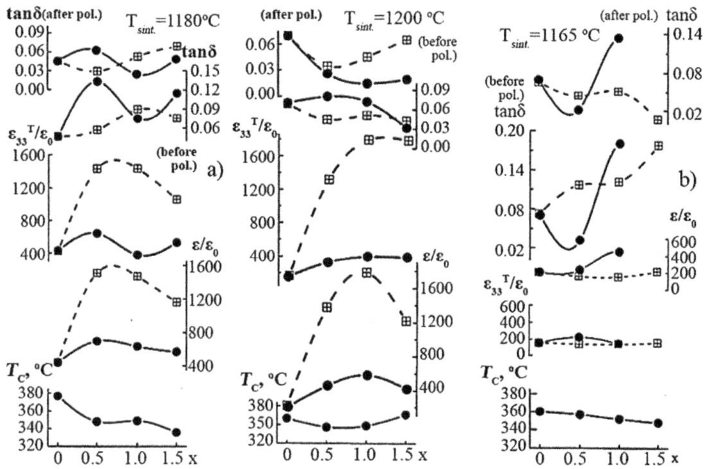

Fig. 5.31 Dependencies of piezoelectric characteristics of SS based on $0.98(Na_{0.54}K_{0.46})(Nb_{0.9}Ta_{0.1})$–$0.02LiSbO_3$, modified by MnO_2 (method of inoculant administration: left **a** super stoichiometric (T_{sint} = 1180 °C, T_{sint} = 1200 °C), right **b** stoichiometric (T_{sint} = 1165 °C)); solid lines define the samples of Ø 10 mm, dotted lines correspond to Ø 20 mm

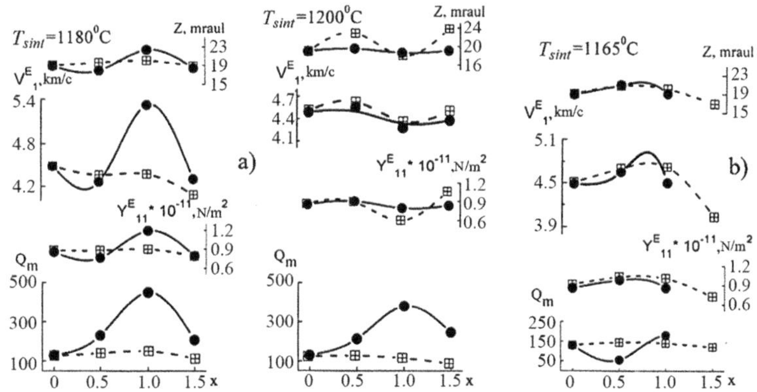

Fig. 5.32 Dependencies of ferroelastic characteristics of SS based on $0.98(Na_{0.54}K_{0.46})(Nb_{0.9}Ta_{0.1})$–$0.02LiSbO_3$, modified by MnO_2 (method of inoculant administration: left **a** super stoichiometric (T_{sint} = 1180 °C, T_{sint} = 1200 °C), right **b** stoichiometric (T_{sint} = 1165 °C)); solid lines define the samples of Ø 10 mm, dotted lines correspond to Ø 20 mm

to those in SS with super stoichiometrically introduced by MnO_2. The influence of technological factors (T_{sint}, diameter of samples) turned out to be ambiguous for each of the parameters regardless of the type of alloying of SS. The observation is undoubtedly related to the phase state of the SS under consideration, as well as to their grain structure (enlarged and more heterogeneous with the stoichiometric MnO_2).

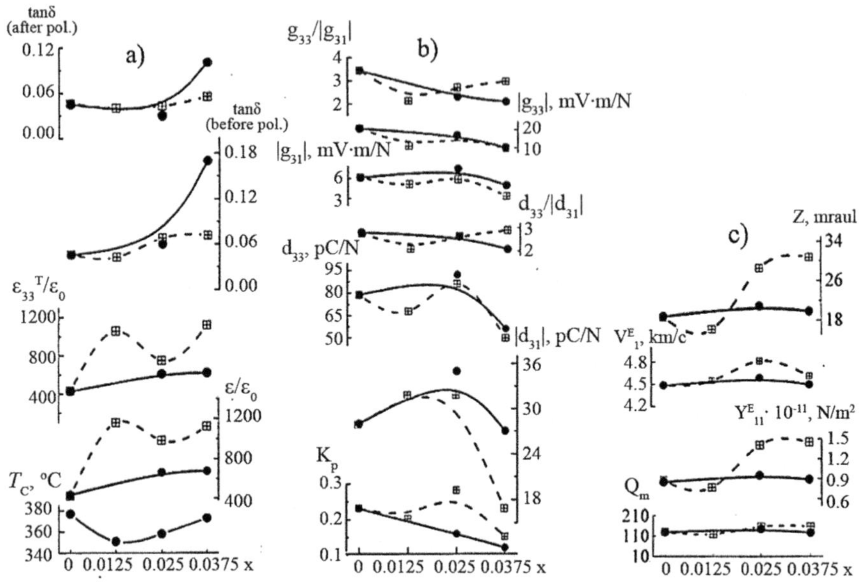

Fig. 5.33 Dependencies of dielectric (**a**), piezoelectric (**b**) and ferroelastic (**c**) characteristics of SS based on $0.98(Na_{0.54}K_{0.46})(Nb_{0.9}Ta_{0.1})$–$0.02LiSbO_3$, modified by a combination of MnO_2 + NiO; solid lines define the samples of Ø 10 mm, dotted lines correspond to Ø 20 mm

The addition of NiO to MnO_2 (Fig. 5.33) does not fundamentally change the dynamics of the behavior of the electrophysical characteristics of SS, which indicates the overwhelming role of MnO_2 in the "tandem" of MnO_2 + NiO, undoubtedly related, among other things, to the crystallochemical features of nickel and to the characteristics of the chemical bonds that form it (less covalent than in the case of MnO_2). The behavior of electrophysical parameters of copper-modified SS is also very similar to the observed behavior (Figs. 5.34, 5.35 and 5.36).

The reason is probably that all the above-described SS crystallize with the participation of the liquid phase, and this has a decisive effect on the formation of macroresponses, leveling the features of each of their modifiers.

However, it should be noted that when administered, MnO_2 and CuO increase sharply Q_m, $\varepsilon_{33}{}^T/\varepsilon_0$, tan$\delta$. At the same time, piezoelectric activity remains almost at the initial level. The increase in Q_m is probably a consequence of the following: manganese and copper cations have small ionic radii and a sufficiently high charge, which suggests that they are partially embedded in the crystal lattice in place of niobium according to the following schemes:

$$(Na_{1-x}K_x)(Nb_{1-y}M_y^{4+})O_{3-y/2}\square_{y/2}(M = \text{Mn};\ \square \text{ is a vacancy})$$

$$(Na_{1-x}K_x)(Nb_{1-y}M_y^{2+})O_{3-3y/2}\square_{3y/2}(M = \text{Cu};\ \square \text{ is a vacancy}).$$

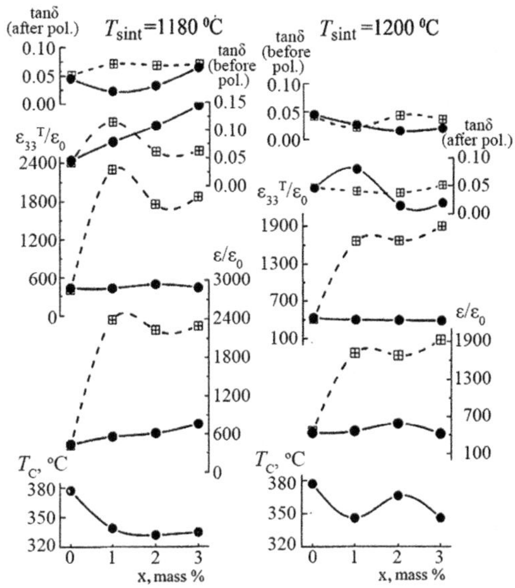

Fig. 5.34 Dependencies of dielectric characteristics of SS based on $0.98(Na_{0.54}K_{0.46})(Nb_{0.9}Ta_{0.1})$–$0.02LiSbO_3$, modified super stoichiometrically by CuO; solid lines define the samples of Ø 10 mm, dotted lines correspond to Ø 20 mm

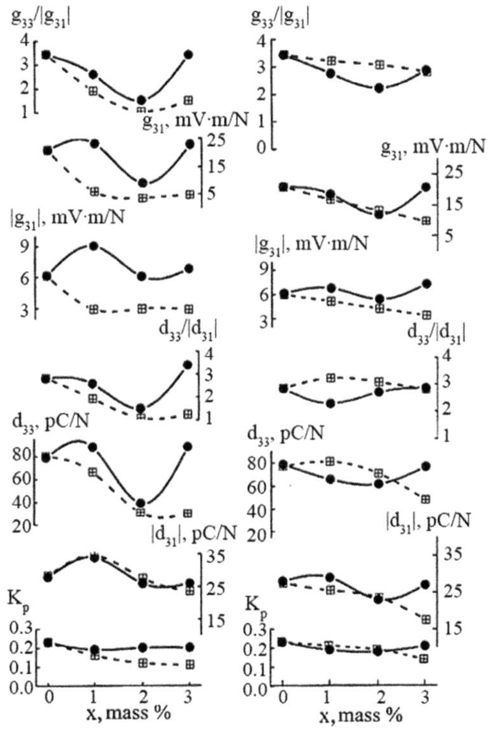

Fig. 5.35 Dependencies of piezoelectric characteristics of SS based on $0.98(Na_{0.54}K_{0.46})(Nb_{0.9}Ta_{0.1})$–$0.02LiSbO_3$, modified super stoichiometrically by CuO; solid lines define the samples of Ø 10 mm, dotted lines correspond to Ø 20 mm

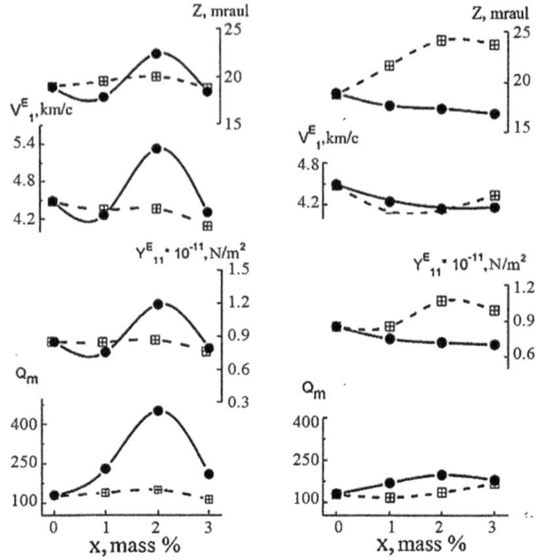

Fig. 5.36 Dependencies of ferroelastic characteristics of SS based on $0.98(Na_{0.54}K_{0.46})(Nb_{0.9}Ta_{0.1})$–$0.02LiSbO_3$, modified super stoichiometrically by CuO; solid lines define the samples of Ø 10 mm, dotted lines correspond to Ø 20 mm

This contributes to the formation of anion-deficient structures, which is accompanied by the formation of vacancies in the oxygen sublattice and the violation of the symmetry of the anionic environment of the *B*-cation, which contributes to an increase in the dipole moment and, as a result, an increase in the HF ceramics. In addition, as is known [6], oxygen vacancies are "traps" or blockers of domain reorientation. This characterizes the introduced MnO_2 as an HF additive. On the other hand, the increase in $\varepsilon_{33}{}^{T}/\varepsilon_0$ and tan δ speaks on the SH of modifiers. Such a "softening" of the SS may be a consequence of the following circumstances.

As evidenced by X-ray data, as the systems are saturated with modifiers, the δ decreases (homogeneous deformation parameter) and naturally [75] increases an average grain size, indicating the implementation of more favorable conditions for diffusion and recrystallization processes [75].

Increase of $\overline{D}$, $\varepsilon_{33}{}^{T}/\varepsilon_0$ at low δ defines the effect of the inverse dependence of the dielectric macroresponse on spontaneous polarization (spontaneous deformation, expressed through the structural parameter δ). A decrease in δ is associated with an increase in the mobility of domain walls and, as a result, an increase in $\varepsilon_{33}{}^{T}/\varepsilon_0$ [88]. Thus, just as in [78, 89], we noted an interesting feature of the result of the action of Mn (and in our case, Cu) on the initial SS: the modifiers simultaneously exhibit both HF and SF qualities, which is due to their crystallochemical features.

The stiffening effect of Mn in [78, 90] is explained by the high degree of covalence of the Mn–O bond, as evidenced by the increased values of electronegativity (EN) of Mn in the corresponding oxidation states, especially Mn (III) and Mn (IV) (see Table 5.2). Following [33], the cations Mn (II), Mn (III), Mn (IV) can be incorporated into the regular structures of double and triple perovskite-like oxides, occupying only the *B*-position. Mn (III) and Mn (IV) with high EN values will indeed significantly increase the degree of covalence of *B–O* bonds and, as a result, increase the HF SS.

Table 5.2 Crystallochemical parameters of cations that are part of solid solutions based on alkali metal niobates and chemical bond characteristics

Cation	R^a, Å [14]	Position of cations in PTS[b]	Electron polarizability			Electronic configurations of the outer layers [33]	EN[a], kcal/g at [94]	ΔEN[a], kcal/g·at [94]	The degree of covalence of the bond, % [94]
			α, Å^3 [95]	α, Å^3 [95]	α × 10^{24}, см^3 [96]				
1	2	3	4	5	6	7	8	9	10
K^{1+}	1.33	*A* and *B*	1.2–1.6	0.988	0.879	s^1	100	430	12
Na^{1+}	0.98	*A* and *B*	0.27–0.76	0.309	0.197	s^1	118	412	16
Li^{1+}	0.68	*B*	0.15–0.44	0.19–0.40	0.079	s^1	125	415	15
Nb^{5+}	0.66	*B*	5.78	3.10	–	ds^1	240	290	46
Nb^{4+}	0.67	*B*	–	–	–	–	215	315	39
Ta^{5+}	0.66	*B*	5.20	2.82	–	d^3s^2	210	320	38
Sb^{5+}	0.52	*B*	3.1	–	–	s^2p^3	270	260	54
Sb^{3+}	0.90	*B*	–	3.101	–	$(s^2)p^3$	195	335	34
Ni^{2+}	0.74	*B*	1.71	1.017	–	$(d^8)s^2$	220	310	40.7
Mn^{4+}	0.52	*B*	–	-	–	$(d^3)d^2s^2$	310	220	64.7
Mn^{3+}	0.70	*B*	3.83–4.05	2.14	–	$(d^4)ds^2$	260	270	51.4
Mn^{2+}	0.91	*B*	1.95–2.35	1.39	–	$(d^5)s^2$	170	360	27.5
Cu^{2+}	0.80	*B*	2.23–2.30	1.23	0.67	$(d^9)ds$	235	295	45
Cu^{1+}	0.98	*A* and *B*	4.42	2.272	1.81	–	177	353	29
Fe^{3+}	0.67	*B*	3.85–3.95	2.14	–	$(d^5)ds^2$	245	285	47
Fe^{2+}	0.80	*B*	1.97–2.30	1.339	–	$(d^6)s^2$	185	345	31
Bi^{3+}	1.2	*A* and *B*	–	3.949	–	$(s^2)p^3$	195	335	33
Bi^{5+}	0.74	*B*	–	–	–	s^2p^3	275	255	55

R is the radius of cations, EN is the electronegativity, ΔEN is the difference between electronegativities of oxygen and cation

[a] Boundaries of the oxide field with the perovskite structure (PTS) $\overline{R}_A \geq 0.90$ Å; 0.51 Å $\leq \overline{R}_B \leq 1.1$ Å; $\overline{R}_A \geq \overline{R}_B$; isomorphism: Δ*R* not more than 10–15%

[b] Placement of cations according to the *A* and *B* positions of the PTS

Ground state Mn (III) in octahedral field is $^5E_g(t^3{}_{2g}e_g)$ [91]. Due to the odd number of e_g electrons (unpaired electrons), Mn (III) has degenerate orbital electronic states, which, in accordance with the Jahn–Teller theorem [92], in a nonlinear polyatomic system leads to the instability of the initial full-symmetric configuration of nuclei quite large. It consists in a noticeable elongation of two bonds with a slight difference in the other four bonds [91].

In [91] it is noted that the spinel structure of Mn_3O_4 ($Mn_2O_3 \cdot MnO$) is also distorted: Mn (II) ions occupy tetrahedral voids, and Mn (III) ions occupy octahedral voids, and due to the distortion of octahedra, the lattice of this compound eventually turns from cubic to elongated tetragonal. Thus, the BO_6 octahedrons of the high-spin ion Mn (III) with the configuration $t^3{}_{2g}e_g$ should also be distorted by increasing the interatomic distances along one of the octahedron axes [91]. As a result, Mn (III) ions are displaced from the centers of the octahedrons they occupy. This provides an additional contribution to the overall distortion of the original unit cell [93]. By analogy with copper-containing perovskites, in which the Jahn–Teller effect determines the tetragonal nature of the crystal lattice distortion, it can be assumed that in the case of Mn (III), the manifestation of this effect in complex multicomponent SS will, if not lead, then significantly enhance the tetragonality of the cell. This is supported by the fact that the T-phase appears in modified SS (see Sect. 5.2).

The contribution of the Jahn–Teller effect to the total cell strain should result in the measured values of the strain parameter $\delta_{\text{mes.}}$ in the corresponding SS, consisting of the sum: $\delta_{\text{mes.}} = \delta + \delta_{\text{J-T}}$, where δ is a homogeneous strain parameter due to spontaneous polarization, $\delta_{\text{J-T}}$ is a strain parameter due to the Jahn–Teller effect. In [2], it has been shown that for different materials, $\delta_{\text{J-T}}$ can fluctuate from ~ 5 to ~ 15% of δ.

Unfortunately, even though the intensities of the multiple lines (200) indicate the presence of a T-phase in the SS, it is impossible to unambiguously determine the angular position of the reflection 002 of this phase. Therefore, its parameters and δ_{T} in Sect. 5.2 are not given.

However, it can be assumed that as the system is enriched with the introduced manganese, the δ_{T} of the emerging T-phase will increase, but at a rate less than the δ_{R} of the disappearing R-phase decreases. As a result, we can only record a sufficiently rapid fall in δ_{R}, which determines the softening of SS.

Thus, the hardening of the SS with Mn in regular positions is associated with an increase in the degree of covalence of the *B*–O bonds due to the higher EN Mn (III) and Mn (IV), and their softening with the manifestation of the Mn (III) Jahn–Teller effect.

It should be noted that in lead-containing materials based on the PZT [Pb(Ti, Zr)O_3] system, the occurrence of the Jahn–Teller effect does lead to a sharp increase in the consolidated δ [2] and, as a result, to a significant hardening of the materials. Even notes that the fact that many HF materials do not differ in a variety of properties (e.g., in size of $\varepsilon^{\text{T}}{}_{33}/\varepsilon_0$), that may be a consequence of the overwhelming effect of Mn, especially Mn (III) exhibiting the Jahn–Teller effect. Its contribution to the occurrence of the dipole moment turns out to be significant, and against this background, the specific features of the complex oxides of the form $\text{Pb}B'_{1-\beta}B''_{\beta}\text{O}_3$ are significantly leveled.

In our case, paradoxically, the manifestation of this effect was expressed in the softening of the SS due to a decrease in the basic R(M) phase (with the appearance of the T-phase), a decrease in δM and an increase in $\overline{D}$ (see Sect. 5.2).

Let us recall that as a measure of HF, it is convenient to take spontaneous deformation, expressed in the form of a homogeneous parameter of δ deformation (which characterizes the degree of deviation of an elementary crystal cell from a cubic cell [33]), since the magnitude of stresses arising during the reorientation of domains depends on it [97]. Indeed, the farther the shape of an elementary cell from a cubic cell, the more energy is spent on mechanical deformations that occur during the reorientation of domains. i.e. with an increase in the δ of HF, the SS increases. At the same time, the coercive field grows E_c, that is the domains' susceptibility to external influences decreases. Most of the parameters of piezoceramics can be associated with these two (δ, E_c) and the HF of materials can be evaluated based on them. Thus, with an increase in HF, in addition to δ, Q_m (mechanical Q-factor), V_R (speed of sound), T_C (Curie temperature), and $\varepsilon^T{}_{33}/\varepsilon_0$ (dielectric constant of polarized ceramics) and tanδ (dielectric loss tangent) decrease [97].

In view of the commensurability of the dimensions of Mn (IV) to the radius (0.56 Å [33]) of octahedral internodes in the densest cubic packing, it is quite likely that it will be in these positions. Such an assumption was made in [67] and was confirmed by the fact of an increase in the electrical resistivity ρ_v of Mn-containing ceramics.

As follows from Table 5.2, it may also be permissible to have Mn in the tetrahedral positions: Mn (II) occupies these positions in the oxide Mn_3O_4 [91], which presents always in the charges, considered by the SS.

The placement of large-sized Mn (II) in tetrahedral voids is probably due to its belonging to a group of ions with an incomplete outer layer, transitional from 8 to 18 electrons, which are characterized by a strong polarizing effect.

The general trend of polarization is a decrease in the number of oxygen frequencies from 6 to 4 for Mn (IV), namely with a ratio of its radius to the radius of oxygen favorable for the tetrahedral configuration.[1] Such a localization is generally common in oxygen compounds with the densest packing of atoms [100]. Additional factors contributing to the formation of the Mn (II) cation not by the substitution SS (Mn (II) in the octahedral positions), but by the insertion SS (Mn (II) in the tetrahedral positions) are the greatly different energy characteristics of the cations, determined by their charges [101].

The possibilities of implementing these processes are facilitated in multicomponent systems, due to the variety of structure-forming cations and their mutual influence. Tetrahedral voids in alkali metal niobates can be formed at the joints of blocks of a regular structure, following niobium oxides [102]. Filling such vacancies can lead to a decrease in the mobility of interblock boundaries, and, therefore, to the stabilization of the structure. It should be noted that the presence of Mn cations in different valence states in SS compositions can be a favorable prerequisite for their ordering and, as a result, cooperative effect on the properties of objects [103].

In Sect. 5.2, it was noted that the introduction of Mn leads to the appearance of LF, which facilitates recrystallization processes and, as a result, leads to the enlargement of the grain structure. The latter circumstance is associated with an increase in the number of domains, an increase in their mobility due to the action of a smaller bulk charge field [104], an increase in the degree of domain reorientation in the process of polarization of ceramics, and, as a result, a softening of SS.

[1] The $r_{Mn(IV)}$, corrected for tetrahedral coordination, is 0.50 Å, $r_{Mn(IV)}/r_O \approx 0.37$, which is included in the range of 0.225–0.415 characteristic of tetrahedral coordination [98, 99]. The degree of correspondence of the structures to the calculated ratio of ionic radii in the case of perovskites is high. Thus, from the geometric conditions of the existence of the perovskite structure, it follows that 0.51 Å $\leq r_B \leq$ 1.1 Å [33], that is the ratio r_B/r_O should be in the range of 0.38–0.81. This is even broader than allowed for octahedral coordination (0.415–0.732) by the Magnus-Bragg rule [98, 99].

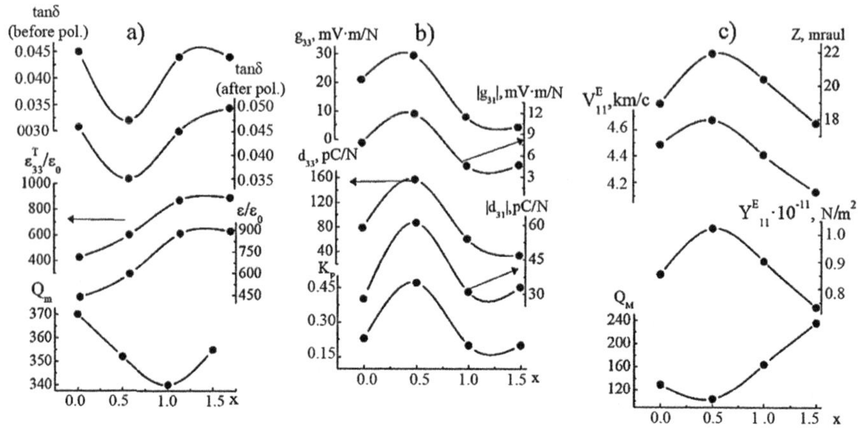

Fig. 5.37 Dependencies of dielectric (**a**), piezoelectric (**b**) and ferroelastic (**c**) characteristics of SS based on $0.98(Na_{0.54}K_{0.46})(Nb_{0.9}Ta_{0.1})$–$0.02LiSbO_3$, modified super stoichiometrically by Bi_2O_3 + Fe_2O_3 (T_{sint} = 1180 °C); solid lines define the samples of Ø 10 mm, dotted lines correspond to Ø 20 mm

The described "dualism" of the introduced manganese, manifested in the simultaneous stiffness and SF of the properties of SS, formed by it, makes it unique in comparison with other elements that act as modifiers.

Thus, the analysis of the crystallochemical characteristics and features of the properties of manganese ions in various valence states made it possible to present a possible scheme of its entry into the structure of the initial SS and, in accordance with it, to clarify the mechanism of influence on the SF of the modified compositions.

Figure 5.37 shows the dependencies of the electrophysical characteristics of SS, modified with the combined modifier $Bi_2O_3 + Fe_2O_3$.

It is clearly seen that with an increase in the content of the modifier (x = 1.0 mass%) the piezoelectric parameters increase sharply, then, with a further increase in the modifier concentration to x = 2.0 mass%, there is a decrease in piezoelectric activity and its slight increase at x = 3.0 mass%. Electromechanical characteristics change as follows: $V^E{}_1$, z, $Y^E{}_{11}$ first increase (x = 1.0 mass%) and then decrease at x = (2.0–3.0) mass%; Q_m, on the contrary, decreases at x = 1.0 mass% and increases in the range of x = (2.0–3.0) mass%. Such behavior is characteristic of ferroelectric SS in the region of the morphotropic phase boundary and, obviously, is a consequence of the tetragonal-monoclinic transition [105].

Comparison of the results of the effect of all the above additives on the matrix base (Fig. 5.38) showed that the simple modifier MnO_2 and the combined $Bi_2O_3 + Fe_2O_3$ are the most effective. Introduced super stoichiometrically, these additives enhance the piezoelectric activity of the initial compositions, their mechanical Q-factor, which is important for the use of the developed materials in devices operating in power modes (piezotransformers, piezomotors, high-voltage generators). The manifestation of SF properties in MnO_2 can also be useful for the use of such media in low frequency receiving devices (hydrophones, microphones, seismic receivers).

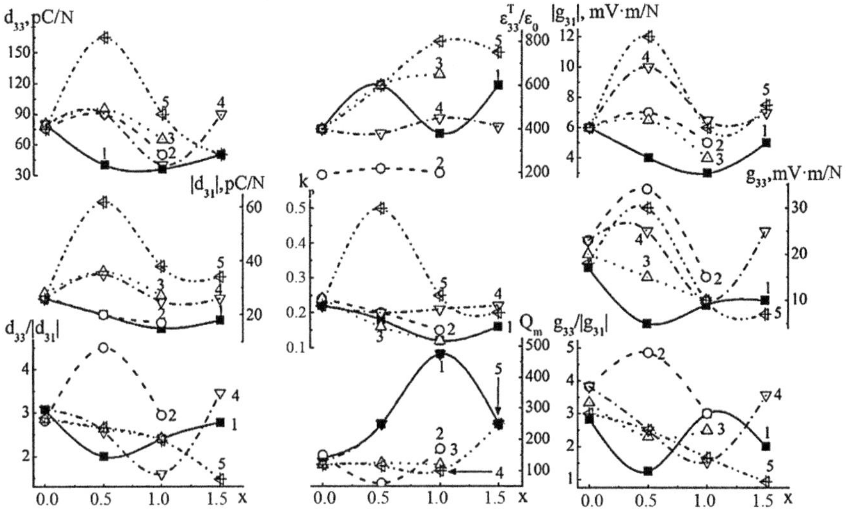

Fig. 5.38 Dependencies of dielectric, piezoelectric and ferroelastic characteristics of SS based on $0.98(Na_{0.54}K_{0.46})(Nb_{0.9}Ta_{0.1})$–$0.02LiSbO_3$, modified (1) super stoichiometrically by MnO_2, (2) stoichiometrically by MnO_2, (3) by combination of $MnO_2 + NiO$, (4) super stoichiometrically by CuO, (5) super stoichiometrically by $Bi_2O_3 + Fe_2O_3$

5.4 Conclusions

The fifth chapter has presented the studies of multielement compositions, based on solid solutions sodium–potassium niobates of compositions $(Na_{0.46}K_{0.54})_{1-x}Li_x(Nb_{0.9}Ta_{0.1})_{1-x}Sb_xO_3$ ($x = 0.02$–0.06). The compositions were modified super stoichiometrically and stoichiometrically with MnO_2, CuO, NiO, Bi_2O_3, Fe_2O_3, and combined additives. The results of studies, investigating the kinetics of the formation (synthesis and sintering) of ceramics, based on alkali and alkaline-earth metal niobates, were discussed. It was demonstrated that sintering invariably occurs in conjunction with the liquid phase, which exerts a profound impact on the fundamental stages of diffusion, polymorphic transformations, recrystallization, dissociation, and other processes. The impacts of each additive, the alloying process, the preparation environment, and the morphological characteristics of the modified samples on the structure and behavior of the dielectric, piezoelectric, and ferroelectric properties of multielement alkaline niobates were intensively researched.

References

1. Danciger, A.Ya., et al: Multicomponent Systems of Ferroelectric Complex Oxides: Physics, Crystal Chemistry, Technology. Design Aspects of Ferro-piezoelectric Materials. Rostov State University Press, Rostov-on-Don, vol. 1, (2001), vol. 2 (2002) (In Russian)

2. Gavrilyatchenko, S.V., et al.: The stabilization of the properties of ferro-piezoelectric ceramics for filters. Ferroelectrics **214**, 273–274 (1998)
3. Sadykov, H.A., et al.: Effect of combined modification on the structure, microstructure and piezo response of lead-free materials based on the (Na, K)NbO_3 system. In: Proceedings of Fourth International Youth Symposium on Physics of Lead-Free Piezoactive and Related Materials (Analysis of the Current State and Development Prospects) (LFPM-2015), 2–6 Sept 2015, Rostov-on-Don, Tuapse, vol. 2, pp. 184–189 (2015) (in Russian)
4. Weissflog, A.: Circuit Theory and Technique of Measurements in the Decimeter and Centimeter Ranges. Soviet Radio, Moscow (1961). (in Russian)
5. Ananyeva, A.A., et al.: Piezo- and dielectric properties of modified potassium-sodium niobate ceramics. In: Methods for Obtaining and Studying Piezo- and Ferroelectric Materials and Feedstock for Them, pp. 89–95. MDNTP, Moscow (1966) (in Russian)
6. Ananyeva, A.A., Ugryumova, M.A.: Development and research of ferroelectric ceramics for electroacoustic transducers. In: Barium Titanate, pp. 201–207. Science, Moscow (1973) (in Russian)
7. Chernyshkov, V.A.: Physical properties of piezoceramics of lithium metaniobate and solid solutions based on it. Ph.D. Thesis (Physical and Mathematical Sciences), Rostov State University Press, Rostov-on-Don (1990) (in Russian)
8. Ugryumova, M.A.: Research and development of new piezoceramics based on solid solutions of barium and lead niobate, potassium and sodium niobate. Ph.D. Thesis (Technical Sciences), Moscow (1968) (in Russian)
9. Dambekalne, M.Ya., et al.: The influence of modification on the piezoelectric properties and texture of ceramics of solid solutions of potassium and sodium niobate. Electron. Technol. Ser. 14. Mater. (Inorganic Dielectrics) **8**, 33–35 (1969) (in Russian)
10. Freidenfeld, E.Zh.: Physico-chemical principles of synthesis and electrical properties of ceramic and glass-crystalline piezoelectrics. DrSc Thesis (Technical Sciences), Moscow (1972) (in Russian)
11. Ugryumova, M.A.: Piezoceramics based on (K, Na)NbO_3 solid solutions. Electron. Technol. Ser. 14. Mater. (Inorganic Dielectrics) **14**(20), 130–138 (1969) (in Russian)
12. Yanson, G.D., et al.: Kinetics of sintering of ferroelectric ceramics based on alkali niobates modified with oxides of alkaline earth elements. In: Proceedings of IV Intersectoral Meeting, Moscow, pp. 40–45 (1972) (in Russian)
13. Grineva, L.D., et al.: Study of the patterns of changes in physical properties when modifying solid solutions based on alkali metal niobates. In: Crystallization and Properties of Crystals, vol. 6, pp. 57–65. Novocherkassk (1979) (in Russian)
14. Yanson, G.D., et al.: Ceramic solid solutions of potassium and sodium metaniobates modified with alkaline earth metal ions. In: Piezo- and Ferroelectric Materials and Their Application, pp. 36–38. MDNTP, Moscow (1972) (in Russian)
15. Nassan, K., et al.: J. Appl. Phys. **41**, 533–537 (1970)
16. Kingery, W.D.: Introduction to Ceramics, 2nd edn. Stroyizdat, Moscow (1970) (in Russian)
17. Okazaki, K.: Technology of Ceramic Dielectrics. Energia, Moscow (1976). (in Russian)
18. Cross, E.: Lead-free at last. Nature **432**(4), 24–25 (2004)
19. Fuyuno, I.: Toyota's production line leads from lab to road. Nature **435**(7045), 1026–1027 (2005)
20. Reznichenko, L.A., et al.: Development of lead-free ferropiezoelectric materials science at the turn of the millennium. In: Mater. First International Interdisciplinary Symposium on Lead-free Ferroelectric Piezoelectric Ceramics and Related Materials: Production, Properties, Applications (Retrospective—Modernity—Forecasts) (LFFC—2012), Rostov-on-Don, pp. 9–19 (2012) (in Russian)
21. Sadykov, Kh.A., et al.: Effects of simple and combined modification of niobate materials that do not contain toxic elements. Constr. Compos. Mater. **3**, 45–55 (2013). (in Russian)
22. Saito, Y., et al.: Lead-free piezoceramics. Nature **432**, 84–87 (2004)
23. Verbenko, I.A., et al.: Preparation and dielectric properties of lead-free ceramics of the composition $[(Na_{0.5}K_{0.5})_{1-x}Li_x](Nb_{1-y-z}Ta_ySb_z)O_3$. Inorganic Mater. **45**(6), 762–768 (2009)

24. Danciger, A.Ya., et al.: Highly Efficient Piezoceramic Materials, vol. 30. Directory, Rostov-on-Don (1994) (in Russian)
25. Danciger, A.Ya., et al.: Highly Efficient Piezoceramic Materials. Search Optimization. Pike Publishing House, Rostov-on-Don (1995) (in Russian)
26. Chen, X., et al.: Phase transitional behavior and dielectric properties of (Ba, Ca) (Ti, Zr)O_3–Bi(Zn, Zr)O_3 lead-free ceramics. Ferroelectrics **488**, 104–111 (2015)
27. Hsiao, Y.-J., et al.: Growth and characterization of $NaNbO_3$ synthesized using reaction-sintering method. Mater. Sci. Eng. B **136**(2–3), 129–133 (2007)
28. Lu, G., et al.: Dielectric properties in the microwave range of $K_{0.5}Na_{0.5}NbO_3$ ceramics. J. Electron. Mater. **46**(1), 123–129 (2017)
29. Santos, I.C.M.S., et al.: Studies on the hydrothermal synthesis of niobium oxides. Polyhedron **21**(20), 2009–2015 (2002)
30. Verbenko, I.A., et al.: Lead-free piezoelectric ceramics and environmentally friendly technology for its production. Ecol. Ind. Prod. **4**, 45–47 (2007). (in Russian)
31. Yin, J., Wu, J., Wang, H.: Composition dependence of electrical properties in (1–x)$KNbO_3$–x$NaNbO_3$ lead-free ceramics. J. Mater. Sci. Mater. Electron. **28**(6), 4828–4838 (2017)
32. Stutes, A.L., Cui, K.: Influence of technological factors on the ferrite sintering process. In: Ceramics (Proceedings First and Second Conference British and Dutch Ceramic Societies), Metallurgy, Moscow, pp. 18–121 (1967) (in Russian)
33. Fesenko, E.G.: The Perovskite Family and Ferroelectricity. Atomizdat, Moscow (1972). (In Russian)
34. Cui, K.: Excessive anisotropic growth of grains and sintering in $MnFe_2O_4$ and $Y_3Fe_5O_{12}$. In: Ceramics (Proceedings of First and Second Conference on British and Dutch Ceramic Societies), Metallurgy, Moscow, pp. 22–32 (1967)
35. Hornstra, J.: The role of grain boundary motion in the last stage of sintering. Physica **27**(3), 342–350 (1961)
36. Reznichenko, L.A., et al.: The phenomenon of secondary discontinuous recrystallization in niobate ceramics. In: Proceedings of the 7th International Conference on Physics of Ferroelectric Semiconductors (IMFS-7), Rostov-on-Don, pp. 170–171 (1996) (in Russian)
37. Akçan, E., Yılmaz, H.: Sintering behavior and electrical properties of $K_4CuNb_8O_{23}$ modified $K_{0.5}Na_{0.5}NbO_3$ ceramics with SnO_2, ZnO or Yb_2O_3 doping. Ceram. Int. **41**, 3659–3667 (2015)
38. Reznichenko, L.A., et al.: Liquid phase in alkali metal niobates. In: Proceedings of the 7th International Conference on Physics of Ferroelectric Semiconductors (IMFS-7), Rostov-on-Don, pp. 149–151 (1996) (in Russian)
39. Wu, L., Ning, H.: Preparation and piezoelectric properties of CuO-added $(Ag_{0.75}Li_{0.1}Na_{0.1}K_{0.05})NbO_3$ lead-free ceramics. J. Wuhan Univ. Technol. Mater. Sci. Ed. **30**(4), 724–728 (2015)
40. Zheng, T., Zu, Y., Wu, J.: Composition dependence of phase structure and electrical properties in lead-free $(1\text{-}x)(K_{0.42}Na_{0.585})(Nb_{1\text{-}y}Sb_y)O_3$–$xBi_{0.5}K_{0.5}ZrO_3$ piezoceramics. J. Alloys Compd. **98**(6), 1829–1835 (2015)
41. Akhnazarova, V.V., et al.: Features of the grain structure of polycrystalline $NaNbO_3$ with connectivity of 3–0. Physics **52**(9), 63–69 (2009)
42. Andryushina, I.N., et al.: The PZT system ($PbZr_{1-x}Ti_xO_3$, $0.0 \leq x \leq 1.0$): specific features of recrystallization sintering and microstructures of solid solutions (Part 1). Ceram. Int. **39**, 753–761 (2013)
43. Reznichenko, L.A., et al.: Variations in the microstructure and properties of multicomponent ferroelectric ceramics as a result of its modification by barium. Ceram. Int. **40**, 15089–15095 (2014)
44. Tian, A., et al.: Effects of Na excess on the preparation and characterization of $(K_{0.5}Na_{0.5})NbO_3$ ceramics. Ferroelectrics **491**, 87–94 (2016)
45. Verbenko, I.A., et al.: Dielectric spectra and behavior when heated under a constant (polarization) electric field of lead-free ferroelectric ceramics of the multicomponent

system $[(K_{0.5}Na_{0.5})_{1-x}Li_x](Nb_{1-y-z}Ta_ySb_z)O_3$, modified with oxides of calcium, strontium, cadmium, copper, titanium. In: Proceedings of the 11th International Conference on Physics of Dielectrics (Dielectrics-2008), Saint Petersburg, vol. 1, pp. 194–197 (2008) (in Russian)
46. Matsuzuki, Y., et al.: AFM observation of the surface morphology and impurity effects on orthorhombic hen egg-white lysozyme crystals. J. Cryst. GrowthCryst. Growth **242**, 199–208 (2002)
47. Reznichenko, L.A., et al.: Influence of modification on the properties of multi-element composite piezoceramics. Constr. Compos. Mater. **2**, 61–67 (2014). (in Russian)
48. Plomp, M., McPherson, A., Malkin, A.J.: Crystal growth of macromolecular crystals: correlation between crystal symmetry and growth mechanisms. J. Cryst. GrowthCryst. Growth **242**, 306–311 (2002)
49. Zhang, K., et al.: Phase transition and piezoelectric properties of dense $(K_{0.48}, Na_{0.52})_{0.95}Li_{0.05}Sb_xNb(1\text{-}x)O_3$–$0.03Ca_{0.5}(Bi_{0.5}, Na_{0.5})_{0.5}ZrO_3$ lead free ceramics. J. Alloys Compd. **664**, 503–509 (2016)
50. Verbenko, I.A.: Multicomponent multifunctional electroactive media with different thermodynamic background. Ph.D. Thesis (Physical and Mathematical Sciences), Southern Federal University Press, Rostov-on-Don (2009) (in Russian)
51. Turova, N.Y.: Inorganic Chemistry in Tables. High Chemical College, RAS, Moscow, (1997) (in Russian)
52. Kuzminov, Yu.S.: Electro-optical and nonlinear-optical crystal of lithium niobate. Science, Moscow (1987). (in Russian)
53. Urusov, V.S.: Theory of Isomorphic Misceousness. Nauka, Moscow (1997). (in Russian)
54. Dambekalne, M.Y., et al.: Effect of modification on piezoelectric properties and texture of potassium and sodium niobate solid solution ceramics. Electron Technique, Ser. 14. Mater. **8**, 33–35 (1970) (in Russian)
55. Lv, X., et al.: Lead-free $KNbO_3$: ZnO composite ceramics. ACS Appl. Mater. Interfaces **8**(44), 30304–30311 (2016)
56. Urusov, V.S.: Theoretical Crystal Chemistry. Moscow State University Press, Moscow (1987) (in Russian)
57. Bahrevar, M.A., et al.: Microstructure and dielectric studies on polymorphic phase transition boundary in the KNN-LST piezoceramics synthesized by aqueous colloidal coating method. Ceram. Int. **41**, 11402–11408 (2015)
58. Razumovskaya, O.N., et al.: Thermogravimetric study of Pb_2NbMnO_6 and Pb_3WMnO_6. Acad. Sci. USSR. Inorganic Mater. **12**(3), 471–473 (1977)
59. Bykov, I.P., et al.: Effect of manganese ion recharging on the dielectric properties of solid solutions based on PZT. In: All-Union Science and Technology Conference on Actual Problems of Technology of Composite Materials and Radio Components in Microelectronic Information Systems, Yalta, vol. I, p. 42 (1990) (in Russian)
60. Reznichenko, L.A., Razumovskaya, O.N., Gavrilyachenko, S.V.: Role of manganese in the "tightening" of ferrotopyezoelectric materials. In: Proceedings of International Scientific Practical Conference on Fundamental Problems of Piezoelectric Instrumentation (Piezotechnics-99), Rostov-on-Don, vol. 1, pp. 98–108 (1999) (in Russian)
61. Geguzin, Y.E.: Physics of Sintering. Nauka, Moscow (1984). (in Russian)
62. Kainarsky, I.S.: Processes of Refractories Technology. Metallurgy, Moscow (1969). (in Russian)
63. Naidich, Yu.V., Lavrinenko, I.A., Eremenko, V.N.: The role of capillary phenomena in the process of compaction during sintering in the presence of a liquid phase. Powder Metall. **1**, 5–11 (1964). (in Russian)
64. Naidich, Yu.V., Lavrinenko, I.A.: Powder Metall. **9**, 32–35 (1967) (in Russian)
65. Naidich, Yu.V., Lavrinenko, I.A.: Sintering in the Presence of a Liquid Metallic Phase. Naukova Dumka, Kiev (1968). (in Russian)
66. Skorokhod, V.V.: Rheological Bases of the Theory of Sintering. Naukova Dumka, Kiev (1972). (in Russian)

67. Prisedskiy, V.V., Klimov, V.V., Klevtsova, L.G.: In: Proceedings of IV Intersectoral Conference on Methods for Obtaining and Analyzing Ferrite, Ferro-Piezoelectric and Capacitor Materials and Raw materials for Them, Moscow, pp. 72–73 (1973) (in Russian)
68. Freidenfeld, E.Zh.: Selection of glass for modification of PZT piezoceramics. In: Proceedings of V Interbranch Conference on Methods of Obtaining and Analyzing Ferrite, Ferroelectric, Piezoelectric and Capacitor Materials and Raw Materials for Them, Moscow, pp. 46–50 (1973) (in Russian)
69. Tien, T.X., Subbarao, E.C., Hrizo, Y.: Ferroelectric phase transitions in the system $PbTiO_3$–$KNbO_3$. J. Amer. Ceram. Soc. **45**(12), 572–575 (1962)
70. Orowan, E.: Fracture and strength of solids. Rep. Prog. Phys.Prog. Phys. **12**, 185–232 (1949)
71. Krasovsky, A.Y.: Physical Foundations of Strength. Naukova Dumka, Kiev (1977). (in Russian)
72. Ugryumova, M.A.: Influence of glass-forming oxides on the mechanical strength of piezoceramics PZT-13. In: Proceedings of the Piezo- and Ferroelectric Materials and Their Application. MDSTP, Moscow (1971) (in Russian).
73. Kleine, R.Z., Freidenfeld, E.J.: Some regularities of modification of ferroelectric ceramic materials by glass. Proc. Piezo- and Ferroelectric Materials and Their Application. MDSTP, Moscow, 95–97 (1975) (in Russian).
74. Zatsarinniy, V.P.: Strength of Piezoceramics, p. 208. Rostov State University Press, Rostov-on-Don (1978) (in Russian)
75. Smeltere, I., et al.: Lead-free ferroelectric ceramics based on sodium potassium niobate. In: Proceedings of the First International Interdisciplinary Symposium on Lead-free Ferroelectric Ceramics and Related Materials: Production, Properties, Applications (Retrospective—Modernity—Forecasts) (LFFC—2012), Rostov-on-Don, Loo, pp. 355–358 (2012) (in Russian)
76. Aminoff, G.: On the crystal structure of Hausmannite (Mn_2O_4). J. Crystallogr. Cryst. Mater. **64**, 475–490 (1926)
77. Chen, Q.Y.: Local equilibrium method in studying phase diagram of Mn-MnO system. Acta Metallurgica Sinica B **24**(6), 440–442 (1988)
78. Fesenko, E.G., et al: Piezoelectric ceramic material. USSR Patent No. 1008198, 01.12.82 (1982) (in Russian)
79. Haumont, R., et al.: Phase stability and structural temperature dependence in powdered multiferroic $BiFeO_3$. Phys. Rev. B **78**, 134108 (2008)
80. Jay, A.H., Andrews, K.W.: Constitution of some binary oxide systems. Nature **154**(3899), 116 (1944)
81. Lyakishev, N.P.: State Diagrams of Double Metal Systems. Mashinostroenie, Moscow (1996–2001) (in Russian)
82. Maitre, A., Francois, M., Gachon, J.C.: Experimental study of the Bi_2O_3-Fe_2O_3 pseudo-binary system. J. Phase Equilib. Diffus.Equilib. Diffus. **25**, 59–67 (2004)
83. Palai, R., et al.: β-phase and β-γ metal-insulator transition in multiferroic $BiFeO_3$. Phys. Rev. B **77**, 014110 (2008)
84. Risold, D., et al.: The bismuth-oxygen system. J. Phase Equilibria **16**(3), 223–234 (1995)
85. Tromel, G., et al.: The phase diagram of the manganese-oxygen system. World of Metall. Erzmetall **29**, 234–237 (1976)
86. Arnold, D.C., et al.: Ferroelectric—paraelectric transition in $BiFeO_3$: crystal structure of the orthorhombic beta phase. Phys. Rev. Lett. **102**, 027602 (2009)
87. Fuentes, J., et al.: Dielectric and piezoelectric properties of the KNN ceramic compound doped with Li, La and Ta. Appl. Phys. A **118**, 709–715 (2015)
88. Reznichenko, L.A., et al.: Features of morphotropic phase transition in the system of $(1-x)NaNbO_3$– $xPbTiO_3$, vol. 2. Kniga, Rostov-on-Don (1995) (in Russian)
89. Gusakova, L.G., et al.: Production and electrophysical properties of lead-free piezoceramics based on modified potassium-sodium niobate. Sci. Works Donetsk Natl. Tech. Univ. Chem. Chem. Technol. **15**, 86–92 (2010) (in Russian)

90. Grineva, L.D.: Study of the patterns of changes in physical properties when modifying ferroelectric solid solutions of complex oxides with a perovskite-type structure. Ph.D. Thesis (Physical and Mathematical Sciences), Rostov State University Press, Rostov-on-Don (1975) (in Russian)
91. Cotton, F., Wilkinson, J.: Modern Inorganic Chemistry, vol. 3. Mir, Moscow (1969) (in Russian)
92. Bersuker, I.B., Polinger, V.Z. (eds.): Vibronic Interactions in Molecules and Crystals. Springer Ser. Chemical Physics (1989)
93. Venevtsev, Y.N., Politova, E.D., Ivanov, S.A.: Ferro-antiferroelectrics of the Barium Titanate Family. Khimiya, Moscow (1954). (in Russian)
94. Povarnykh, A.S.: On the use of electronegative elements in crystal chemistry and mineralogy. Notes Ukrain. Branch All-Union Mineral. Soc. UkSSR AN, Kiev **3**(Pt. 1), 3–27 (1962) (in Russian)
95. Shannon, R.D., Fischer, R.X.: Empirical electronic polarizabilities of ions for the prediction and interpretation of refractive indices: oxides and oxysalts. Am. Miner. **101**, 2288–2300 (2016)
96. Skanavi, G.I.: Physics of Dielectrics (Area of Weak Fields). State Publishing House of Technical and Theoretical Literature, Moscow, Leningrad (1949). (in Russian)
97. Fesenko, E.G., et al.: New Piezoceramic Materials. Rostov State University Press, Rostov-on-Don (1983). (In Russian)
98. Narai-Szabo, I.: Inorganic Crystal Chemistry. Hungarian Academy of Sciences Press, Budapest (1969)
99. Ormont, B.F.: Introduction to Physical Chemistry and Crystal Chemistry of Semiconductors. High School, Moscow (1973). (in Russian)
100. Poray-Koshitz, M.A.: Practical Course of X-ray Structural Analysis. Moscow State University Press, Moscow (1960). (in Russian)
101. Evans, R.: Introduction to Crystal Chemistry. Goshimizdat, Moscow, Leningrad (1948) (in Russian)
102. Prisedsky, V.V., et al.: Extended defects in non-stoichiometric $PbZrO_3$. In: Proceedings of the II Meeting on Solid State Chemistry, Sverdlovsk, Pt. 1, p. 17 (1978) (in Russian)
103. Tretyakov, Yu.D.: Solid-Phase Reactions. Khimiya, Moscow (1979). (in Russian)
104. Trofimov, A.I.: Ultrasonic Systems for Monitoring Curvatures of Technological Channels of Nuclear Reactors. Energoatomizdat, Moscow (1994). (In Russian)
105. Gusakova, L.G., et al.: Lead-free nanostructured piezoceramic material based on (K, Na)NbO_3. Nanosyst. Nanomater. Nanotechnol. **10**, 303 (2012) (in Russian)

Chapter 6
Thermal Frequency Behavior of Multi-element Compositions Based on Alkali Metal Niobates

6.1 Features of Thermal Frequency Behavior of Modified Niobium Solid Solutions

The information gathered so far indicates that in the family of ferroelectric (FE) ceramic materials, perhaps the most complex are niobium-containing systems, and among them the compounds, based on the presence of alkali metals.

On the one hand, the mismatch between the radii of the alkali cations and the parameters of the interoctahedral cavities occupied by them in the perovskite structure causes its "looseness" and, consequently, the easy loss of stability, associated with the occurrence of a series of successive phase transformations and the formation of states that "respond" differently to external influences. To some extent, this is related to the manifestation of three properties of complex niobium oxides: compound polymorphism, morphotropy and non-stoichiometry of composition [1, 2].

Conversely, the crystallochemical characteristics of the structure contribute to the hydrolysis of initial niobium substances during synthesis and the narrowing of the interval of its optimal temperatures. Additionally, the increased reactivity of reagents and high volatility of alkali metals at relatively low temperatures, as well as the excessive anisotropic growth of giant ideomorphic grains during recrystallization, self-destruction of ceramics, and their polarization electrolysis, are also influenced by these features. Furthermore, these factors, along with numerous others, render the dependence of niobate materials' properties on their thermodynamic prehistory (conditions of obtaining) critical.

The known determining influence on the macro-properties of niobate media of the phase, particle size, defect, and rheological state of niobium pentaoxide, the main initial reagent involved in the synthesis of alkaline niobates, as well as the problems, associated with the variable valence of niobium, make the interpretation of many scientific results ambiguous. Furthermore, there are grounds for considering the kinetics of formation of this class of substances from the perspective of self-organization effects, which were previously highlighted in Ref. [3].

I. A. Verbenko et al., *Environmentally Friendly Smart Materials with Special Electrical and Magnetic Properties*, Engineering Materials,
https://doi.org/10.1007/978-3-031-90387-8_6

The objective of this chapter is to examine the distinctive characteristics of the thermal-frequency behavior, exhibited by niobium solid solutions, and to provide a scientific interpretation of these observations.

The findings of the investigation into the initial and modified SS based on AMN in response to a range of external influences, including temperature and frequency of alternating electric field, are presented in Figs. 6.1, 6.2, 6.3, 6.4, 6.5, 6.6, 6.7, 6.8, 6.9, 6.10, 6.11, 6.12, 6.13, 6.14, 6.15 and 6.16.

A common characteristic of all curves is the behavior of $\varepsilon'/\varepsilon_0$ ($\varepsilon_0 = 8.854 \times 10^{-12}$ F/m; the characteristic of ferroelectrics based on AMN, exhibits a somewhat blurred maximum at temperatures approaching the Curie point. Additionally, an

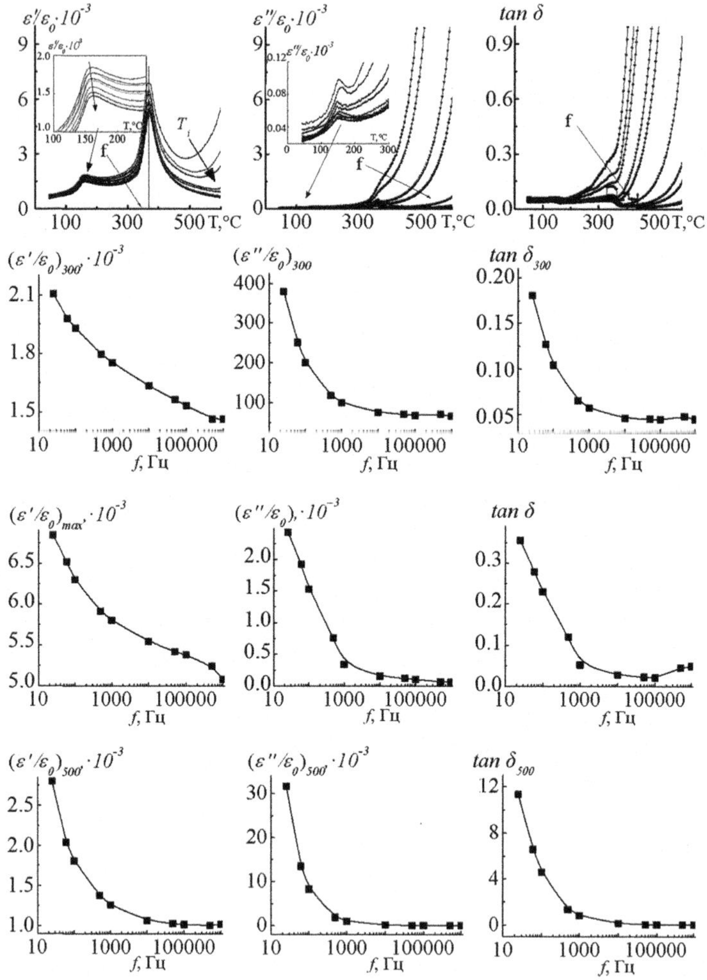

Fig. 6.1 Thermal frequency dependencies of dielectric characteristics of solid solutions of composition $(1 - y)(Na_{0.54}K_{0.46})(Nb_{0.9}Ta_{0.1})O_3 - yLiSbO_3$ ($y = 0.02$), ($T_{sint.} = 1180$ °C)

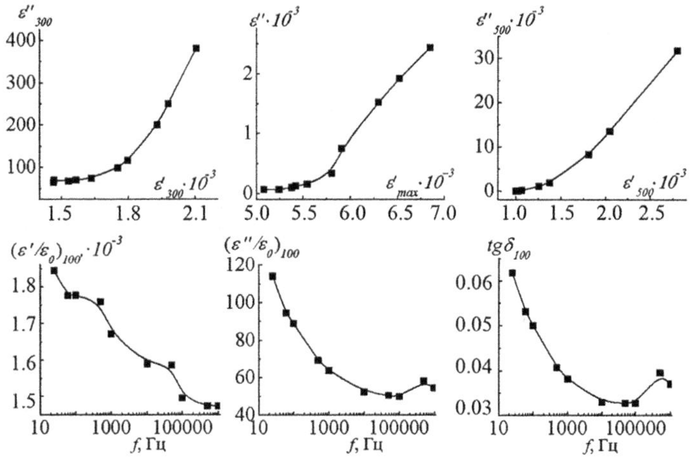

Fig. 6.1 (continued)

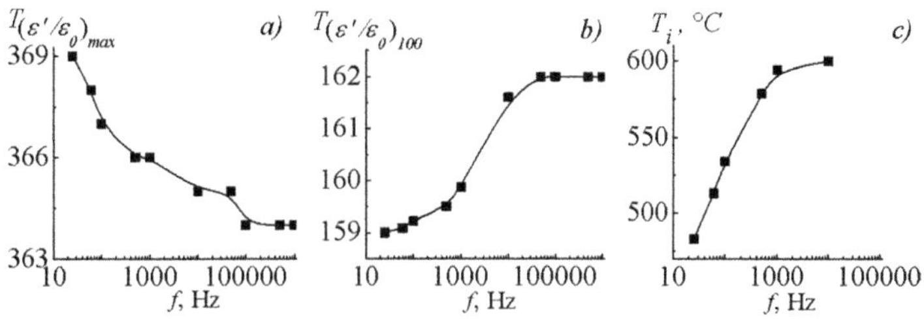

Fig. 6.2 Dependencies on the frequency f of the alternating electric field on: **a** the temperatures of maxima $(\varepsilon'/\varepsilon_0)$ near the Curie temperature, $T_{(\varepsilon'/\varepsilon_0)_{\max}}$; **b** in the vicinity of 100 °C, $T_{(\varepsilon'/\varepsilon_0)_{100}}$ and **c** the temperature of the beginning of the sharp growth of $\varepsilon'/\varepsilon_0$ in the paraelectric region, T_i

anomaly, or low relaxation maximum $(\varepsilon'/\varepsilon_0)$ is observed in the vicinity of 100 °C. A rapid increase in $\varepsilon'/\varepsilon_0$ is observed at temperatures T_i that exceed the temperature, corresponding to the peak value of $(\varepsilon'/\varepsilon_0)_{\max}$, and this increase is more pronounced when the frequency of the alternating electric field is higher. The formation of a relatively weak dispersion in the FE region and a strong one at the temperature, corresponding to the $\varepsilon'/\varepsilon_0$ maximum and in the paraelectric region, is observed, with an increase in strength as the temperature increases and the frequency decreases. A slight shift in the temperature of the $\varepsilon'/\varepsilon_0$ maximum, $\varepsilon'/\varepsilon_0$, $T_{(\varepsilon'/\varepsilon_0)_{\max}}$ towards lower temperatures and a decrease in the peak $(\varepsilon'/\varepsilon_0)_{\max}$ are observed as the frequency increases.

We will commence with an examination of the relaxation phenomena that are manifested in the temperature-frequency dependencies of the complex dielectric permittivity at temperatures $T \approx 100$ °C and $T > 400$ °C. The presence of slow polarization mechanisms in dielectrics gives rise to a time lag in the establishment

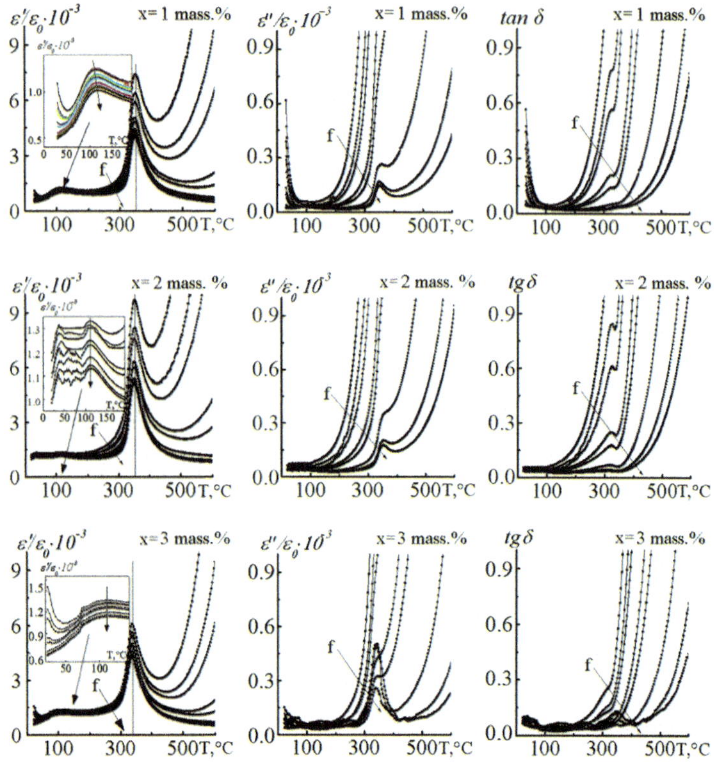

Fig. 6.3 Dependencies of $\varepsilon'/\varepsilon_0$, $\varepsilon''/\varepsilon_0$ and tan δ on temperature at different frequencies of solid solutions based on $(1-y)(Na_{0.54}K_{0.46})(Nb_{0.9}Ta_{0.1})O_3 - yLiSbO_3$ ($y = 0.02$), modified super stoichiometrically with MnO_2 (x) ($T_{\text{sint.}} = 1180$ °C)

of polarization P, at the application of a constant electric field E to the dielectric. In alternating electric fields, dielectric losses occur, which have a significant impact on dielectric spectra [3, 4].

The available experimental data indicates that dielectrics exhibiting Debye spectra are relatively uncommon. In contrast, a continuous distribution of relaxation times within an interval $\tau_1 \leq \tau \leq \tau_2$ is observed with greater frequency. The probability of a given relaxation time (τ) occurring is determined by the distribution function of relaxation times ($f(\tau)$), resulting in non-Debye dielectric spectra [4–9].

The formation of the low-temperature maximum $(\varepsilon'/\varepsilon_0)_{100}$ is likely attributable to the processes, associated with the adsorption and desorption of water by the surface of ceramic materials, which subsequently diffuses deep into the samples and the dissociation of both the "atmospheric" water and the "technological" water remaining in the samples after synthesis and sintering may also contribute to this phenomenon.

As the frequency increases, the value $(\varepsilon'/\varepsilon_0)_{100}$ decreases, the maximum itself becomes blurred and shifts to the right (on the abscissa axis) in the direction on the

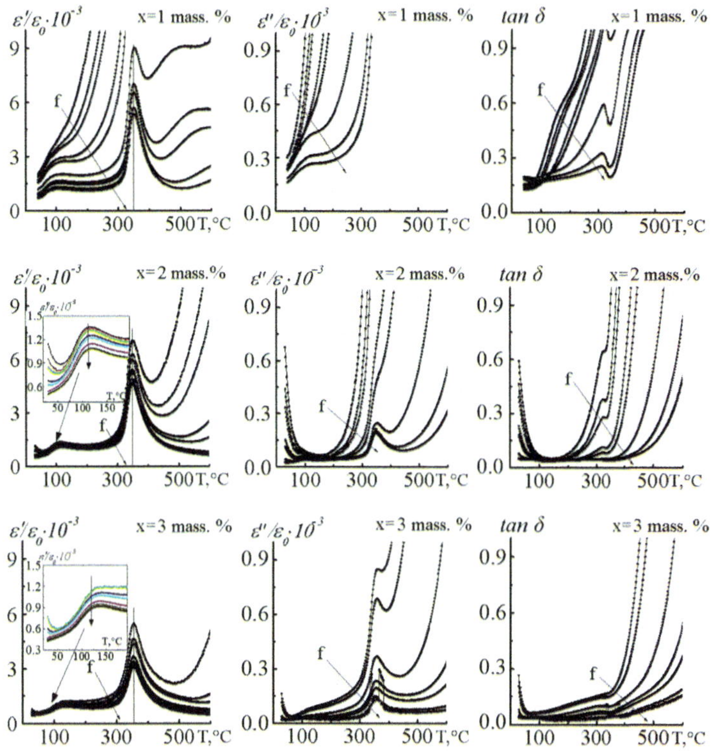

Fig. 6.4 Dependencies of $\varepsilon'/\varepsilon_0$, $\varepsilon''/\varepsilon_0$ and tan δ on temperature at different frequencies of solid solutions based on $(1-y)(Na_{0.54}K_{0.46})(Nb_{0.9}Ta_{0.1})O_3 - yLiSbO_3$ ($y = 0.02$), modified super stoichiometrically with MnO_2 (x) ($T_{\text{sint.}} = 1200$ °C)

temperature increase, indicative of a typical relaxation process. As previously indicated in Ref. [5], within the relaxation region, the polarization charges are unable to respond to the applied electric field in a timely manner. The contribution of the relaxants to the total polarization of the dielectric decreases, which results in a reduction in the $\varepsilon/\varepsilon_0$. In this instance, the frequency dependence $\varepsilon'/\varepsilon_0$ may exhibit multiple such declines, or 'steps', each of which corresponds to the 'switching off' of the subsequent polarization mechanism. This phenomenon can be clearly observed in Fig. 6.1 (curve $(\varepsilon'/\varepsilon_0)_{100}(f)$).

The physical interpretation of the rightward shift of the temperature maximum $(\varepsilon'/\varepsilon_0)_{100}$ at increasing frequency of the alternating electric field can be reduced to the following. A change in temperature has a significant impact on the relaxation time of the polarization establishment process in a dielectric. At low temperatures, the relaxation time is prolonged, resulting in incomplete establishment of the polarization process within a single half-period of applied voltage. This phenomenon contributes to the observed reduction in the dielectric constant. As the temperature rises, the time constant of the relaxation process declines markedly, allowing the process of

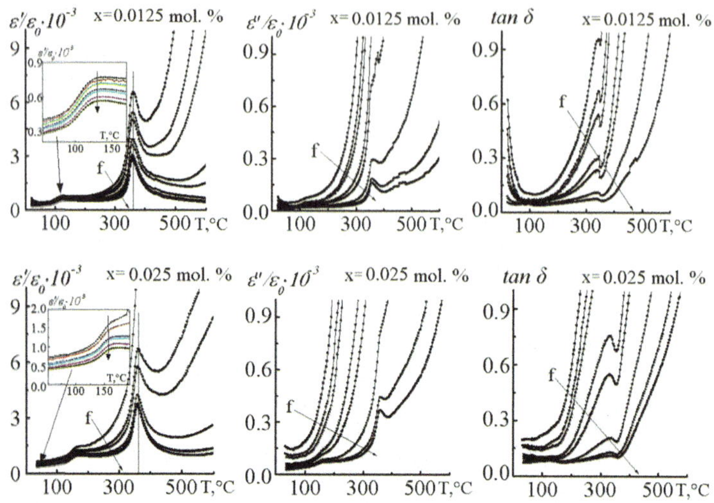

Fig. 6.5 Dependencies of $\varepsilon'/\varepsilon_0$, $\varepsilon''/\varepsilon_0$ and tan δ on temperature at different frequencies of solid solutions based on $(1-y)(Na_{0.54}K_{0.46})(Nb_{0.9}Ta_{0.1})O_3 - yLiSbO_3$ ($y = 0.02$), modified by Mn (x) stoichiometrically ($T_{sint.} = 1165$ °C)

polarization establishment during the half-period of the applied voltage to develop more extensively. This results in an increase in $\varepsilon'/\varepsilon_0$ (at AC voltage) with rising temperature. In the high frequency region, polarization will only have time to set up at very small relaxation times, that is, at elevated temperatures. This phenomenon is the underlying cause of the shift of $T_{(\varepsilon'/\varepsilon_0)100}$ to the high-temperature region. In other words, for the relaxors, lagging of the field requires a higher temperature the higher the frequency of the alternating electric field. In a certain sense, temperature can be considered the 'whip' that 'forces' the relaxors to contribute to the dielectric constant.

At temperatures exceeding 400 °C, relaxation phenomena exhibit distinctive characteristics associated with Maxwell–Wagner relaxation, which is contingent upon the disparity in dielectric permittivity and electrical conductivity between crystallites and intergranular interlayers [10]. The rapid growth of the real part of the dielectric permittivity $\left(\varepsilon'/\varepsilon_0\right)$ with temperature and the shift of the minima $\left(\varepsilon'/\varepsilon_0\right)$ to the high-temperature region with increasing frequency are both characteristic of this phenomenon.

The increase of $\varepsilon'/\varepsilon_0$ in the paraelectric region is a consequence of the reduction of $Nb^{5+} \rightarrow Nb^{4+}$ and, as a result, the formation of anion-deficient non-stoichiometricity with the emergence of oxygen vacancies. These vacancies have time at low frequencies to contribute to the dielectric permittivity by increasing the conductivity of ceramic grains (process known as interlayer Maxwell–Wagner polarization [11–13]). The reduction temperature of niobium in Nb_2O_5 monoxide is approximately 450 °C, with the reduction process continuing to at least 740 °C [14]. In multielement compositions, the process can commence at lower temperatures, even decreasing against

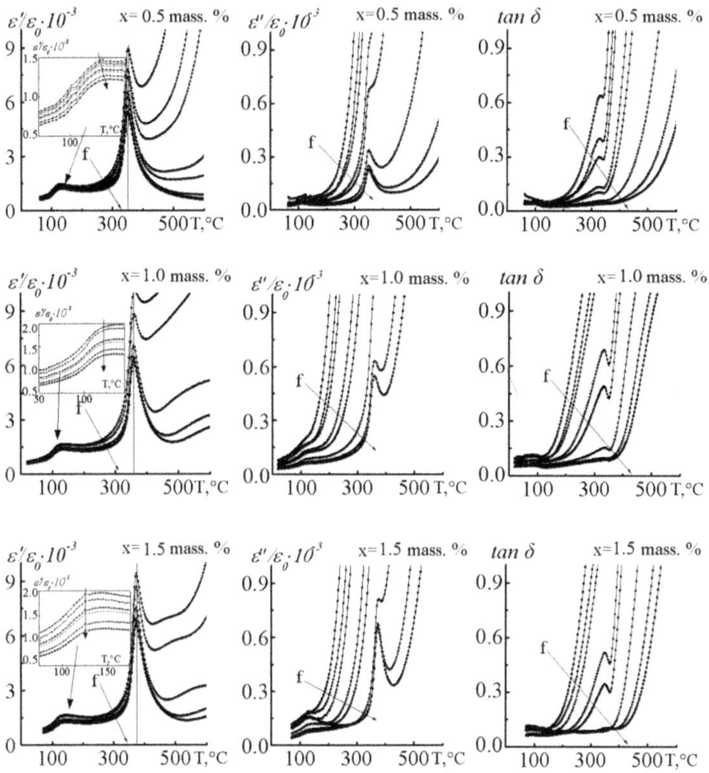

Fig. 6.6 Dependencies of $\varepsilon'/\varepsilon_0$, $\varepsilon''/\varepsilon_0$ and tan δ on temperature at different frequencies of solid solutions based on $(1 - y)(Na_{0.54}K_{0.46})(Nb_{0.9}Ta_{0.1})O_3 - yLiSbO_3$ ($y = 0.02$), modified super stoichiometrically by $NiO + MnO_2$ (x) (2 mass% NiO) ($T_{sint.} = 1165$ °C)

the background of structural rearrangements occurring 'nearby' (FE-PE transitions) due to the Headwall effect [14].

In relation to the shift of $T_{(\varepsilon'/\varepsilon_0)_{max}}$ towards lower temperatures as the frequency of the alternating electric field increases, it is necessary to perform several considerations.

As previously stated, at temperatures proximate to the FE-PE transition, oxygen vacancies emerge in niobate materials because of reductive processes, namely the conversion of Nb^{5+} to Nb^{4+}. In multielement compositions, due to the temperature decrease of Nb reduction, their 'throwing' into the FE region is possible.

The oxygen vacancies that emerge at this point begin to interact with the domain walls, thereby inhibiting their motion [15]. As the frequency increases, the number of slow-moving oxygen vacancies that become 'traps' (blockers) of domains decreases. At lower temperatures, the FE-PE transition and the formation of the maximum $(\varepsilon'/\varepsilon_0)_{max}$ are possible, which predetermines a leftward shift $T_{(\varepsilon'/\varepsilon_0)_{max}}$.

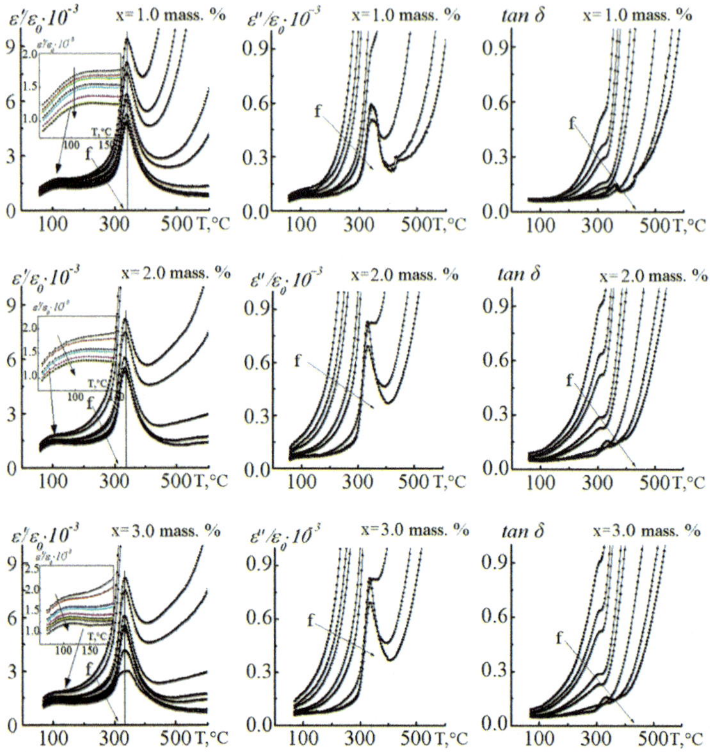

Fig. 6.7 Dependencies of $\varepsilon'/\varepsilon_0$, $\varepsilon''/\varepsilon_0$ and tan δ on temperature at different frequencies of solid solutions based on $(1-y)(Na_{0.54}K_{0.46})(Nb_{0.9}Ta_{0.1})O_3 - yLiSbO_3$ ($y = 0.02$), modified super stoichiometrically with CuO (x) ($T_{sint.} = 1180$ °C)

To substantiate this hypothesis, an experiment was conducted to ascertain whether the anomalous thermal-frequency behavior of $T_{(\varepsilon'/\varepsilon_0)_{max}}$ can be attributed to oxygen vacancies.

It should be noted that for complex oxides, including those with a perovskite-type structure (ABO_3), a slight deviation from the ideal stoichiometry is typical, particularly in the case of the AO component and oxygen. The composition of a non-stoichiometric binary perovskite-type oxide can typically be represented by the formula $A_{1-\gamma}BO_{3-\gamma-\delta}$, where the subscript γ reflects the deviation from stoichiometry in the content of the AO component, and δ is an additional independent non-stoichiometry in oxygen at a fixed value of γ, which varies at a fixed ratio of cations A and B in the solid phase. Each temperature is associated with the equilibrium state:

$$ABO_3 \rightleftarrows A_{1-\gamma}BO_{3-\gamma-\delta} + \gamma AO + \delta/2O_2, \tag{6.1}$$

which exhibits a specific non-stoichiometry. Additionally, as observed in [15, 16], the PTS is distinguished by a notable degree of γ-non-stoichiometry, indicating a

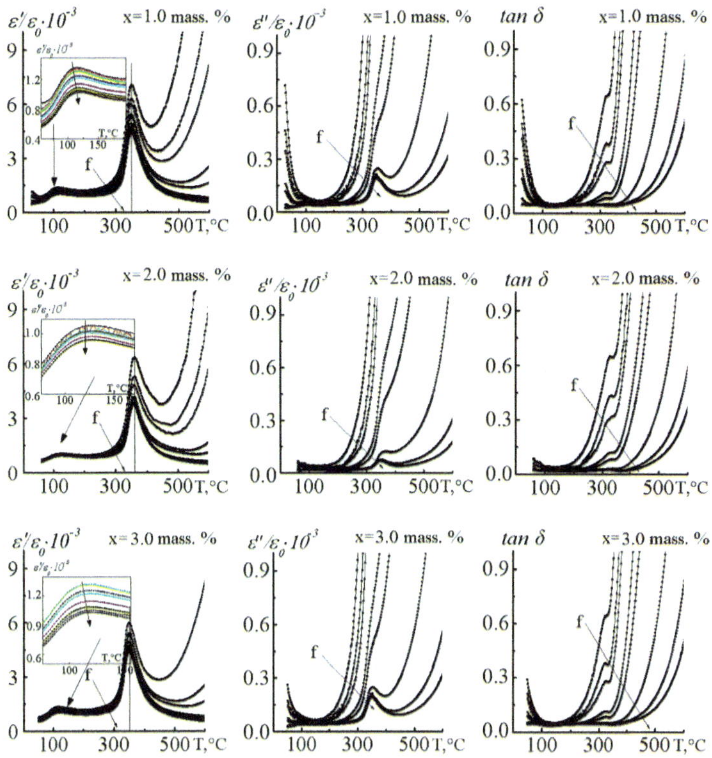

Fig. 6.8 Dependencies of $\varepsilon'/\varepsilon_0$, $\varepsilon''/\varepsilon_0$ and tan δ on temperature at different frequencies of solid solutions based on $(1 - y)(Na_{0.54}K_{0.46})(Nb_{0.9}Ta_{0.1})O_3 - yLiSbO_3$ ($y = 0.02$), modified super stoichiometrically with CuO (x) ($T_{sint.} = 1200$ °C)

deficiency in the *A*O component, while the δ-non-stoichiometry is considerably less pronounced. The equilibrium partial pressure of oxygen is dependent upon the conditions under which the materials are obtained, the type and quantity of modifying additives introduced. An increase in δ-non-stoichiometry, and consequently the number of oxygen vacancies, results in a shift of equilibrium (6.1) to the right. This is achieved by reducing the equilibrium partial pressure of oxygen.

The experiment was discerned through the application of a heat treatment to pre-sintered ceramics, conducted within a reducing atmosphere generated by the combustion of graphite within the furnace working chamber. The experimental design and a photograph of the furnace in which the ceramics were fired are presented in Fig. 6.17. It should be noted that the furnace was designed by Yu. M. Popov, a prominent technologist at the Research Institute of Physics of SFedU.

The furnace elements comprise a one-meter-long ceramic tube with an inner diameter of 0.022 m (1), equipped with metal flanges at either end (5). Two thermal chambers (2-A and 2-B) are positioned at 0.20 m from one another. The flanges are equipped with apertures into which thermocouples (3) are positioned; the extremities

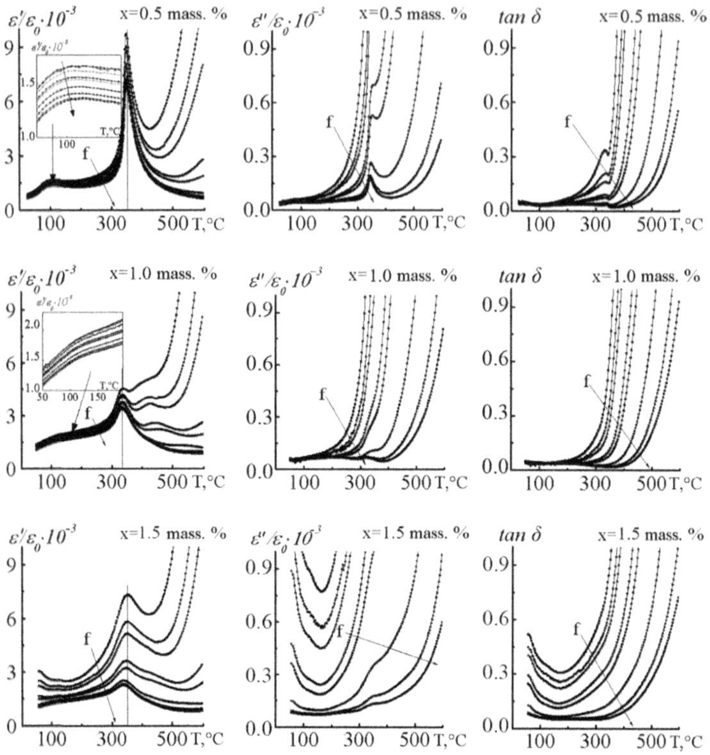

Fig. 6.9 Dependencies of $\varepsilon'/\varepsilon_0$, $\varepsilon''/\varepsilon_0$ and $\tan\delta$ on temperature at different frequencies of solid solutions based on $(1-y)(Na_{0.54}K_{0.46})(Nb_{0.9}Ta_{0.1})O_3 - yLiSbO_3$ ($y = 0.02$), modified super stoichiometrically with $Bi_2O_3 + Fe_2O_3$ (x) ($T_{sint.} = 1180$ °C)

of these thermocouples are connected to the core of each thermal chambers (the thermal chambers volume is 0.57 m^3). A boat constructed with graphite (4) is located at the center of thermo-chamber A. Another boat, containing a sample (7), is placed near the boat. Initially, thermochamber B is activated. When it 'reaches' a temperature of 1000 °C, the thermo-chamber A, in which the sample is placed, is switched on. We heat up to 840 °C with a dwell time of 1 h. Subsequently, thermocell A is deactivated, while thermocell B maintains the temperature, thereby sustaining the oxygen deficiency. Once the furnace B has cooled to 200 °C, the sample is removed. The combustion of graphite at temperatures exceeding 800 °C facilitates the binding of oxygen present in the surrounding working volume and the formation of a reducing atmosphere. The outcome is a ceramic material with diminished electrical conductivity and a high concentration of oxygen vacancies.

The mechanism of the process can be expressed by the following equations:

$$ABO_3 - \mathrm{xO} \rightarrow AB_{1-2x}(\mathrm{B\bar{e}})_{2x}\mathrm{O}^{2-}_{3-x}\square, \text{ for AMN}$$

$$A^{1+}\mathrm{Nb}^{5+}\mathrm{O}_3^{-2} - \mathrm{xO} \rightarrow A^{1+}\mathrm{Nb}^{5+}_{1-2x}\left(\mathrm{Nb}^{5+}\bar{\mathrm{e}}\right)_{2x}\mathrm{O}^{2-}_{3-x}.$$

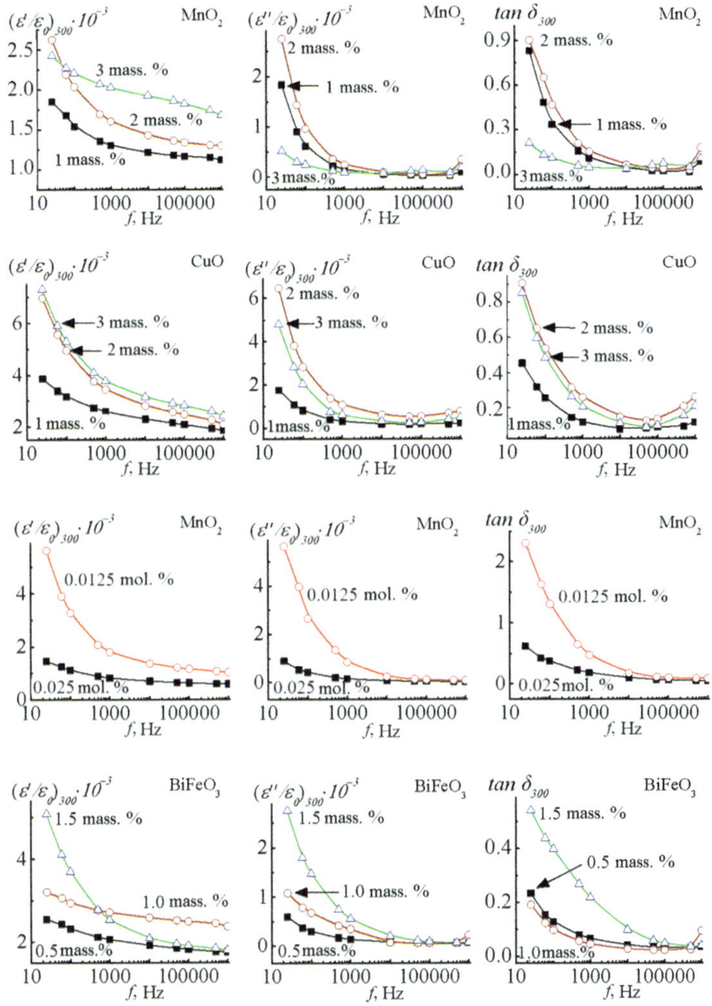

Fig. 6.10 Frequency f of the alternating electric field dependencies of the values $\varepsilon'/\varepsilon_0$, $\varepsilon''/\varepsilon_0$ and tan δ in the ferroelectric region (at 300 °C) of modified solid solutions with different additive contents (logarithmic scale on the abscissa axis)

In this instance, partial capture of electrons by niobium atoms results in the formation of tetravalent ions, thereby maintaining the electrical neutrality of the samples: $A^{1+}Nb^{5+}O_3^{-2} - xO \rightarrow A^{1+}Nb^{5+}_{1-x}Nb_x{}^{4+}O^{2-}_{3-x/2}\square_{x/2}$ (where $\square$ is a vacancy). The application of an electric field results in the movement of the electrons, thereby increasing the electrical conductivity of the material. The binding of oxygen by graphite is carried out until, at a certain concentration, a complex oxide with the highest possible oxygen non-stoichiometry is formed. As the sample is further

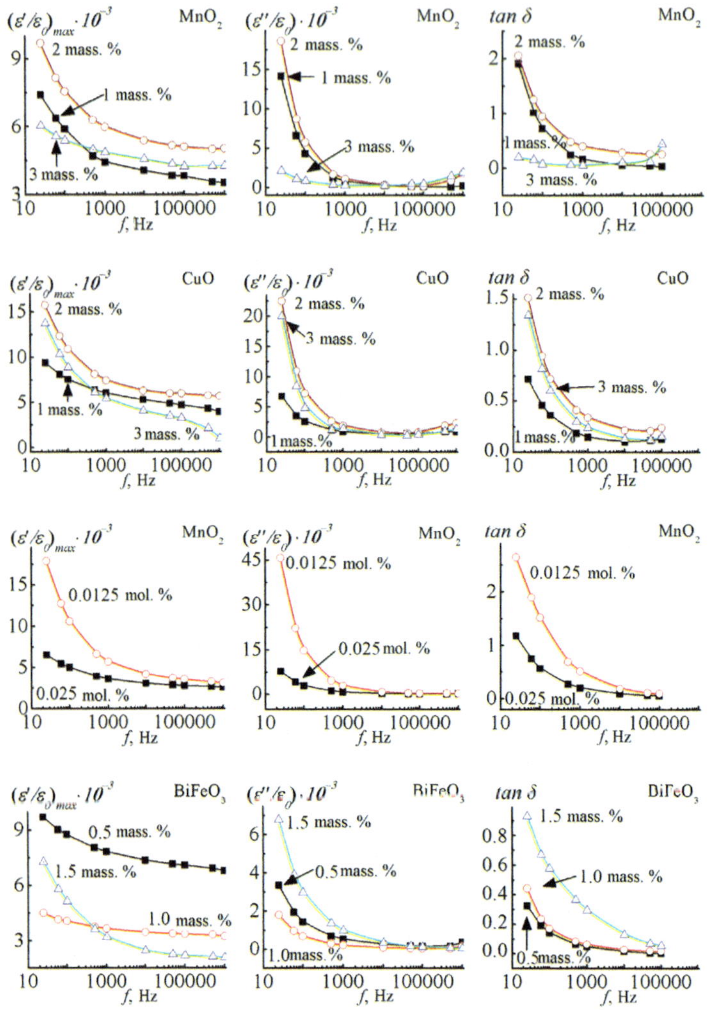

Fig. 6.11 Frequency f of the alternating electric field dependencies of the maximum values of $\varepsilon'/\varepsilon_0$, $\varepsilon''/\varepsilon_0$ and $\tan\delta$, corresponding to $(\varepsilon'/\varepsilon_0)_{max}$, of modified solid solutions with different additive contents (logarithmic scale on the abscissa axis)

reduced, degradation may occur due to the interaction of vacancies, formed in the *A*- and O-sublattices, which could result in the precipitation of impurity phases.

Thus, in the case of AMN, the reasons for the formation of oxygen vacancies are:

(i) the oxygen non-stoichiometry observed in all PTS;
(ii) the natural reduction of $Nb^{5+} \rightarrow Nb^{4+}$ (niobium is an ion with variable valence) at temperatures close to 400 °C (in Nb_2O_5) and decreasing in multielement compositions against the background of some structural transformations (Hedwal effect [17]); we have near the temperature of the FE-PE transition;

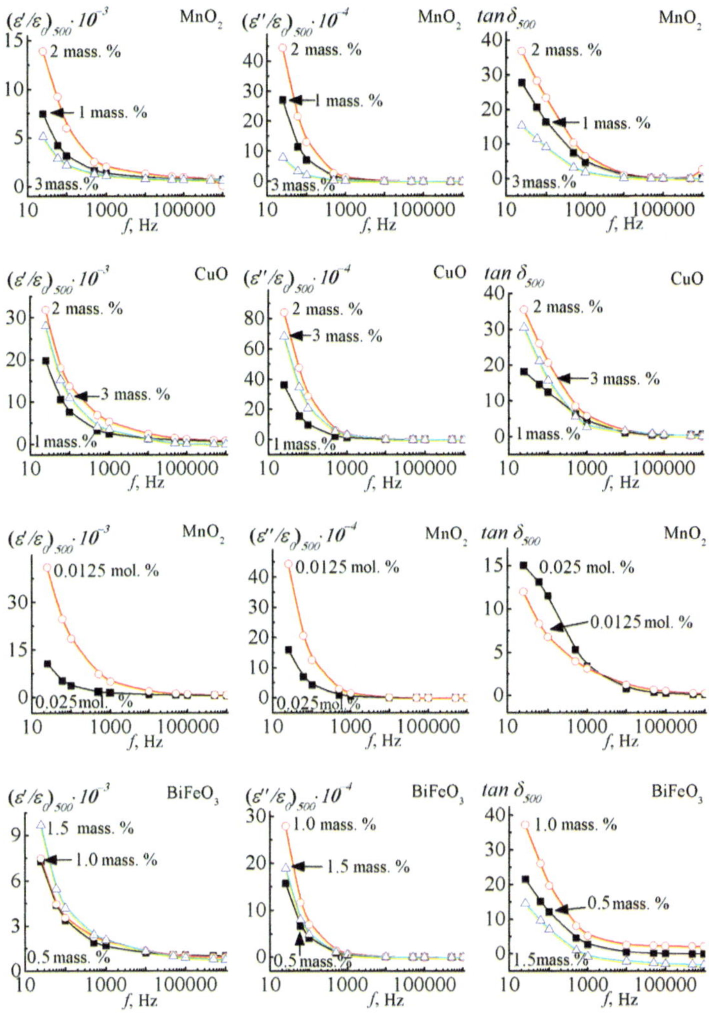

Fig. 6.12 Frequency dependencies f of the alternating electric field values of $\varepsilon'/\varepsilon_0$, $\varepsilon''/\varepsilon_0$ and tan δ, in the ferroelectric region (at 500 °C) of modified solid solutions with different additive contents (logarithmic scale on the abscissa axis)

(iii) the artificial reduction of niobium due to the creation of an oxygen-poor atmosphere, both through technical and technological methods and because of modification effects resulting from the introduction of heterovalent additives with uncompensated electric charge into main bases.

The additional introduction of oxygen vacancies resulted in a further reduction in $T_{(\varepsilon'/\varepsilon_0)_{\max}}$ at elevated frequencies (Fig. 6.18d), compared to the pre-heat treatment state of the samples. The observed increase in $T_{(\varepsilon'/\varepsilon_0)_{\max}}$ at low frequencies is likely attributable to the formation of impurities, as will be discussed below.

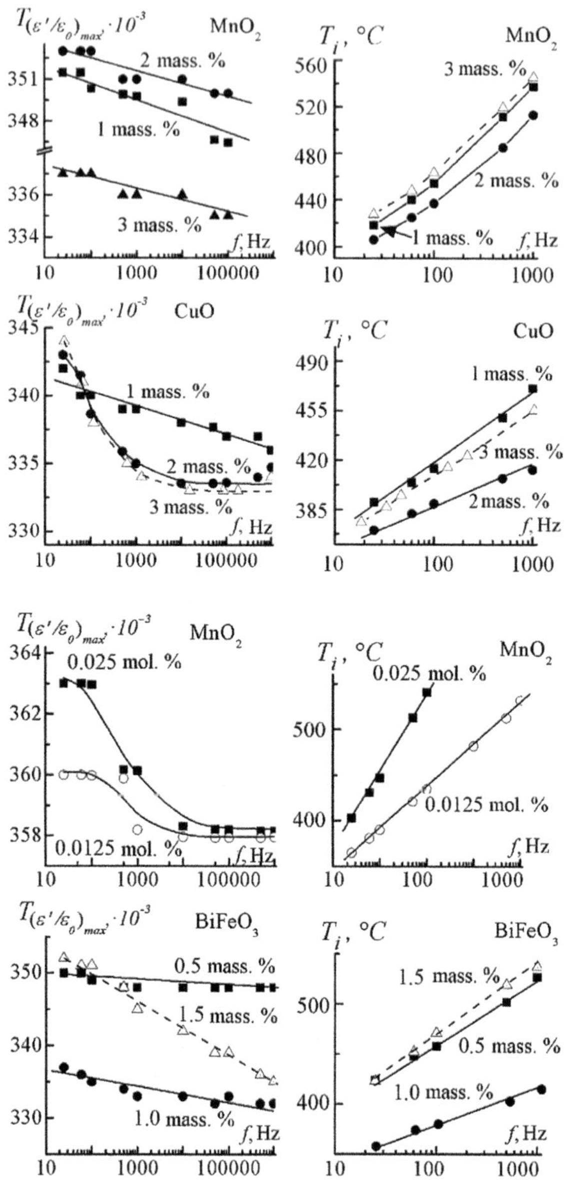

Fig. 6.13 Frequency dependencies f of the alternating electric field, maximum temperature $(\varepsilon'/\varepsilon_0)_{max}$, $T(\varepsilon'/\varepsilon_0)_{max}$, and onset temperature of $(\varepsilon'/\varepsilon_0)$ growth in the paraelectric region, T_i, of modified solid solutions with different additive contents (logarithmic scale on the abscissa axis)

The behavior of $T_{(\varepsilon'/\varepsilon_0)_{100}}$ at 100 °C is an indicative phenomenon. Prior to treatment, the temperature exhibited a tendency to shift towards higher temperatures as the frequency increased (Fig. 6.18a, e), which is characteristic of a normal relaxation process (at 100 °C, the interactions of the ceramics with water, both in its molecular and dissociated forms, are the only factors that bear relevance). There are no oxygen vacancies dealing with the reduction of Nb because of the low temperature and the

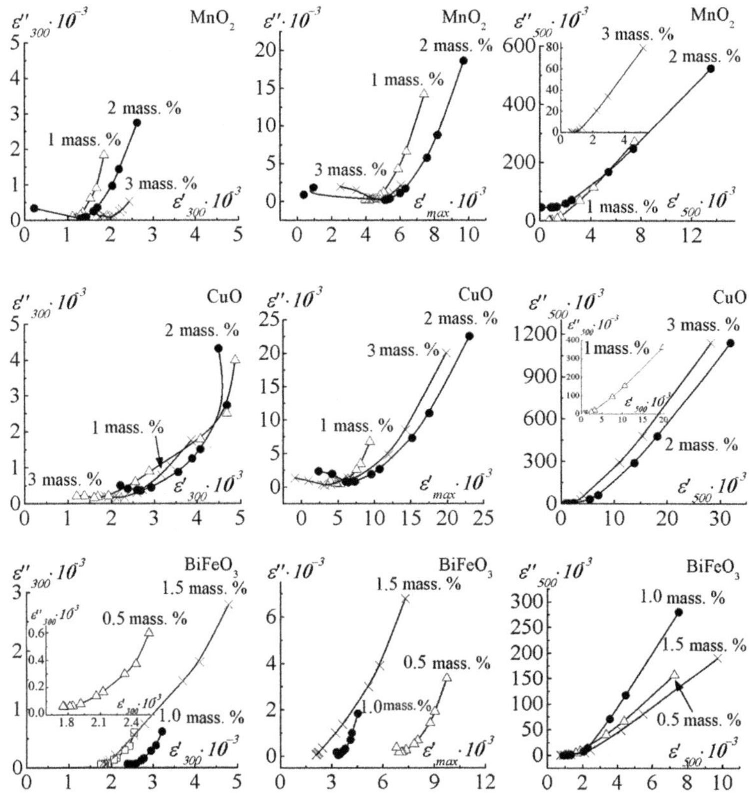

Fig. 6.14 Dependencies of $(\varepsilon'/\varepsilon_0)_{300}/(\varepsilon'/\varepsilon_0)_{300}$, $(\varepsilon'/\varepsilon_0)_{max}/(\varepsilon'/\varepsilon_0)_{max}$, $(\varepsilon'/\varepsilon_0)_{500}/(\varepsilon'/\varepsilon_0)_{500}$ in modified ceramics

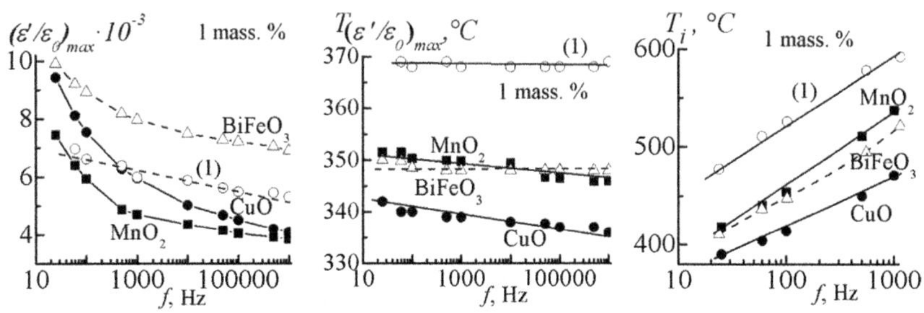

Fig. 6.15 Dependencies on the frequency f of alternating electric field for the peak value of relative permittivity, $\varepsilon'_{max}/\varepsilon_0$, temperature $T(\varepsilon'/\varepsilon_0)_{max}$, temperature of the onset of growth $\varepsilon'/\varepsilon_0$ in the paraelectric region T_i of initial SS (1) and modified solid solutions with 1 mass% additives (logarithmic scale on the abscissa axis)

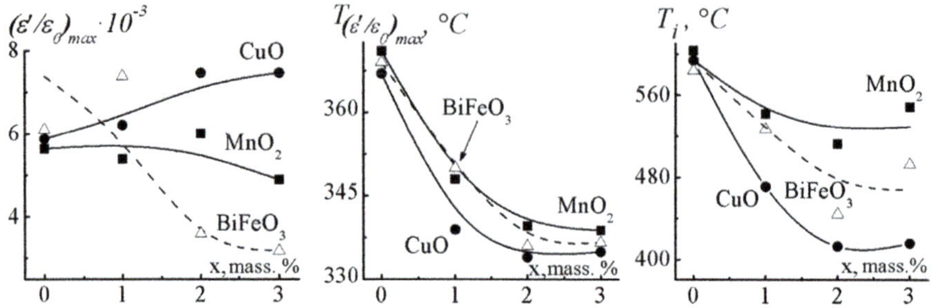

Fig. 6.16 Dependencies of the peak value of relative permittivity $\varepsilon'_{\mathrm{max}}/\varepsilon_0$, temperature $T(\varepsilon'/\varepsilon_0)_{\mathrm{max}}$, temperature T_i of the onset of $\varepsilon'/\varepsilon_0$ growth in the paraelectric region of modified solid solutions ($f = 1$ kHz) on the concentration of the introduced additives

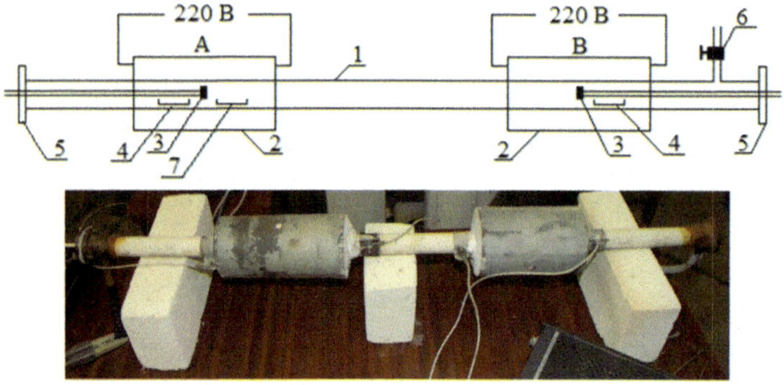

Fig. 6.17 Schematic (top) and furnace (bottom) for heat treatment of ceramics in a reducing atmosphere

fact that it is distant from the FE-PE transition. After heat treatment, only at low frequencies the behavior of $T_{(\varepsilon'/\varepsilon_0)_{100}}$ repeated, but with a slower rate, and at high frequencies, the temperature of the maximum $T_{(\varepsilon'/\varepsilon_0)_{100}}$ began to shift to the left, as it is characteristic of the main peak $\varepsilon'/\varepsilon_0$. It is important to consider that oxygen vacancies move at a relatively slow rate [18], and it is likely that their diffusion is only significantly activated at low frequencies.

A comparative analysis of the radiographic data, obtained on the untreated sample (1) and the heat-treated sample (2), revealed the following differences (Table 6.1):

- in sample 2, extraneous phase $Na_5Ta_{11}O_{30}$ appears, clearly visible on the diffractogram taken from the polished surface of the sample (in Fig. 6.19a marked with stars);
- homogeneity of the chemical composition of sample 2 is worse than that of sample 1, which is manifested in a more complex profile and an increase in the half-width of the multiplet lines 200, 020 (Fig. 6.19b);
- the volume of the perovskite monoclinic cell of sample 2 is larger than that of sample 1, which corroborates the occurrence of oxygen vacancies [19, 20] in the

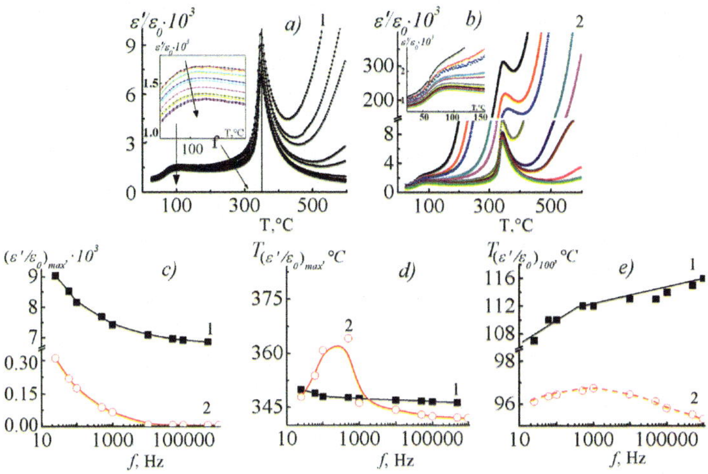

Fig. 6.18 Thermal frequency dependencies $(\varepsilon'/\varepsilon_0)$, $(\varepsilon'/\varepsilon_0)_{max}$ (**a–c**) and frequency values of $T_{(\varepsilon'/\varepsilon_0)max}$, $T_{(\varepsilon'/\varepsilon_0)100}$ (**d, e**) of solid solutions based on $(1-y)(Na_{0.54}K_{0.46})(Nb_{0.9}Ta_{0.1})O_3 - yLiSbO_3$ ($y = 0.02$), modified super stoichiometrically with 0.5 mass% $Bi_2O_3 + Fe_2O_3$ (x): (1) before heat treatment in reducing atmosphere, (2) after treatment

Table 6.1 Phase composition and structural characteristics of untreated and heat treated media

Type of sample	Phase composition (PS—perovskite)	Parameters of unit cell, Å Degree and volume, Å^3
Untreated (1)	PS	$a = 3.991$; $b = 3.952$; $\beta = 90.26$; $V = 62.96$
Heat treated (2), ceramic	PS + $Na_5Ta_{11}O_{30}$	The parameters are not given due to the incorrectness of their determination from the surface of ceramics because of the occurrence of textures
Heat treated (2), powder (crushed ceramics)	PS + traces of $Na_5Ta_{11}O_{30}$	$a = 3.992$; $b = 3.958$; $\beta = 90.21$; $V = 63.06$

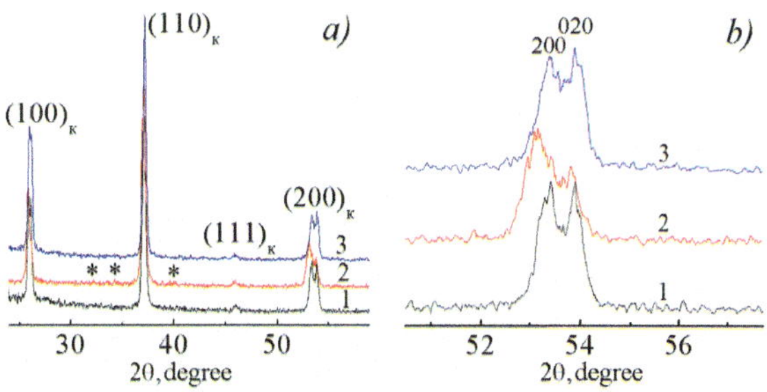

Fig. 6.19 Diffractograms of samples (left) and X-ray lines 200, 020 of samples 1 and 2: 1—before heat treatment, 2—after heat treatment (ceramics), 3—after heat treatment (powder); asterisks indicate the lines of impurity for phase $Na_5Ta_{11}O_{30}$

result of reductive processes that result in the emergence of ions with a diminished oxidation degree Nb^{4+}, exceeding Nb^{5+}. Furthermore, the formation of vacancies in the anionic sublattice results in the phenomenon of electrostatic repulsion of positively charged cations that are localized in the vicinity of the oxygen vacancy [19], which also contributes to the expansion of the cell, increasing its volume, as has been observed in practice. It should be noted that the data presented in [21] revealed a direct proportional relationship between the thermal expansion of the samples and the oxygen non-stoichiometry.

The results obtained indicate that oxygen vacancies play a crucial role in the anomalous behavior of $T_{(\varepsilon'/\varepsilon_0)_{\max}}$. This is also confirmed by the greater rate of decrease of $T_{(\varepsilon'/\varepsilon_0)_{\max}}$ in CuO-modified ceramics, compared to the similar dependence on materials with MnO_2 (Fig. 6.15), which exhibit fewer oxygen vacancies. Vacancy (□) generation scheme: $Na^{1+}Nb^{5+}_{1-x}Mn^{4+}_{x}O^{2-}_{3-x/2}\square_{x/2}$; $Na^{1+}Nb^{5+}_{1-x}Cu^{2+}_{x}O^{2-}_{3-3x/2}\square_{3x/2}$; $Na^{1+}_{1-x}Bi^{3+}_{x}Nb^{5+}_{1-x}Fe^{3+}_{x}O_3$ (there are no vacancies and $T_{(\varepsilon'/\varepsilon_0)_{\max}}$ is independent of f).

The comprehensive analysis of the physicochemical properties of niobium-containing compounds and SS, as presented here and in other sources such as [22–25], has revealed the following findings.

Niobate materials are systems of sufficient complexity, comprising several interacting elements, exhibiting critical coupling parameters and exhibiting relatively high probabilities of fluctuations. The latter increase over time, accumulate, and exert an influence on the collective behavior of elements and subsystems, which can result in the dissolution of the former structure and the emergence of a new order constituted by the elements of the destroyed environment. It can be concluded that the studied ceramics are open systems, situated far from the thermodynamic equilibrium point and already not at the sintering stage. At this point, oxygen from the surrounding environment diffuses into the near-surface regions and interacts with oxygen vacancies. To describe the phenomena that arise in such systems, the term 'anti-relaxation phenomena' is employed. These phenomena are manifested in the temperature-frequency dependencies of the complex dielectric permittivity at temperatures $T \approx 350$ °C, situated in proximity to the dielectric permittivity maxima. The observed decrease in temperature of these maxima with increasing frequency cannot be described within the framework of equilibrium thermodynamics. The transition from a valence of Nb^{5+} to Nb^{4+} for niobium, the emergence of numerous oxygen vacancies and the formation of new structures render the system so complex that a synergetic approach is necessary. The competition between equilibrium (the enhanced electrical conductivity and enhanced Maxwell–Wagner relaxation resulting from the heat treatment) and non-equilibrium thermodynamics is clearly visible in curve 2 (Fig. 6.18d) in the form of a maximum $\varepsilon'/\varepsilon_0$.

The tendency of niobium to change valence and the simultaneous occurrence of many oxygen vacancies necessitate the application of non-equilibrium thermodynamics to niobium-containing solid solutions. In addition to the relaxation processes, described in the context of equilibrium thermodynamics, an 'anti-relaxation' phenomenon emerges, necessitating a synergetic approach for its accurate description and interpretation.

6.2 Effect of Through Conductivity on Dielectric Spectra of Objects

In real dielectrics with relaxation processes, the through conductivity γ is a common phenomenon. It would be remiss not to include ferroactive ceramics based on AMN within this category. Furthermore, the crystallochemical characteristics of this category of substances (including the increased volatility of alkali metal niobates, the ease with which the valence state of Nb can be altered, the propensity to undergo hydrolysis and polarization electrolysis, and so forth) give rise to the emergence of through conductivity that can transform dielectric spectra.

The objective of the present section is to establish the patterns of γ influence on the thermal frequency behavior of an AMN-based multielement composite in the paraelectric region. In accordance with the hypothesis proposed in [22], it is assumed that at the micro level, there is no interaction between relaxants and the charge carriers responsible for the electrical conductivity [5, 26, 27].

Figures 6.20 and 6.21 illustrate the temperature-dependent behavior of the $\varepsilon'/\varepsilon_0(T)$ and $\varepsilon''/\varepsilon_0(T)$ of $0.98(Na_{0.54}K_{0.46})(Nb_{0.9}Ta_{0.1})O_3 - 0.02LiSbO_3 + 1.0$ mass% MnO_2 ceramics within temperatures and frequencies of the specified ranges, determined conducted in the cooling mode.

It can be observed that the $\varepsilon'/\varepsilon_0(T)$ and $\varepsilon''/\varepsilon_0(T)$ dependencies exhibit a complex character, which is a consequence of the complex sequence of phase transformations occurring in the ceramics $0.98(Na_{0.54}K_{0.46})(Nb_{0.9}Ta_{0.1})O_3 - 0.02LiSbO_3 +$ 1.0 mass% MnO_2 at these temperatures. Herewith the Curie region (the temperature range of transition from the FE to PE) is situated in the vicinity of 340–360 °C. As the Curie–Weiss law is only fulfilled from $T \sim 380$ °C (Fig. 6.21), the clusters of the FE phase in this material emerge at this temperature.

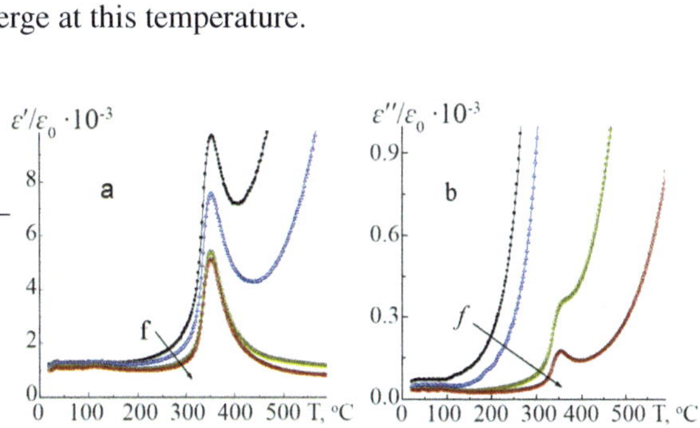

Fig. 6.20 Dependencies of $\varepsilon'/\varepsilon_0$ (**a**), $\varepsilon''/\varepsilon_0$ (**b**) on temperature of ceramics $0.98(Na_{0.54}K_{0.46})(Nb_{0.9}Ta_{0.1})O_3 - 0.02LiSbO_3 + 1.0$ mass% MnO_2; the arrows indicate the direction of growth of the frequency f

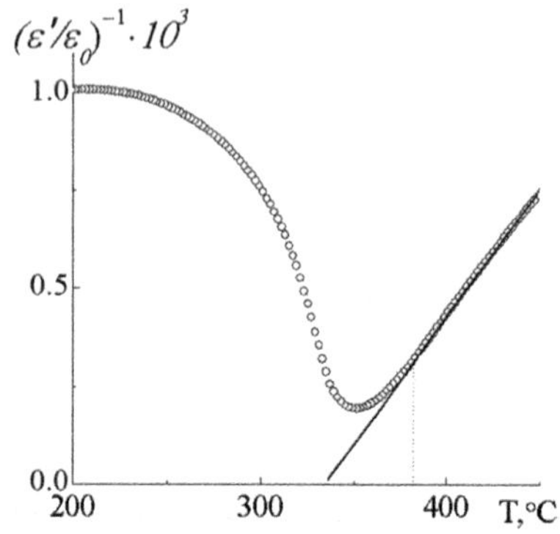

Fig. 6.21 Dependencies of $(\varepsilon'/\varepsilon_0)^{-1}$ on temperature of ceramics $0.98(Na_{0.54}K_{0.46})(Nb_{0.9}Ta_{0.1})O_3 - 0.02LiSbO_3 + 1.0$ mass% MnO_2

A salient feature of the dependencies $\varepsilon''/\varepsilon_0(T)$ is the abrupt increase at temperatures approaching 200 °C, which can be attributed to the enhanced electrical conductivity of the object. It is established that the through-conductivity at direct current contributes $\gamma_{dc}/(2\pi f \varepsilon_0)$ to the imaginary part of the complex dielectric permittivity, thereby significantly distorting the character, nature and shape of dielectric spectra, including relaxation spectra. To study the contribution of γ_{dc}, we have conducted a series of measurements on the DC conductivity and the values of $\varepsilon'/\varepsilon_0$ and $\varepsilon''/\varepsilon_0$ in the frequency range of $(1–10^5)$ Hz of $0.98(Na_{0.54}K_{0.46})(Nb_{0.9}Ta_{0.1})O_3 - 0.02LiSbO_3 + 1.0$ mass% MnO_2 ceramics in the paraelectric region (at $T = 400$ °C). This was done to exclude the contribution of the segmental electric subsystem.

To achieve the desired result, the experimental sample was heated to 430 °C and then cooled at a gradual rate (within approximately 20 min) to 400 °C. The temperature of the sample was then maintained at a constant level throughout the measurement period. Figure 6.22 illustrates the real part of the complex electrical conductivity $\gamma(f)$ (where $\gamma = 2\pi f \varepsilon'' \varepsilon_0$) of the investigated ceramics at $T = 400$ °C.

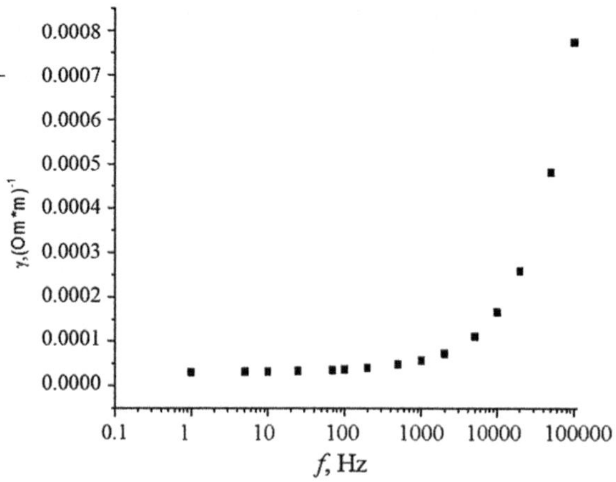

Fig. 6.22 Dependencies of $\gamma(f)$ of the ceramics $0.98(Na_{0.54}K_{0.46})(Nb_{0.9}Ta_{0.1})O_3 - 0.02LiSbO_3 + 1.0$ mass% MnO_2 at $T = 400$ °C

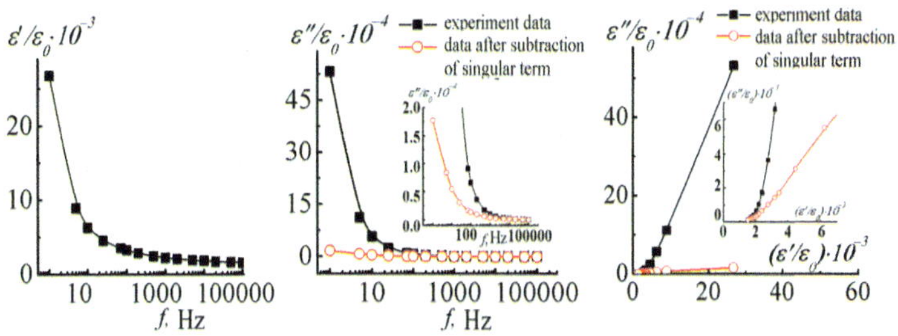

Fig. 6.23 Dependencies of $\varepsilon'/\varepsilon_0(f)$, $\varepsilon''/\varepsilon_0(f)$ and $\varepsilon''/\varepsilon_0(\varepsilon'/\varepsilon_0)$ of $0.98(Na_{0.54}K_{0.46})(Nb_{0.9}Ta_{0.1})O_3 - 0.02LiSbO_3 + 1.0$ mass% MnO_2 ceramics at $T = 400$ °C; red markers show the values of $\varepsilon''/\varepsilon_0$ after subtraction of singular terms

It can be observed that at frequencies below 100 Hz, the dependence $\gamma(f)$ reaches saturation, and the values of γ agree (within the limits of experimental error) with the value of γ_{dc}, measured at DC current. This demonstrates that the primary contribution to the dielectric response is through conductivity, even at frequencies below 100 Hz. The experimental dependencies of the $\varepsilon'/\varepsilon_0(f)$, $\varepsilon''/\varepsilon_0(f)$ and $\varepsilon''/\varepsilon_0(\varepsilon'/\varepsilon_0)$ of $0.98(Na_{0.54}K_{0.46})(Nb_{0.9}Ta_{0.1})O_3 - 0.02LiSbO_3 + 1.0$ mass% MnO_2 ceramic at $T = 400$ °C are shown in Fig. 6.23. It can be observed that following the subtraction of the singular term $\gamma_{dc}/(2\pi f \varepsilon_0)$ from the values of $\varepsilon''/\varepsilon_0$, the form of the dependencies $\varepsilon''/\varepsilon_0(f)$ and $\varepsilon''/\varepsilon_0(\varepsilon'/\varepsilon_0)$ has undergone a significant alteration, and there are indications of at least two relaxation processes occurring within the object.

It was demonstrated that optimal results (Fig. 6.23) can be achieved by simultaneously approximating the dependencies $\varepsilon'/\varepsilon_0(f)$, $\varepsilon''/\varepsilon_0(f)$ and $\varepsilon''/\varepsilon_0\left(\varepsilon'/\varepsilon_0\right)$ within the framework of the dielectric model. This model assumes that two non-interacting relaxation processes contribute to the dielectric response within the analyzed frequency range (Fig. 6.24, denoted as low-frequency (LF) and high-frequency (HF)), the each of which was described by the Cole–Cole formula:

$$\varepsilon^* = \varepsilon_\infty + \frac{\varepsilon_S - \varepsilon_\infty}{1 + (i\omega\tau)^{1-\alpha}},$$

where ε_s is the static dielectric permittivity, ε_∞ is the high-frequency dielectric permittivity, τ is the most probable relaxation time, and α is the distribution parameter.

This evidence suggests that in this material, within the paraelectric region, the dielectric response, and the analyzed frequency range, there are contributions from two distinct types of non-Debye relaxors, exhibiting markedly disparate distribution parameters and the most probable relaxation times (frequencies). The analysis of measurements at $T = 400$ °C does not provide sufficient information to determine the nature of the substances in question. Further measurements will be conducted at a broader range of temperatures in the future. Nevertheless, in consideration of the range of localizations of τ and the substantial values of α, it is reasonable to conclude that the nature of the detected relaxation processes can be attributed to alterations and rearrangements in the real (defective) subsystem of a given material.

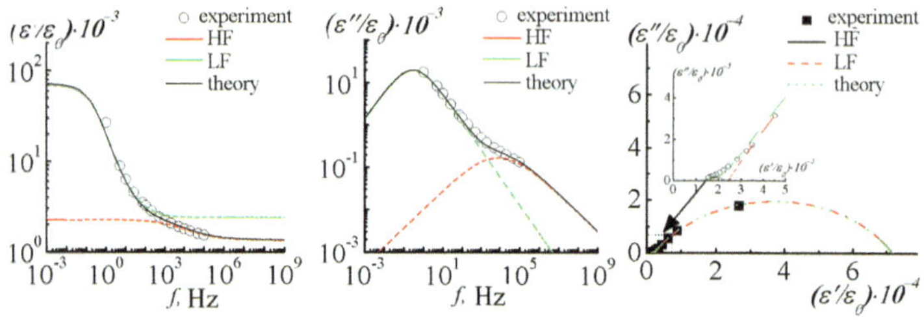

Fig. 6.24 Dependencies of $\varepsilon'/\varepsilon_0(f)$, $\varepsilon''/\varepsilon_0(f)$ and $\varepsilon''/\varepsilon_0(\varepsilon'/\varepsilon_0)$ of $0.98(Na_{0.54}K_{0.46})(Nb_{0.9}Ta_{0.1})O_3 - 0.02LiSbO_3 + 1.0$ mass% MnO_2 ceramics at $T = 400$ °C in the range of $f = (1\text{–}10^5)$ Hz (empty circle); lines are the results of the approximation: HF fitting—Cole–Cole ($\varepsilon_s = 72{,}000$, $\varepsilon_\infty = 2300$, $\tau = 0.55$, $\alpha = 0.34$), HF fitting—Cole–Cole ($\varepsilon_s = 2300$, $\varepsilon_\infty = 1350$, $\tau = 0.00002$, $\alpha = 0.55$)

6.3 Conclusions

The final chapter has been dedicated to the investigation of the thermal-frequency characteristics of modified solid solutions, based on alkali metal niobates in the presence of a wide range of external influences, including temperature and the frequency of alternating electric fields. The causes and mechanisms of oxygen non-stoichiometry in complex compounds with perovskite-type structure were described in detail. The pivotal role of oxygen non-stoichiometry in the genesis of thermal-frequency behavior for such structures is demonstrated. Attempts are made to determine the nature of relaxation processes, occurring in multielement solid solutions, based on sodium–potassium niobates.

References

1. Sadykov H.A., et al.: Effect of combined modification on the structure, microstructure and piezo response of lead-free materials based on the (Na, K)NbO_3 system. In: Proceedings of the Fourth International Youth Symposium on Physics of Lead-Free Piezoactive and Related Materials (Analysis of the Current State and Development Prospects) (LFPM-2015), 2–6 Sept 2015, Rostov-on-Don—Tuapse, vol. 2, pp. 184–189 (2015) (in Russian)
2. Wang, K., Li, J.F.: (K, Na)NbO_3-based lead-free piezoceramics: phase transition, sintering and property enhancement. J. Adv. Ceram. **1**, 24–37 (2012)
3. Voloshin, A.V.: Tantalo-Niobates: Systematics, Crystal Chemistry and Evolution of Mineral Formation in Granitic Pegmatites. Nauka, Saint Petersburg (1993) (in Russian)
4. Reznichenko, L.A., et al.: Secondary discontinuous recrystallization as a manifestation of the effect of self-organization in niobate ferroelectric ceramics. Ferroelectrics **247**(1–3), 95–105 (2000)
5. Bogatin, A.S.: Determining the relaxation polarization parameters in dielectrics with high electrical conductivity. Bull. Russ. Acad. Sci. Phys. **79**, 729–731 (2015)
6. Brown, W.: Dielectrics, p. 328. Foreign Literature, Moscow (1961) (in Russian)
7. Fröhlich, H.: Theory of Dielectrics. Foreign Literature, Moscow (1960). (in Russian)
8. Turik, A.V., et al.: Relaxation processes in dielectrics with non-Debye spectra. Solid State Phys. **53**(12), 2299–2301 (2011)

9. Turik, A.V., Bogatin, A.C: Relaxation Polarisation in Non-Debye Dielectrics. Lambert Academic Publishing, Saarbrücken, Deutschland (2013)
10. Skanavi, G.I.: Physics of Dielectrics (Area of Weak Fields). State Publishing House of Technical and Theoretical Literature, Moscow, Leningrad (1949). (in Russian)
11. Hippel, A.R.: Dielectrics and Waves. Chapman & Hall, New York; Wiley, London (1954)
12. Turik, A.V., et al.: Disordered ferroelectric systems: giant dielectric enhancement, Maxwell-Wagner relaxations and conductor—insulator transition. Ferroelectrics **307**(1), 171–176 (2004)
13. Turik, A.V., Radchenko, G.S.: Maxwell-Wagner relaxation in piezoactive media. J. Phys. D **35**(11), 1188–1192 (2002)
14. Vezzoli, G.C.: Electrical properties of NbO_2 and Nb_2O_5 at elevated temperature in air and flowing argon. Phys. Rev. B **26**(7), 3957–3963 (1982)
15. Sadykov, Kh.A., et al.: Dependence of dielectric properties of lead-free ferropiezoceramic materials on modification with 3D-metal oxides. Phase Transitions, Ordered States New Mater., 32–34 (2018) (in Russian)
16. Fesenko, E.G., et al.: New Piezoceramic Materials. Rostov State University Press, Rostov-on-Don (1983). (In Russian)
17. Titov, S.V.: Effects of modification in alkali metal niobates, lead titanate, lead zirconate and their solid solutions. Ph.D. Thesis (Physical and Mathematical Sciences), Rostov State University Press, Rostov-on-Don (2001) (in Russian)
18. Tretyakov, Yu.D., Lepis, H.: Chemistry and Technology of Solid-phase Materials. Moscow State University Press, Moscow (1985). (in Russian)
19. Arkharov, V.I.: Mesoscopic phenomena in solids and their mesostructure. In: Problems of Modern Physics: Collection of Articles for the 100th Anniversary of A. F. Ioffe's Birth, pp. 384–409. Nauka, Leningrad (1980) (in Russian)
20. Powder Diffraction File. Data Card. Inorganic Section. Set 34. Card 1228. JCPDS. Swarthmore. Pennsylvania. USA (1948)
21. Tikhonova, L.A., et al.: Physicochemical properties of electrode materials based on $La_2NiO_{4\pm\delta}$. Bull. BSU **2**(1), 14–16 (2010) (in Russian)
22. Aschroft, W.N., Mermin, N.D.: Solid State Physics. Brooks Cole, San Diego, CA, USA (1976)
23. Danciger, A.Ya., et al: Multicomponent Systems of Ferroelectric Complex Oxides: Physics, Crystal Chemistry, Technology. Design Aspects of Ferro-Piezoelectric Materials. Rostov State University Press, Rostov-on-Don, vol. 1, (2001), vol. 2 (2002) (In Russian)
24. Verbenko, I.A.: Phase diagrams of state and macro responses of multicomponent mesoscopically inhomogeneous media with special electric and magnetic states. DrSc Thesis (Physical and Mathematical Sciences), Southern Federal University Press, Rostov-on-Don (2016) (in Russian)
25. Verbenko, I.A., Reznichenko, L.A. Lead-free ferroelectric ceramics based on alkali metal niobates: history, technology, prospects. In: Proceedings of the Second International Youth Symposium on Physics of Lead-free Piezoactive and Related Materials. (Analysis of the Current State and Development Prospects) (LFPM-2013), Rostov-on-Don—Tuapse, vol. I, no. 2, pp. 52–64 (2013) (In Russian)
26. Bogatin, A.S., et al.: The reason for division of polarization relaxation processes into strong and weak. Bull. Russ. Acad. Sci. Phys. **74**, 1066–1068 (2010)
27. Bogatin, A.S., Bulanova, A.L., Andreev, E.V., et al.: Identifying dielectric spectra: Cole-Cole distribution, constant phase element, and lognormal distribution. Bull. Russ. Acad. Sci. Phys. **84**, 1425–1426 (2020)

Index

I. A. Verbenko et al., *Environmentally Friendly Smart Materials with Special Electrical and Magnetic Properties*, Engineering Materials,
https://doi.org/10.1007/978-3-031-90387-8

The manufacturer's authorised representative in the EU is Springer Nature Customer Service Centre GmbH, Europaplatz 3, 69115 Heidelberg, Germany. If you have any concerns regarding our products, please contact ProductSafety@springernature.com

Printed and bound by CPI Group (UK) Ltd, Croydon, CR0 4YY

01/10/2026

02225866-0003